# From Local to Global Optimization

## Nonconvex Optimization and Its Applications

Volume 53



*The titles published in this series are listed at the end of this volume.*

# From Local to Global Optimization

Edited by

Athanasios Migdalas
*Technical University of Crete,*
*Chania, Crete, Greece*

Panos M. Pardalos
*University of Florida,*
*Gainesville, Florida, U.S.A.*

and

Peter Värbrand
*Linköping University,*
*Norrköping, Sweden*

KLUWER ACADEMIC PUBLISHERS
DORDRECHT / BOSTON / LONDON

A C.I.P. Catalogue record for this book is available from the Library of Congress.

ISBN 0-7923-6883-5

Published by Kluwer Academic Publishers,
P.O. Box 17, 3300 AA Dordrecht, The Netherlands.

Sold and distributed in North, Central and South America
by Kluwer Academic Publishers,
101 Philip Drive, Norwell, MA 02061, U.S.A.

In all other countries, sold and distributed
by Kluwer Academic Publishers,
P.O. Box 322, 3300 AH Dordrecht, The Netherlands.

*Printed on acid-free paper*



Printed in the Netherlands.

# Contents

Professor Hoang Tuy

**This volume is dedicated to Professor Hoang Tuy on the occasion of his $70^{th}$ birthday.**

# Preface

On August 1997 a conference titled "*From Local to Global Optimization*" was held at Storgården in Rimforsa near the Linköping Institute of Technology, Sweden. The conference gave us the opportunity to celebrate Hoang Tuy's achievements in Optimization during his 70 years of life. This book consists of a collection of research papers based on results presented during the conference and are dedicated to Professor Hoang Tuy on the occasion of his 70th birthday.

The papers cover a wide range of recent results in Mathematical Programming. The work of Hoang Tuy, in particular in Global Optimization, has provided directions for new algorithmic developments in the field.

We are indebted to the Kluwer Academic Publishers for inviting us to publish this volume, and the Center for Industrial Information Transfer (CENIIT) for financial support.

We wish to thank the referees for their help and the authors for their papers. We also wish to join all contributors of this book in expressing birthday wishes and gratitude to Hoang Tuy for his inspiration, support, and friendship to all of us.

Athanasios Migdalas, Panos M. Pardalos, and Peter Värbrand

November 1998

## Hoàng Tuy: An Appreciation

Its a pleasure for me as colleague and friend to take this opportunity to celebrate Hoàng Tuy's numerous contributions to the field of mathematical programming.

Hoàng Tuy is perhaps most famous for his contributions to global optimization. Already in 1964 he suggested the pioneering "Tuy"-cut in concave minimization, and since then he has written numerous papers and several books. He is one of the founding fathers of global optimization. His scientific contributions cover a broad area outside of global optimization as well including minimax problems, inconsistent convex inequalities and stability, and fixed point computational theory.

Hoàng Tuy started in real analysis. He grew up and studied under difficult conditions - Vietnam has been at war most of his life time. He was first known as an outstanding high school teacher. He wrote a text book on geometry, published by the guerilla movement in 1949, that was in use for a long time. He followed the Soviet university program in mathematics on his own, and later, in 1959, he received his Candidate's Degree (= U.S. Ph.D.) at Moscow University after one year and a half as a student under the supervision of Menshov and Shilov. He published 5 papers in real analysis in Russian, a language which he learned the hard way by studying Russian mathematical books.

Then he became interested in Operations Research and visited Kantorovich in Novosibirsk 1962 and 1964. To the mathematical programming community this change of interest turned out to be decisive and led to many substantial and important contributions. Convex optimization was fairly well understood, when Hoàng Tuy opened up the new field of concave and global optimization, which up til then had been looked upon as an area of research that was almost hopelessly difficult. It is worth noting that Hoàng Tuy's interest grow out of a practical problem in transportation - reorganizing the logistics of trucking so as to reduce the distance that the trucks travel empty.

Hoàng Tuy had a strong desire to make use of mathematics in solving practical problems and helping to develop the economy of Vietnam. Due to the war conditions this turned out to be more difficult than he had expected. In stead his major contributions came to fall in theoretical mathematical programming but also as educator and administrator. He became vice chairman of the Institute of Mathematics in Hanoi when it was started in 1970 and served as chairman from 1980. In the same year he became General Secretary of the Vietnamese Mathematical Society. He has also been a member of various governmental committees.

Hoàng Tuy has spent considerable time as invited guest professor at leading universities all over the world. In 1989 he began a long period of fruitful cooperation with Optimization at Linköping University. We are very proud and deeply grateful for the inspiration and knowledge that his regular visits to Linköping has meant for the group as a whole and for the Ph.D. students in particular. In 1995 he accepted an honorary doctor's degree from the Technical Faculty at Linköping University. I am glad that this occurred during my last term as President of Linköping University, so that I could be among the first to congratulate him.

Hoàng Tuy has been recognized as a leading scientist in Vietnam. In 1996 he was, as one of the first, awarded the Hô Chi Minh Prize, Vietnam's highest national reward. This closed the circle in a way. In 1951 Hoàng Tuy made a long walk north. The only possible route was to use what was later to be known as the Hô Chi Minh trail. The walking took 3 months under severe conditions. There were three major dangers; the French, malaria and the tigers. He carried only rice, salt and his math books, from which he had removed the covers and cut out the margins on every page in order to make them lighter.

I have learned to know Hoàng Tuy as a wise man with an extra ordinary background and life experience. There does not seem to be any difficulty or challenge - may it be in mathematics, education, administration or human life in general - that he will not turn into a successful human endeavor.

Sven Erlander
Linköping, August 24, 1998

Chapter 1

# CONVEX GLOBAL UNDERESTIMATION FOR MOLECULAR STRUCTURE PREDICTION

A.T. Phillips
J.B. Rosen
and K.A. Dill

**Abstract** Key problems in computational biology, including protein and RNA folding and drug docking, involve conformational searching. Current search methods – Monte Carlo, Molecular Dynamics, Simulated Annealing, and Genetic Algorithms – are too slow for protein folding by many orders of magnitude. They get stuck in kinetic traps. We describe a global optimization method, the CGU method, which appears to be very promising. We know the method always finds the same conformation from 100 different starting points, indicating that it finds the unique global minimum for the many different sequences we have tried. We know the CGU doesn't get stuck in kinetic traps because the search time is independent of the shapes of the landscapes (amino acid sequence and composition). We know that the method is much faster than a standard Simulated Annealing algorithm that we have tested: the SA method doesn't find global minima for chains longer than 10 residues, and the performance advantage of the CGU method increases with chain length. And computational results show that the computer time scales with $n^4$ where $n$ is the number of degrees of freedom, and we consistently reach the global minimum of the model energy function for PPT, a 36-amino acid peptide ($n = 72$), in less than 3 hours on a 32 processor Cray T3E.

**Keywords**: Convex Global Underestimation, Protein Folding, Simulated Annealing, Computational Biology, Molecular Structure



*A. Migdalas et al. (eds.), From Local to Global Optimization,* 1–18.

## 1. Introduction

A major goal of computational biology is to develop a computer algorithm for ab initio protein folding to predict native structures from amino acid sequences without other information (4). Such a folding algorithm would be an important complement to experiments. For determining protein structures, crystallography and NMR are slow in some cases and impossible in others. The number of known amino acid sequences is nearly $10^6$ and growing exponentially, while the number of known native structures is only in the thousands. And some amino acid sequences are too unrecognizable to benefit from homology modeling. Despite a number of important advances in the computational area ( (3; 5; 6; 12; 15; 19; 21; 23; 30; 31; 38; 39; 40)), no general algorithm yet exists.

Protein folding is a difficult interdisciplinary problem, requiring a combination of protein biochemistry, global optimization, and high performance computing. A fast global optimization search method, detailed protein-like models, and accurate potential energy functions are essential to the successful development of a computational protein folding algorithm. Attempting to develop all of these at once is very difficult; such approaches have not yet folded proteins successfully. Instead, our approach is to concentrate on the global optimization search algorithm, and then we and others can use it to improve models. With a fast global search, we can quickly test any energy function model change to determine if it is indeed an improvement.

Obviously, we need some energy function model to serve as the test function for developing the search strategy. For this purpose we use the simple energy function of Sun (36), which is protein-like but flawed. We use it to test and improve the conformational search method. The obvious criticism is: will a search method developed with an imperfect energy function also ultimately work on the true energy function? We believe there are three bases to justify our approach.

First, our strategy requires a model that is at least a reasonable first approximation. It must serve our needs for development, in this case having relatively few energy parameters and degrees of freedom, but it also must have protein-like properties. As noted below, the model of Sun suits these purposes. Second, we will indeed refine our models and refine the CGU method to work with them as well. Third, more detailed energy functions are not necessarily more desirable, at the present stage. More detail does not yet imply better folded structures. Atomic resolution force fields are extensively tested only on small perturbation problems, and none of them has yet successfully folded a protein from open conformations. Furthermore, with more detailed models, developing new search strategies takes much longer because they waste time exploring details. For folding proteins, details may be important, but for developing early-stage search strategies, details are just an obstacle to development. The key point is that for the initial objective of developing a fast global search algorithm, the Sun model will be an excellent choice.

The Sun model is a better testbed than some other model problems used to test conformational search methods. Search strategies are sometimes tested on van der Waals clusters to prove they reach the global energy minimum. The Sun model has the advantage that it is a problem of folding, not clustering: it has chain connectivity, specific monomer sequences, chain steric constraints, the ability of a chain to collapse and fold into compact unique native states with a hydrophobic core and hydrogen bonded secondary structure. Moreover, when the native secondary structures are given, the Sun function has been used to find native-like structures of about 7 small proteins (36). Hence, since a perfect protein model does not yet exist, we consider the Sun model a good starting point for developing a search strategy, and then improving the energy function.

## 2. Overview of the CGU Search Method

Our global optimization algorithm, the CGU (Convex Global Underestimator) algorithm (26), is very different than Molecular Dynamics (MD), Monte Carlo (MC), Simulated Annealing (SA), or Genetic Algorithms (GA). The CGU method does not search the tops of energy landscapes, does not get caught in kinetic traps, and its speed does not depend on the shapes of energy landscapes (amino acid sequences), but only on the sizes of landscapes (for a chain having $n$ degrees of freedom, the computer time is on the order of $n^4$).

Using hundreds of different starting points, for chains of lengths up to 36 monomers, we know that the CGU finds the same unique structure in every run, indicating: (1) that it is probably finding global optima, and (2) that it is doing so with great reliability. Having tried this for many different sequences of different chain lengths, we know that finding global minima is not restricted to certain sequences or specific native structures. Our tests, summarized later, show that the performance of the CGU method is much better than Simulated Annealing (14) on these problems, and the advantage grows with chain length. SA seldom finds global optima, so it is not even known how the computer time for SA scales with chain length. Using 32 nodes of a Cray T3E, the CGU finds global minima for 50-mers in 9 hours.

In addition to the global minimum, the CGU method also computes a large number of low energy local minima. This is useful for learning the shapes of energy landscapes and the nature of energy gaps, which is helpful in computational studies of folding kinetics. We can characterize the low energy regions of landscapes of the Sun model and others, which are much more realistic than lattice models. We have also devised a novel way to display this $n$-dimensional landscape, where $n$ is the number of degrees of freedom, in a simple graph (see Figure 1.7) that shows the distribution of local minima.

## 3. Summary of Recent Results

### 3.1. Current Search Methods are Much Too Slow

Conformational searching is often done by Monte Carlo/Simulated Annealing ((22; 25; 27; 29)), Genetic Algorithms ((35; 37)), Molecular Dynamics ((2; 24)), and by transformation methods, for example based on the diffusion equation ((16; 17; 18; 33)). SA, GA, and MD are much too slow, arguably by 3-10 orders of magnitude. What is the basis for this estimate? First, search methods need to provide some assurance they reach global, not local, minima. Biomolecules often achieve thermodynamic stability, implying that they usually do not get stuck in kinetic traps. Proteins fold in milliseconds to seconds, while MD all-atom simulations are limited to nanoseconds. Second, search methods must be fast enough to reach the large sizes of real molecules. Current methods often aim to fold crambin (46 amino acids), but most real proteins of interest have hundreds of monomers or more. Search methods must scale gracefully with chain length. Third, search methods must be fast enough to deal with the many degrees of freedom in realistic chain representations in continuum space; lattice models are not sufficient. Current methods are not close to satisfying these requirements, and factor-of-2 type fixes are insufficient.

## 3.2. Current Search Methods Get Stuck in Kinetic Traps

What is wrong with current search methods? They get stuck in kinetic traps. MC, MD, SA, and GA methods all search the tops of conformational energy landscapes, which have deep valleys separated by hills. When searches fall into local minima, it can take a long time to climb over barriers to proceed toward global minima. The symptom of kinetic trapping is that search times depend more strongly on the shape of the landscape (amino acid sequence) than on its size (number of amino acids). Because it is seldom known whether a search method reaches a true global minimum, or just local minima, remarkably little is known about the speed properties of standard search methods (1). The only scaling law that is known for conformational searching is the exact enumeration time, which scales exponentially with chain length ((4; 28)).

Our CGU method is very different. It is an ab initio method which searches under the landscape, not on top of it. We have shown that it does not get caught in traps: search time depends very little on the amino acid sequence or the native structure of the fold. Search time depends only on the chain length, and the scaling is on the order of $n^4$ (on average) where $n$ is the number of degrees of freedom.

## 3.3. How does the CGU Method Work?

The CGU method searches for the globally optimal lowest energy conformation $\phi_G$. Our approach involves the construction of a convex function which underestimates all known local minima and does so by the least possible amount (Figure 1.1).

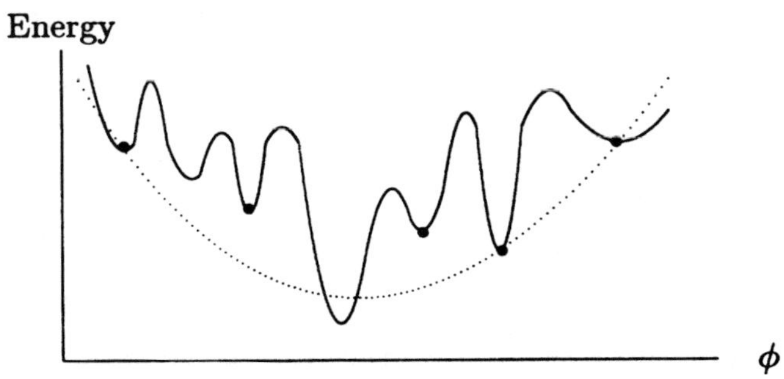


*Figure 1.1.* The Convex Global Underestimator (CGU)

Based on the premise that protein folding energy landscapes are funnels with bumps ((8; 20; 32; 41)), there is information about where to find the native structure distributed everywhere throughout the landscape. To find the bottom of a funnel, even a bumpy one, you need to head generally downhill. In general, we know that most well-understood proteins must have such landscapes. The idea of a funnel is nothing more than the statement that proteins fold much faster than the "Levinthal time", the exhaustive search time, and always to the same unique state.

The CGU simply makes use of the overall funnel-like shape to guide and localize the search to regions that are estimated to be near the native structure. This is done iteratively. Specifically, we are given a primary sequence of amino acids with $n$ degrees of freedom $\phi \in \Re^n$, and a potential energy function $F(\phi)$. Typically the $n$ degrees of freedom will include at least the backbone dihedral angles $\phi/\psi$, and possibly others, such as the sidechain $\chi$ angles, so that $n \geq 2*$ (the number of residues). Our strategy for searching for the global energy minimum $F_G \equiv F(\phi_G)$ of $F(\phi)$ involves an iterative process of three phases during each iteration: (I) sampling the landscape, (II) forming the convex global underestimator surface, a parabolic surface under the lowest minima found so far, and (III) finding the minimum on this underestimator surface.

**Phase I**: $k \geq 2n + 1$ local energy minimum conformations $\phi^{(j)}$ are generated in the search region of interest (a minimum of $2n + 1$ conformations are required for construction of the convex underestimator in $n$ dimensions as described below). These conformations may be generated in many ways, but presently they are sampled from a uniform random distribution (over the desired search region) and then relaxed to a local energy minimum state by a Quasi-Newton (QN) continuous minimization technique. This QN approach is well known as a robust general continuous minimization method for this type of problem, and it has the additional benefit of providing an approximation to the Hessian of $F(\phi)$ at each local solution.

**Phase II**: The CGU function $U(\phi)$ is then constructed as a more global surface to "fit" these $k$ local minima by underestimating all of them in the least possible amount (i.e. the L1 norm) by solving the optimization problem:

$$min \sum_{j=1}^{k} \delta_j \qquad (1)$$

where $\delta_j = F(\phi^{(j)}) - U(\phi^{(j)}) \geq 0$ is required for all conformations $j = 1, \ldots, k$. While many choices for the convex underestimating function $U(\phi)$ are possible, our approach is to use a separable quadratic function of the form

$$U(\phi) = c_0 + \sum_{i=1}^{n} (c_i \phi_i + \frac{1}{2} d_i \phi_i^2). \tag{2}$$

This choice is not essential but has many important benefits. First, convexity of $U(\phi)$ is easily guaranteed by simply requiring $d_i \geq 0$ in Eq 2. Second, since $c_i$ and $d_i$ appear only linearly in the constraints of Eq 2, the solution to Eq 1 can be computed by a simple linear programming technique, the complete details of which are given in (26). Third, the minimum energy conformation of $U(\phi)$, denoted $\phi_{PRED}$, is very easily computed by $(\phi_{PRED})_i = -c_i/d_i$. This conformation then serves as a prediction for $\phi_G$. In this way the CGU searches under the landscape of $F(\phi)$ and provides a prediction $\phi_{PRED}$ which can then be used in Phase III.

**Phase III**: Given the predicted structure $\phi_{PRED}$ and the best known local minimum structure computed so far, denoted $\phi_L$, the search region is now localized around $\phi_{PRED}$ while also including $\phi_L$ (the boxed region in Figure 1.2).

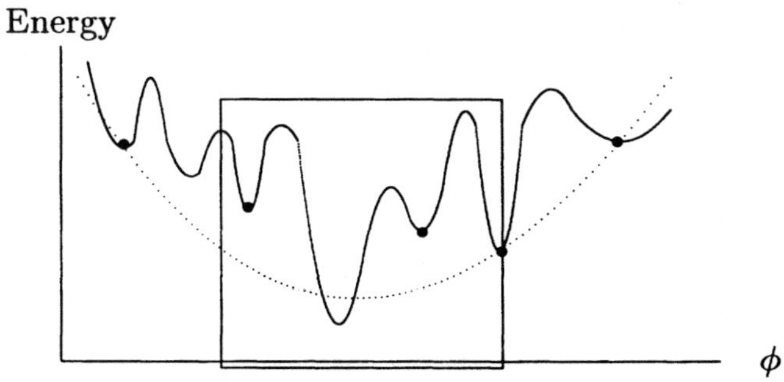


*Figure 1.2.* The Reduced Search Region

Phases I-III are repeated over the continually reduced search regions until $\phi_{PRED} = \phi_L$. That is, when the CGU predicts $\phi_{PRED} = \phi_L$, then the method terminates, and $\phi_L$ is declared the global minimum energy conformation, denoted $\phi_{CGU}$.

This CGU method is clearly not *guaranteed* to find the global minimum $\phi_G$ of the potential function $F(\phi)$. We are aware of no practical off-lattice method that makes such a claim. In fact, we can construct examples for which the CGU method finds $\phi_{CGU} \neq \phi_G$. Nevertheless we believe the CGU method may quite robustly find global minima of reasonable models of proteins, based on growing evidence that protein energy landscapes are "folding funnels" with bumps ((8; 20; 32; 41)). Since the lateral area of an energy landscape at a given depth represents the number of conformations having the same internal free energy, the funnel idea is simply that as folding progresses toward lower energies, the chain's conformational options become increasingly narrowed, ultimately resulting in the one native structure. This is fundamentally a consequence of the fact that proteins are heteropolymers. Such landscapes are ideal for the CGU method since the convex quadratic underestimator closely approximates the funnel, and ignores the bumps, as the algorithm narrows its search region. In these cases the CGU method is very likely to succeed and find $\phi_{CGU} = \phi_G$.

## 3.4. Evidence that the CGU Method Does Work

We have evidence that the CGU method does in fact find the global minimum conformations, $\phi_G$, in the model, for different sequences. Our current tests involve the simple Sun model, which has the form shown in Figure 1.3 where each sidechain $C_s$ is represented by a single united atom (the size for which is dictated by the specific amino acid) and is classified as either hydrophobic or polar. The only degrees of freedom are the pairs of $\phi/\psi$ backbone dihedral angles. The potential function used is of the form

$$F(\phi) = \sum \alpha_i F_i(\phi) \tag{3}$$

where $\sum \alpha_i = 1$, and $\alpha_i > 0$ are energy function parameters determining the relative weight of each energy term $F_i(\phi)$. In our tests, we chose a modified version of the Sun energy function (36) which includes $F_i$ terms representing hydrogen bond formation, hydrophobic attraction, steric repulsion, and $\phi/\psi$ restrictions based on the Ramachandran maps of the twenty individual amino acids, parameterized from the Protein Databank. Since the CGU method requires a differentiable potential function (in order to apply our continuous minimization method), we approximate the discrete Ramachandran map data by a continuous func-

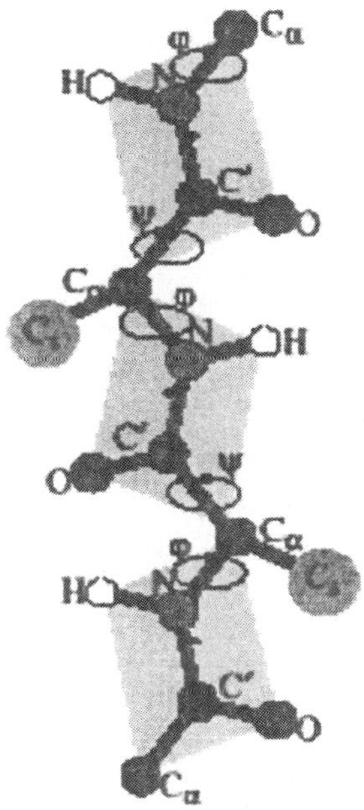

*Figure 1.3.* The Simplified Molecular Model

tion of $\phi$ and $\psi$ (see (11) for complete details). However, the CGU search method is still quite general; it depends neither on the level of detail of the molecular model nor on the functional forms of the energy terms $F_i(\phi)$.

The following evidence suggests that the CGU method is a successful search strategy:

1 Having completed more than 100 trials for each sequence tested (including a 36 residue avian pancreatic polypeptide, a 30 residue zinc finger motif, a 23 residue beta-beta-alpha motif (34), 9 residue oxytocin, and 5 residue met-enkephalin), the CGU method consistently finds the same global solution $\phi_{CGU}$ for each one, from any starting point. The CGU method is not deterministic; that is, the local minimum conformation generation phase (Phase I) is "randomized" so that every trial of the method will sample a different set of starting conformations. Although this is not proof, we take this as good evidence that the solution $\phi_{CGU}$ is also the global one $\phi_G$.

2 The native structures found by the CGU, $\phi_{CGU}$, have lower potential energies than the best ones found by simulated annealing. As the chain length increases, the energy discrepancy, $(F_{SA} - F_{CGU})/|F_{CGU}|$, between the best SA structure, $\phi_{SA}$, and the best CGU structure, $\phi_{CGU}$, typically widens as shown in Figure 1.4. Our tests were based on a widely available simulated annealing code, ASA (13). As a fair test, we provided the SA method with the same number of starting conformations as used in all of the conformation generation phases of the CGU method.

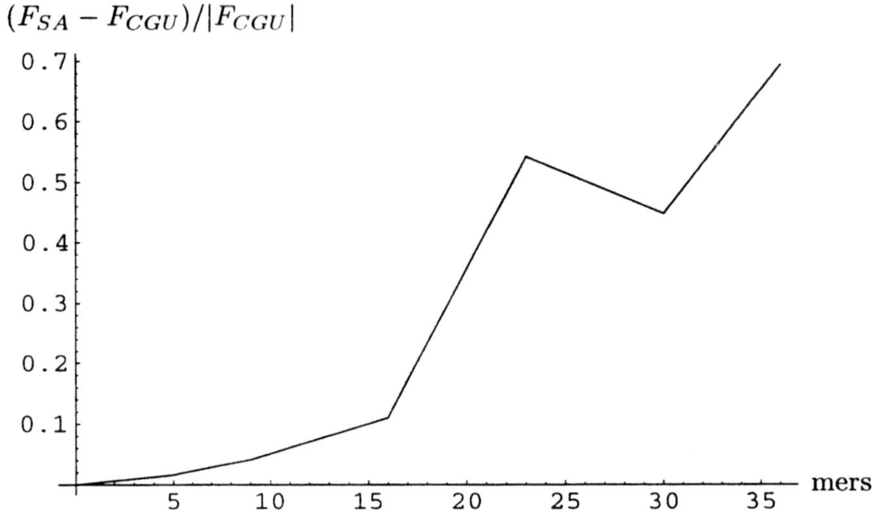


*Figure 1.4.* Energy Gap Between SA and CGU

Our model natives (in the Sun model) always have lower energies than the true natives (PDB). In the context of this work, this is good news because it shows that the flaws are in the energy function, not the search strategy. The model native structures are not correct, but they are protein-like (7): they are compact, with hydrophobic cores and hydrogen-bonded secondary structure (Figure 1.5). Hence, if we can push the search speeds still higher, we could then refine energy functions, or include more degrees of freedom, to improve the model.

3 The computer time required to find the native state is independent of monomer sequence, monomer composition, and native chain fold. The CGU search time doesn't depend on the shape of the landscape, and the CGU method does not get caught in kinetic traps. As shown in (9), the running time of the CGU algorithm

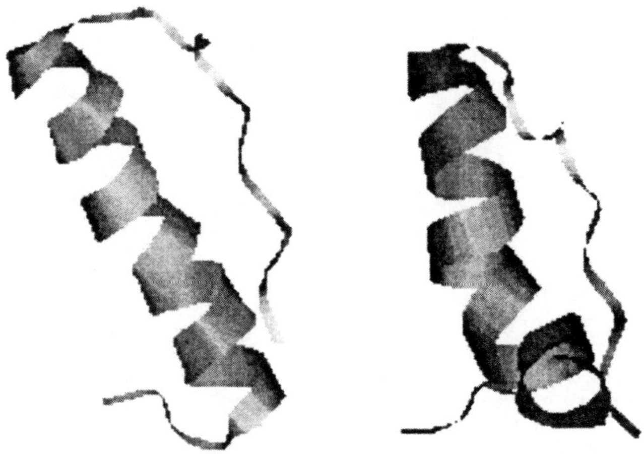

*Figure 1.5.* True Native $\phi_N$ (left) vs Model Native $\phi_{CGU}$ (right) Structures for PPT

is insensitive to the monomer sequence. For example, permutations of a 30 residue sequence (in which the percent hydrophobicity remained fixed across all cases) resulted in various model native structures. These computations also showed that the computation time required by the CGU method is approximately independent of monomer sequence. Furthermore, for a given monomer composition, i.e. sequence of hydrophobic (H) and polar (P) type residues, the CGU algorithm is also time invariant (see (9)) with respect to the specific monomer sequence, even though different sequences fold to very different model native structures.

4 The overall running times are quite reasonable at this stage: on a 32-node Cray T3E, a 30-residue sequence requires one hour and a 50-residue sequence requires nine hours. Unlike other search methods, the CGU search time has a well-understood dependence on the number of degrees of freedom, scaling on average by $n^4$ (26). Figure 1.6 shows a plot of the average computation time $T(r)$ in minutes, on a 32 node Cray T3E, as a function of the chain length $r$ (thus, the number of degrees of freedom $n = 2r$ in our present model) for a number of small peptide sequences.

5 In practice, we observe that the CGU algorithm terminates in no more than 10 major iterations on the small proteins tested. Furthermore, the number of major iterations appears to be independent of the number of degrees of freedom $n$. Therefore, the dependence on $n$ is determined essentially by the energy function (and

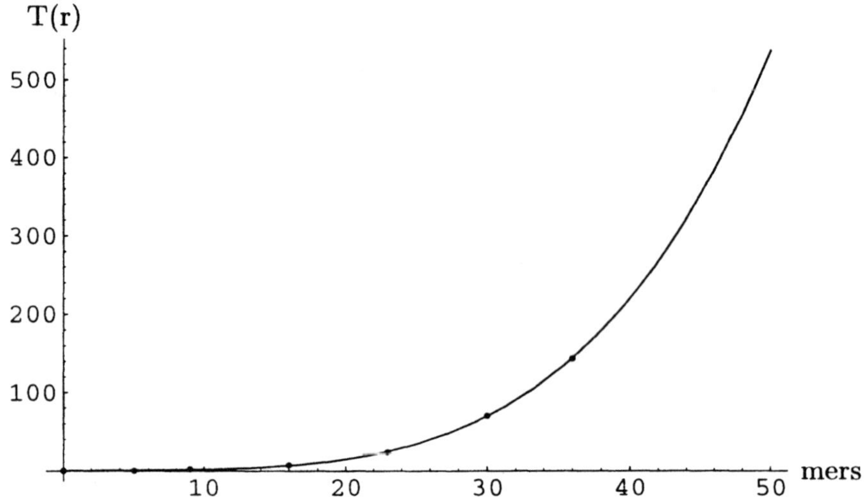


*Figure 1.6.* CGU Solution Time (Minutes) vs Mers

its gradient) evaluation $O(n^3)$, to give an $O(n^4)$ dependence for computation of the global minimum.

6 The CGU method scales linearly with the number of processors. That is, since the local minimization phase (Phase I) can be performed independently and in parallel, and this phase accounts for approximately 99% of the total computation time (when the number of residues $r \leq 100$), a linear increase in the number of processors results in a linear decrease in overall computation time required.

## 3.5. Energy Landscape Information Provided by the CGU Method

The successes we have observed are based, in large part, on the choice of the underestimating function $U(\phi)$ as a separable quadratic in Eq 2. This choice not only makes the solution of Eq 1 simple and efficient (it accounts for less than 1% of the total computation time), but it also provides important insight into the form and features of the energy landscape. This property also appears to be unique to our approach.

For each degree of freedom $\phi_i$, the CGU we have chosen associates a coefficient $d_i > 0$. We have shown in (10) that, based on the Boltzmann distribution law and the form of the CGU given in Eq 2, we can interpret

$(\phi_G)_i$ as the mean value of $\phi_i$ and $k_B T/d_i$ as the variance $\sigma_i^2$. Hence, large $d_i$ indicate a small variance in $\phi_i$ from its global minimum/mean value $(\phi_G)_i$.

Also, since the true energy landscape of $F(\phi)$ can be thought of as a surface above an $n$-dimensional horizontal hyperplane, with each point in the hyperplane representing a conformation $\phi$, the distribution of local minima, provided by repeated iterations of the CGU method, in effect represents the energy surface $F(\phi)$. We have found a simple way to visualize this high dimension landscape. Upon completion of the CGU method, we have available a large set of local minimum conformations (isomers having been removed during each iteration), among which $\phi_G$ is energetically best. We then compute a "landscape" CGU which underestimates, in the minimum L1 norm, this entire set of local conformations in such a way that $\phi_G$ remains the global minimum. This is done by defining this new landscape CGU $U_L(\phi)$ by

$$U_L(\phi) = F_G + \frac{1}{2}(\phi - \phi_G)^T D(\phi - \phi_G) \tag{4}$$

where $D = diag(d_1, d_2, \ldots, d_n)$. Thus the landscape CGU $U_L(\phi)$ depends only on $\phi_G$ and on the set of "landscape coefficients" $d_i$. These are easily obtained by solving the linear program given in Eq 1, but with $U(\phi)$ in Eq 2 replaced by $U_L(\phi)$ given by Eq 4, and with the extra constraint $0 \leq d_i \leq d_{max}$ for $i = 1, \ldots, n$ (see (9) for complete details). In this formulation, $d_{max}$ is a large specified upper bound which prevents the underestimating function from increasing too rapidly as a function of the deviation of any torsion angle $\phi_i$ from its global minimum value $(\phi_G)_i$. Having solved this linear program for $d_i$, if we define

$$\overline{\Delta\phi} = \sqrt{(\phi - \phi_G)^T D(\phi - \phi_G)}, \tag{5}$$

then this Root Mean Square Weighted Deviation (RMSWD) provides a simple and convenient means for plotting the energy difference $U_L(\phi) - F_G$ for any conformation $\phi$. In fact, directly from Eq 4 and Eq 5, we have

$$U_L(\phi) - F_G = \frac{1}{2}(\overline{\Delta\phi})^2. \tag{6}$$

Figure 1.7 shows this two dimensional visualization of the energy landscape for the case of the 36 residue avian pancreatic polypeptide. This figure plots the normalized energy gap for each of the local minima, and shows their relationship to the landscape CGU energy surface. We are aware of no other conformational search strategy that provides this level of energy landscape information for realistic 3D models.

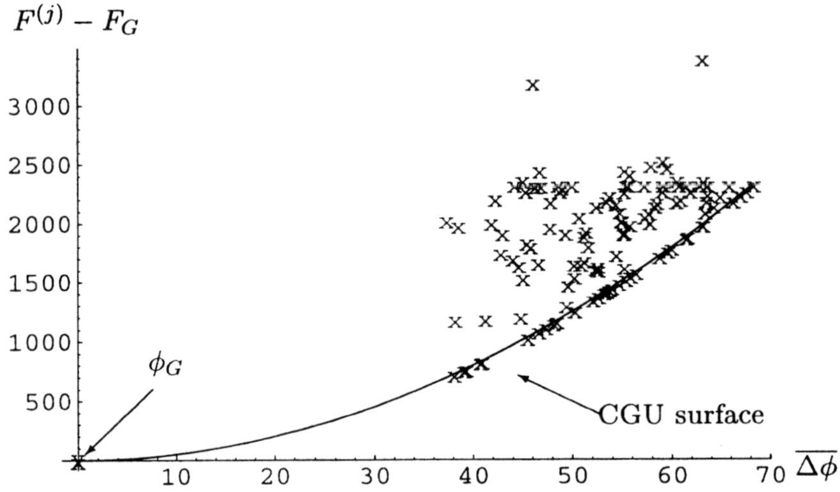


*Figure 1.7.* RMSWD Energy Landscape Projection Obtained from PPT

## 4. Summary

The CGU global optimization search method is a promising alternative to other search stragies including Monte Carlo, Molecular Dynamics, Simulated Annealing, and Genetic Algorithms. Those methods get stuck in kinetic traps. We know the CGU doesn't get stuck in kinetic traps because the search time is independent of the shapes of the landscapes (amino acid sequence and composition). We know the CGU method always finds the same conformation from 100 different starting points, indicating that it finds the unique global minimum for the many different sequences we have tried. We know that the method is much faster than a standard Simulated Annealing algorithm that we have tested: the SA method doesn't find global minima for chains longer than 10 residues, and the performance advantage of the CGU method increases with chain length. And we know that the computer time scales with $n^4$ where $n$ is the number of degrees of freedom.

# References

[1] Beutler, T.C., and K.A. Dill. (1996), A fast comformational search strategy for finding low energy structures of model proteins, *Protein Science* 5:2037-2043.

[2] Bishop, T.C., H. Heller, and K. Schulten (1997), Molecular dynamics on parallel computers: applications for theoretical biophysics, *Toward Teraflop Computing and New Grand Challenge Applications*, 129-138, R.V. Kalia and P. Vashishta (Eds).

[3] Boczko, E.M., and C.L. Brooks (1995), First-principles calculation of the folding free energy of a three-helix bundle protein, *Science* 269:393-396.

[4] Chan, H.S., and K.A. Dill (1993), The protein folding problem, *Physics Today*, February 1993, pp. 24-32.

[5] Covell D.G. (1992), Folding protein alpha-carbon chains into compact forms by Monte Carlo methods, *Proteins: Struct Funct Genet* 14:409-420.

[6] Covell D.G. (1994), Lattice model simulations of polypeptide chain folding, *J Mol Biol* 235:1032-1043.

[7] Dill, K.A. (1990), Dominant forces in protein folding, *Biochemistry* 29(31):7133-7155.

[8] Dill, K.A., and H.S. Chan (1997), From Levinthal to pathways to funnels, *Nature Structural Biology* 4(1):10-19.

[9] Dill, K.A., A.T. Phillips, and J.B. Rosen (1997), Protein structure and energy landscape dependence on sequence using a continuous energy function, *Journal of Computational Biology* 4(3):227-239.

[10] Dill, K.A., A.T. Phillips, and J.B. Rosen (1997), Protein structure prediction and potential energy landscape analysis using continuous

global minimization, *Proceedings of the First Annual International Conference on Computational Molecular Biology* (RECOMB97), pp. 109-117.

[11] Dill, K.A., A.T. Phillips, and J.B. Rosen (1997), Molecular structure prediction by global optimization, *Developments in Global Optimization*, 217-234, I.M. Bomze et al. (Eds).

[12] Hinds D.A., and M. Levitt (1994), Exploring conformational space with a simple lattice model for protein structure, *J Mol Biol* 243:668-682.

[13] Ingber, L. (1989), Very fast simulated re-annealing, *J. Mathl. Comput. Modeling* 12:967-973.

[14] Kirkpatrick, S., C.D. Gelatt, Jr., and M.P. Vecchi, (1983), Optimization by simulated annealing, *Science* 220(4598):671-680.

[15] Kolinski A., and J. Skolnick (1994), Monte Carlo simulations of protein folding. I. Lattice model and interaction scheme, *Proteins: Struct Funct Genet* 18:338-352.

[16] Kostrowicki, J., L. Piela (1991), Diffusion equation method of global minimization: performance for standard test functions, *JOTA* 69:269-284.

[17] Kostrowicki, J., L. Piela, B.J. Cherayil, and H.A. Scheraga (1991), Performance of the diffusion equation method in searches for optimum structures of clusters of Lennard-Jones atoms, *J Phys Chem* 95:4113-4119.

[18] Kostrowicki, J., and H.A. Scheraga (1992), Application of the diffusion equation method for global optimization to oligopeptides, *J Phys Chem* 96:7442-7449.

[19] Kuntz I.D., G.M. Crippen, P.A. Kollman, and D. Kimelman (1976), Calculation of protein tertiary structure, *J Mol Biol* 106:983-994.

[20] Leopold, P.E., M. Montal, and J.N. Onuchic (1992), Protein folding funnels: A kinetic approach to the sequence structure relationship, *Proc Natl Acad Sci USA* 89:8721-8725.

[21] Levitt M., and A. Warshel (1975), Computer simulation of protein folding, Nature 253:694-698.

[22] Li, Z., and H. Scheraga (1987), Monte Carlo minimization approach to the multiple minima problem in protein folding, *Proc Natl Acad Sci USA* 84:6611-6615.

[23] Monge, A., R. Friesner, and B. Honig (1994), An algorithm to generate low-resolution protein tertiary structures from knowledge of secondary structure, *Proc Natl Acad Sci USA* 91:5027-5029.

[24] Nelson, M., W. Humphrey, A. Gursoy, A. Dalke, L. Kalé, R.D. Skeel, and K. Schulten (1997), NAMD – A parallel, object oriented molecular dynamics program, *International Journal of Supercomputing Applications and High Performance Computing* 10:251-268.

[25] O'Toole, E.M., and A.Z. Panagiotopoulos (1992), Monte Carlo simulation of folding transitions of simple model proteins using a chain growth algorithm, *J Chem Phys* 97:8644-8652.

[26] Phillips, A.T., Rosen, J.B., and Walke, V.H. (1995), Molecular structure determination by global optimization, *Dimacs Series in Discrete Mathematics and Theoretical Computer Science* 23:181-198.

[27] Ripoll, D.R., and S.J. Thomas (1990), A parallel Monte Carlo search algorithm for the conformational analysis of proteins, *Proc IEEE/ACM Supercomputing '90*, pp. 94-102.

[28] Shakhnovich, E.I., and A.M. Gutin (1990), Enumeration of all compact conformations of copolymers with random sequence of links, *J Chem Phys* 93:5967-5971.

[29] Shakhnovich, E.I., G. Farztdinov, A.M. Gutin, and M. Karplus (1991), Protein folding bottlenecks: a lattice Monte Carlo simulation, *Phys Rev Lett* 67:1665-1668.

[30] Sippl M., M. Hendlich, and P. Lackner (1992), Assembly of polypeptide and protein backbone conformations from low energy ensembles of short fragments: Development of strategies and construction of models for myoglobin, lysozyme, and thymosin beta 4, *Protein Science* 1:625-640.

[31] Skolnick, J., and A. Kolinski (1990), Simulations of the Folding of a Globular Protein, *Science* 250:1121-1125.

[32] Socci, N.D., and J.N. Onuchic (1994), Folding kinetics of protein-like hetropolymers, *J Chem Phys* 100:1519-1528.

[33] Stillinger, F.H. (1985), Role of potential-energy scaling in the low-temperature relaxation behavior of amorphous materials, *Phys. Rev. B* 32:3134-3141.

[34] Struthers, M.D., R.P. Cheng, and B. Imperiali (1996), Design of a monomeric 23-residue polypeptide with defined tertiary structure, *Science* 271:342-345.

[35] Sun, S. (1993), Reduced representation model of protein structure prediction: statistical potential and genetic algorithms, *Protein Science* 2:762-785.

[36] Sun, S., P.D. Thomas, and K.A. Dill (1995), A simple protein folding algorithm using binary code and secondary structure constraints, *Protein Engineering* 8(8):769-778.

[37] Unger, R., and J. Moult (1993), Genetic algorithms for protein folding simulations, *J Mol Biol* 231:75-81.

[38] Vajda S., M.S. Jafri, O.U. Sezerman, and C. DeLisi (1993), Necessary conditions for avoiding incorrect polypeptide folds in conformational search by energy minimization, *Biopolymers* 33:173-192.

[39] Wallqvist A., M. Ullner, and D.G. Covell (1994), A simplified amino acid potential for use in structure predictions of proteins, *Proteins: Struct Funct Genet* 18:267-280.

[40] Wilson C., and S. Doniach (1989), A computer model to dynamically simulate protein folding – Studies with crambin, *Proteins: Struct Funct Genet* 6:193-209.

[41] Wolynes, P.G., J.N. Onuchic, and D. Thirumalai (1995), Navigating the folding routes, *Science* 267:1619-1620.

# Chapter 2

# BAYESIAN HEURISTIC APPROACH (BHA) AND APPLICATIONS TO DISCRETE OPTIMIZATION

Jonas Mockus

*Institute of Mathematics and Informatics,*
*Akademijos 4, Vilnius, Lithuania*
jonas@optimum.mii.lt

**Abstract** Industrial size discrete optimization problems are most often solved using "heuristics" (expert opinions defining how to solve a family of problems). The paper is concerned about ways to speed up the search in a discrete optimization problem by combining several heuristics involving randomization. Using expert knowledge an *a priori* distribution on a set of heuristic decision rules is defined and is continuously updated while solving a particular problem. This approach (BHA or Bayesian Heuristic Approach) is different from the traditional Bayesian Approach (BA) where the *a priori* distribution is defined on a set of functions to be minimized. The paper focuses on the main objective of BHA that is improving any given heuristic by "mixing" it with other decision rules. In addition to providing almost sure convergence such mixed decision rules often outperform (in terms of speed) even the best heuristics as judged by the three considered examples. However, the final results of BHA depend on the quality of the specific heuristic. That means the BHA should be regarded as a tool for enhancing the best heuristics but not for replacing them.

This way the formal Bayesian Approach (BA) is extended to a semi-formal Bayesian Heuristic Approach (BHA) where heuristics may be included more flexibly. The paper is concluded by an example of Dynamic Visualization Approach (DVA). The goal of DVA is to exploit heuristics directly, bypassing any formal mathematical framework.



*A. Migdalas et al. (eds.), From Local to Global Optimization,* 19–48.

The formal description of BHA and its application is published in a number of books and papers. In this paper the BHA is introduced and explained using three examples as illustrations, namely knapsack, flow-shop, and batch scheduling. This helps to show when and why BHA works more efficiently as compared with the traditional optimization methods. The global optimization software is described in short as a tool needed implementing BHA.

The purpose of the paper is to inform the authors inventing and applying various heuristics about the possibilities and limitations of BHA hoping that they will improve their heuristics using this powerful tool.

**Keywords:** Bayesian heuristics, discrete optimization,dynamic visualization approach.

## 1. Introduction

Many of discrete optimization problems are NP-complete, see [16]. This means in practice that one needs exponential time $T$ to obtain exact solution, for example, $T = C\ 2^n$, where $C > 0$ and $n$ is a number of variables. Therefore approximate algorithms are used if $n$ is large.

A way to get approximate solution of discrete optimization problem is by random generation of large number of decisions and selecting the best one. The randomized techniques converges to the exact solution with probability one if the probability to obtain any feasible decision is positive. This is the advantage of Randomization Approach (RA). The disadvantage is that RA converges very slowly, as usual.

Another way of approximate solution is by application of heuristics including the expert knowledge. This way one gets some "heuristic" solution. This is Heuristic Approach (HA). Usually HA do not guarantee convergence, because the "heuristic" solution can be far away from the exact one.

The convergence of HA can be provided and the RA may be speeded-up by selecting randomization probabilities as a function of the heuristics. This way the Randomized Heuristics Approach (RHA) is defined. There are numerous cases of successful applications of various RHA algorithms and many excellent publications regarding this approach. We shall mention just a few well known examples of RHA such as simulated annealing (see [4, 36]), genetic algorithms (see [12, 1]), RHA (see [10, 11]), GRASP (see [9]).

However, there remains a problem how to select the best randomization function. There are reasons to approach this problem by considering a "mixture" of different randomization functions. By the mixture of functions we denote a set of functions with attached probabilities. The mixture of randomized functions means that decisions are made using a "double-lottery". The first one is a "ticket-lottery" where we win just a "ticket" for a "decision-lottery". The ticket-lottery defines probabilities of different decision-lotteries. The decision-lottery defines probabilities of different decisions. The number of the repetitions $K$ [1] is fixed depending on desired accuracy and the best feasible decision is selected as the final result.

The "quality" of the mixture is estimated by the final results. The problem is how to define the best mixture. By the best mixture we mean the probabilities of different randomization functions such that optimize the mixture quality. The dependence of the mixture quality on randomization parameters is a stochastic and usually multi-modal function. A family of methods designed for such functions is called the Bayesian Approach BA, see [21]. Application of the BA to mixture optimization means that one defines an a priori distribution on a set of functions determining the mixture quality. The Bayesian algorithm (see [21]) performs the next observation (calculation of mixture quality function) minimizing the expected deviation. Therefore this technique is called the Bayesian Heuristic Approach (BHA), see [23] .

The formal description of BHA and its applications is given in two books [21, 23] and a number of papers [24, 22, 19, 25, 27, 30, 28, 18, 26]. In this paper the basic concepts of BHA are introduced and explained using three following examples as illustrations. This helps to clarify the reasons when and why BHA works more efficiently as compared with the traditional optimization methods.

The simple knapsack problem illustrates the main steps of BHA algorithm using the simplest greedy heuristic. The flow-shop problem shows the improvements achieved while "improving"by BHA three different heuristics: the longest -time, the longest-remaining-time and the Gupta (derived while trying to extend the solution of the Johnson's problem to multi-machine case). The batch scheduling problem shows the results of BHA solving an industrial problem by improving the commonly used Material Requirement Planning (MRP) heuristics.

[1] The length of the sequence of ticket-lotteries.

In the last section an example of application of Dynamic Visualization Approach (DVA) for informal interactive optimization is shortly discussed. Such optimization is needed if an optimization problem is not well defined .

## 2. Knapsack Problem

A convenient way to explain the BHA is by applying it to a simple NP-complete problem. A good example[2] is a knapsack problem. The knapsack problem is to maximize the total value of a collection of objects when the total weight $g$ of those objects is limited. We denote value of the object $j$ by $c_j$ and its weight by $g_j$.

$$\max_y \sum_{j=1}^{n} c_j y_j \tag{1}$$

$$\sum_{j=1}^{n} g_j y_j \leq g \tag{2}$$

$$\tag{3}$$

Here the objective depends on $n$ Boolean variables

$$y_j = \begin{cases} 1, & \text{if object } j \text{ is in the collection} \\ 0, & \text{otherwise,} \end{cases} \tag{4}$$

### 2.1. Exact Algorithms

The simplest exact algorithm is exhaustive search of all the vectors $y = (y_1, ..., y_n)$. The exhaustive search needs
$T = C\ 2^n$ time. Here the constant $C$ does not depend on the problem. The search efficiency of exact algorithms is improved by Branch&Bound (B&B) techniques where the time-constant $C_\omega \leq C$ depends on the specific problem $\omega$. The exponential time is the price of guarantee which one has to pay if accuracy is to be provided for all the knapsack problems. If this price is to high then the alternative is approximate algorithms which provide reasonably good solutions for many problems but not for all.

[2]This example is good just for illustration how BHA works but not for showing the advantages of this approach. The BHA works more efficiently in scheduling problems, for example in flow-shop and batch scheduling problems described in this paper.

## 2.2. Approximate Algorithms

The simplest approximate algorithm is Monte-Carlo where the decision $m$ is taken with probability

$$r_m = 1/2^n. \tag{5}$$

Here the decision $m$ means the vector $y^m = (y_1^m, ..., y_n^m)$. The algorithm stops after $K$ repetitions meaning that $1 \leq m \leq K$. The most valuable feasible collection after $K$ repetitions is chosen as the final result

$$C(K) = \min_{1 \leq m \leq K} \sum_{j=1}^{n} c_j y_j^m, \tag{6}$$

$$\sum_{j=1}^{n} g_j y_j^m \leq g$$

The final result converges to the exact solution if the number of repetitions $K \to \infty$, meaning that $C(K) \to C^*$ with probability one, where $C^*$ is the value of the optimal collection. However the convergence is very slow.

**Greedy Heuristic Algorithms.** Define the heuristics $h_i$ as the specific value of object $i$

$$h_i = \frac{c_i}{g_i} \tag{7}$$

This is a well known and widely used greedy heuristics (see [13]). At each step the greedy heuristics includes the feasible[3] object with highest heuristic $h_i$. The algorithm stops when there remain no feasible objects. The greedy heuristic algorithm is very fast. We need only one "repetition" consisting of at most $n$ iterations[4]. Thus the total time $T = Cn^2$. However it may stuck in some non-optimal decision, especially when all the heuristics $h_i$ are almost equal.

**Permutation Heuristic Algorithms.** The greedy heuristics involve the expert knowledge about the problem by defining appropriate

[3]Satisfying inequality (2).

[4]The number of iterations $R$ is defined as a number of calculations of objective function (1) including constraints ( 2)

function $h_i$. There are cases when the expert knowledge is included by fixing some initial decision determining a feasible vector $y^0 = (y_1^0, ..., y_n^0)$. In such cases the permutation heuristics are convenient way of improving the initial decision. For example, one may transform an initial feasible collection $y^0$ by replacing some objects of this collection. Denote by $y^i$, $i = 1, ..., k$ the transformed collection and by $C^i$ its value. In such a case the difference $h_i = C^i - C^0$ may be considered as a heuristic. Here $C^0$ denotes the value of the initial collection. Obviously, the initial collection is updated at each step.

Using permutations the difference of values of genuine objective functions does not define the optimal objective function value in general. For example, an algorithm improving objective at each step may stuck in some non-optimal decision. Therefore the difference $h_i$ should be regarded just as heuristics. The suitable number $k \geq 2$ of permutations considered at each step is defined empirically.

**Randomized Heuristic Algorithms.** Randomization is a simple way to force the heuristic algorithm out of non-optimal decisions. The traditional way of randomization greedy heuristics is by taking feasible decision $i$ with probability

$$r_i = \frac{h_i}{\sum_{j=1}^{k} h_j}. \tag{8}$$

To obtain good approximation large number of repetitions $K$ is needed. This algorithm is better than greedy one, because it converges with probability one if $K \to \infty$. It is better than Monte-Carlo, too, since it includes an expert knowledge by making the decision probabilities dependent on heuristics.

The popular way of randomizing the permutation heuristics is simulated annealing (see [31]), where $k = 2$ and randomization function is exponential

$$r_i = \begin{cases} e^{\frac{h_i}{x}}, & \text{if } h_i < 0 \\ 1, & \text{otherwise,} \end{cases} \tag{9}$$

where $x$ is the "initial temperature"[5].

It is not clear why consider only the linear randomization (8) for greedy heuristics and only the exponential one (9) for permutation heuristics?

[5]Usually the initial temperature is multiplied by a cooling function depending on the repetition number

The following expression define a set of randomization functions similar to corresponding terms of Taylor series

$$r_i = \frac{h_i^l}{\sum_j h_j^l}, \; l = 0, 1, 2, ..., L. \tag{10}$$

and

$$r_i^{\infty} = \begin{cases} 1, & \text{if } h_i = \max_j h_j \\ 0, & \text{otherwise,} \end{cases} \tag{11}$$

Here $l = 0$ denotes Monte-Carlo component, $l = 1$ and $l = 2$ defines linear and quadratic components of randomization. The index $\infty$ denotes the pure heuristics with no randomization.

Applying randomization (10) to permutation heuristics one should use normalization. For example

$$h_i = \frac{d_i - d_{min}}{d_{max} - d_{min}} + a. \tag{12}$$

where $d_i = C^i - C^0$ and

$$d_{min} = \min_{1 \le j \le k} d_j, \; d_{max} = \max_{0 \le j \le k} d_j, \; a > 0. \tag{13}$$

One may define the best randomization function empirically by considering each randomization function separately and estimating its quality by the best decision obtained applying this function $K$ times. That is a traditional way. The main disadvantage is that the best average results often are obtained using a "mixture" of different randomization functions not just the "pure" ones. The mixture means using a randomization function $r_i^l$ with some probability $x(l)$

Therefore one should regard the problem in a more general set-up defining the best mixture of different randomization functions. In such a case the methods of continuous global stochastic optimization are needed since one cannot obtain the optimal solution just by comparing a finite number of randomization functions.

Denote the probability distribution as $x = (x(l), \; l = 0, 1, ...., L, \infty)$. Denote by $f_K(x)$ the best decisions obtained using $K$ times a mixture $x$. This way we extend the set of possible decisions $x$ from the discrete set $x(l) \in \{0, 1\}, \; l = 1, ..., L, \infty$ to a continuous one. The difficulty is that $f_K(x)$ is stochastic and, usually, multi-modal function. A natural way to consider such problems is by regarding a functions $f_K(x)$ as a sample of a stochastic function (defined by some a priori distribution) and obtaining the next observation by minimizing the expected deviation from the exact solution. This technique is called Bayesian Heuristic Approach (BHA), see [23].

# 3. Flow-Shop Problem

The flow-shop problem is a simple example of large and important family of scheduling problems. We denote by $J$ and $S$ the set of jobs and machines. Denote by $\tau_{j,s}$ the duration of operation $(j, s)$, where $j \in J$ denotes a job and $s \in S$ denotes a machine.

Suppose that the sequence of machines $s$ is fixed for each job $j$. One machine can do only one job at a time. Several machines cannot do the same job at the same moment. The decision $d_i(j) \in D_i$ means the start of a job $j \in J_i$ at stage $i$. We define the set of feasible decisions $D_i$ as the set $J_i$ of jobs available at the stage $i$ conforming to the flow-shop rules.

The objective function is the make-span $v$. Denote by $T_j(d)$ the time when we complete job $j$ (including the gaps between operations) using the decision sequence $d$. Then the make-span for $d$ is

$$v(d) = \max_{j \in J} T_j(d). \tag{14}$$

## 3.1. Algorithm

**Permutation Schedule.** We can see that the number of feasible decisions for the flow-shop can be very large. The number can be reduced by considering only smaller subset of schedules, the so-called permutation schedules.

The permutation schedule[6] is a schedule with the same job order on all machines. Such a schedule can be defined by fixing job indices 1,2,...,n. We assume the first operation to be on the first machine, the second on the second, and so on. The schedule is transformed by a single permutation of job indices. It is generally assumed that permutation

[6]The term "permutation schedule" here defines a specific subset of schedules of the flow-shop problem. This term is not directly related to permutation heuristics. In fact, namely the greedy heuristics are considered in this paper while solving the flow-shop problems.

schedules approach the optimal decision sufficiently closely and are easier to implement (see [2]).

Denote

$$\tau_j = \sum_{s=1}^{|S|} \tau_{j,s},$$

where $\mid S \mid$ stands for the number of machines. We define $\tau_j$ as the length of the job $j$.

**Heuristics.** Consider, for comparison, the following three heuristics $h_j$ defining the priorities of jobs $j$

- The Longer-Time (LT) heuristics

$$h_j = \tau_j. \tag{15}$$

- The Longer-Remaining-Time (LRT) heuristics $h(j)$ that depends on the remaining time $\tau_i(j)$ of the job $j$ at the decision stage $i$

$$h(j) = \tau_i(j). \tag{16}$$

Here

$$\tau_i(j) = \sum_{s \in S(i,j)} \tau_{j,s},$$

where $S(i,j)$ is the set of machines to be used after the stage $i$ to complete the job $j$.

- The Gupta heuristics

$$h_j = \frac{t_j - A_i}{A^i - A_i} + a. \tag{17}$$

where

$$A_i = \min_{j \in J_i} t_j, \ A^i = \max_{j \in J_i} t_j, \ a > 0.$$

depends on the Gupta priorities (see [2]).

$$t_j = \frac{e_j}{\min_{1 \le s \le |S-1|}(\tau_{j,s} + \tau_{j,s+1})}.$$

Here

$$e_j = \begin{cases} +1, & \text{if } \tau_{j,1} < \tau_{j,|S|} \\ -1, & \text{otherwise.} \end{cases}$$

The LT, LRT, and Gupta heuristics (17) define job preferences by different priority rules. The priority rule LT (15) prefers longer job. The LRT prefers jobs with longer remaining times. The Gupta priority rule (17) in two-machine case is similar to the well known Johnson's rule using permutation schedule.

In all the three cases a randomized decision function $r_i$ is defined and a stochastic function $f_K(x)$ is optimized. This function defines the shortest make-span (see (14)) found as a result of $K$ repetitions.

## 3.2. Results

Table I is for comparison of the BHA algorithm after 100 iterations using simple LT heuristics (15) and different randomization procedures with the exact Branch-and-Bounds (B&B) algorithm "truncated" after approximately the same time.

Assume that $J = S = O = 10$, where $J$, $S$, $O$ are the number of jobs, machines, and operations, respectively. Lengths and sequences of operations are generated as random numbers uniformly distributed from 0 to 99. The expectations and standard deviations are estimated by repeating optimization of a randomly generated problem 40 times In Table I the

*Table 2.1.* The results of Bayesian methods using Longer-Job heuristics (15) in hundreds of time units

| $R = 100,\ K = 1,\ J = 10,\ S = 10$, and $O = 10$ | | | | | |
|---|---|---|---|---|---|
| Randomization | $f_B$ | $d_B$ | $x_0$ | $x_1$ | $x_2$ |
| Delta | 6.183 | 0.133 | 0.283 | 0.451 | 0.266 |
| Taylor 3 | 6.173 | 0.083 | 0.304 | 0.276 | 0.420 |
| CPLEX | 12.234 | 0.00 | — | — | — |

symbol $f_B$ denotes a mean, and $d_B$ denotes a standard deviation of make-span. The "Delta" denotes randomization including terms $l = 0, 1, 2, \infty$ (see expressions (10) and (11)), the "Taylor 3" denotes randomization (10) where the number of terms is $L = 2$, and "CPLEX" denotes the results of the well known general discrete optimization software after 2000 iterations (one CPLEX iteration is comparable to a Bayesian observation). The bad results of CPLEX show that the standard Mixed Integer Linear Programming (MILP) technique is not efficient in solving this specific problem of discrete optimization. It is not yet clear how much one improve the results using specifically tailored B&B.

Table II shows how the results of BHA after 100 iterations depends on heuristics used. Three heuristics were considered: LT, LRT and Gupta. The results of deterministic versions ("pure" heuristics) are also included, for comparison. The number of jobs $J = 7$, the number of machines $S = 25$, and the number of operations $O = 25$.

We define lengths and sequences of operations generating random numbers uniformly distributed from 0 to 99, and estimate expectations and standard deviations repeating optimization of 10 randomly generated problems 40 times. No comparisons with CPLEX are made in Table II, since the example is too large. In Table II the notation "Taylor 4" means

*Table 2.2.* The results of Bayesian methods using Gupta (17), Longer-Time (15), and LRT (16) heuristics

| $R = 100$, $K = 1$, $J = 7$, $S = 25$, and $O = 25$ | | | | | | | |
|---|---|---|---|---|---|---|---|
| Randomization | $f_B$ | $d_B$ | $T$ | $x_0$ | $x_1$ | $x_2$ | $x_3$ |
| Gupta heuristics (see expression (17)) | | | | | | | |
| Delta | 1861.38 | 16.89 | — | 0.383 | 0.349 | 0.267 | — |
| Taylor 3 | 1857.38 | 11.90 | — | 0.008 | 0.365 | 0.625 | — |
| Taylor 4 | 1857.02 | 16.27 | — | 0.252 | 0.271 | 0.224 | 0.253 |
| Legendre | 1870.21 | 15.16 | — | 0.386 | 0.377 | 0.238 | — |
| Deterministic | 2092 | — | — | — | — | — | — |
| LT heuristics (see expression (15)) | | | | | | | |
| Delta | 1876.81 | 20.68 | — | 0.420 | 0.341 | 0.239 | — |
| Taylor 3 | 1868.67 | 18.13 | 149 | 0.387 | 0.308 | 0.305 | — |
| Taylor 4 | 1874.82 | 16.52 | 199 | 0.277 | 0.263 | 0.231 | 0.229 |
| Legendre | 1856.85 | 17.48 | 149 | 0.311 | 0.326 | 0.363 | — |
| Deterministic | 2261 | — | — | — | — | — | — |
| LRTime heuristics (see expression (16)) | | | | | | | |
| Delta | 2113.50 | 47.59 | — | 0.041 | 0.023 | 0.934 | — |

randomization (10) with the number $L = 3$ , "Deterministic" denotes a choice of the best heuristic, without any randomization, and $T$ denotes the CPU time.

We see that the Bayesian adaptation of heuristics using BHA significantly improved the results in all the cases. The Gupta heuristic (17) provides better results compared to the LT and LRT heuristics. The results of Table I and Table II in general correspond to the theoretical predictions.

## 3.3. Learning

Usually we compare the Bayesian and other methods neglecting the "learning" potential of Bayesian methods. For example, Figure 2.1 estimates the learning effect (see [23]) by showing the density of optimal values of parameters $x_1$ and $x_2$. Those values correspond to 100 different randomly generated flow-shop problems. We see that the maximal density of optimal values of $x_1$ and$x_2$ exceeds the average density several times. According to the results of [23] it means that the optimal parameters of Bayesian methods are rather robust and thus could be applied to other related problems.

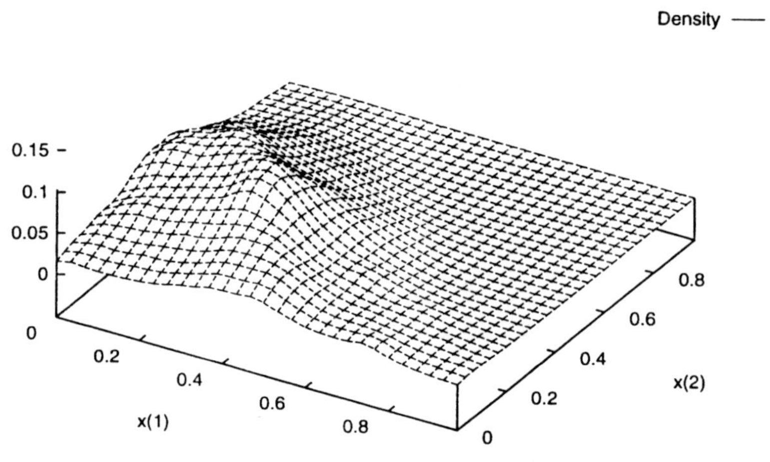


*Figure 2.1.* Density of optimal parameters in solving different randomly generated flow-shop problems

# 4. Batch Process Scheduling by MRP Heuristics

## 4.1. Why Consider Batch Scheduling?

Batch scheduling problem is a good example of real-life mixed-discrete-continuous optimization problem. A wide range of products in the chemical processing industry are produced using the batch mode of production. This mode of production has long been the accepted procedure for the manufacture of many types of chemicals (specialty chemicals, pharmaceuticals, polymers, biochemical, and foods), particularly those which are produced in small quantities and for which the production processes or the demand pattern are likely to change.

The most important feature of batch processes is their flexibility in processing multiple products by accommodating diverse operating conditions associated with each product. Therefore, in spite of the traditional drive towards continuous production, the batch mode remains the only alternative for a number of sectors of the processing industry. As a result, there has been an increasing interest in the development of procedures for scheduling batch process operations.

## 4.2. Heuristic Priority Rules

We next discuss the application of the well known Material Requirement Planning Heuristic (MRP), see [15, 23]. MRP is an inventory management and production planning technique which, given a delivery schedule for a final products, determines the initiating times for all raw materials orders, for the production runs needed to prepare all required intermediate products, as well as the starting times for the production of the final product itself. Given a delivery time for a product shipment, each branch of the product processing tree is traced from the product in question and each component requirement is calculated. If the inventory is inadequate, then a production order is issued for that component. The tracing of each branch of the processing tree continues until all raw materials requirements have either been met or ordered.

For scheduling problems relevant to the chemical industry (see [23, 29, 30, 32]) this heuristic is extended to handle unit and task assignment, batch size determination, and other features. For the purpose of this work three simple and natural heuristic rules are employed:

- *Product selection rule.* The production run of each product in a batch and continuous processing system has a due date. One begin with a product that has the earliest due date. The rationale for this is that later due dates are less certain. When the production of this product is scheduled one recursively schedules the production of intermediates required for producing it.

- *Equipment item selection rule.* To schedule the production of a given product or an intermediate one has to decide a task and to select an equipment unit to process this task. When there are several tasks producing the same product, the quantity of a product to be produced is assigned randomly. An equipment item is selected that is free during the time interval, closest to the deadline.

- *Task start selection rule.* Usually this time interval is longer than the task processing time. Thus one starts a given task so that it ends exactly at the end of this interval.

Using MRP heuristics BHA algorithm is represented by the following steps:

Step 1. Fix parameters $x$ using the global Bayesian method [21];

Step 2. Evaluate the schedule for these parameters;

Step 3. If the number of iterations is greater than the maximal value, then stop;

Step 4. Go to step 1.

We comment the first two steps:

Step 1. The parameter vector $x = (x_0, x_1, x_2)$ is controlled by the global Bayesian method [21]. Note, that any method generating a positive vector $x$ provides the convergence with probability one. Here the BA is applied for both the theoretical and the practical reasons.

Theoretically BHA is interesting because it unites the Bayesian approach and Heuristic programming by defining an a priori distribution on a family of heuristics.

The practical reason is that Bayesian methods are designed for the global stochastic optimization so we may expect good results optimizing a multi-modal and stochastic auxiliary function $f(x)$;

Step 2. The schedule is formed by using randomized MRP heuristics at a fixed parameter vector $x$. The value of this schedule is the profit (for example, the value of products). It is possible that the generated schedule is infeasible, i.e., the capacity for some states is exceeded or some states contain negative amount of material. Then we penalize it;

The results are given in Table III which includes the results of two BHA algorithms: one using the MRP heuristics with polynomial randomization, and the other one using the randomization of simulated annealing type, see [23].

## 5. Computing Results

In Table III various BHA algorithms are compared with the truncated solutions of the MILP uniform discrete-time model using B&B [37]. The results are reported for different test examples, both the batch and the continuous ones.

*Table 2.3.* BHA and B&B comparison

| Problem | Alg. | Profit | CPU time | Numb. Boolean var. | Numb. cont. var. | Numb. constr. |
|---|---|---|---|---|---|---|
| BATCH1 | B&B | 2744.5 | 584.9 | 305 | 860 | 1073 |
| | | same | 1217.5 | same | same | same |
| | | same | 1901.2 | same | same | same |
| | BHA-1 | 2696 | 0.03 | 18 | | |
| | | 2939 | 0.04 | 14 | | |
| | BHA-3 | 3007 | 6.9 | 384 | 376 | 744 |
| | | 3079 | 85.4 | same | same | same |
| | | 3079 | 1396.5 | same | same | same |
| uni-mod | B&B | 3230 | 6 | | | |
| | | 3230 | 11.21 | 36 | 101 | 88 |
| | BHA-1 | 3230 | 0.03 | 24 | | |
| | BHA-2 | 3230 | 0.9 | | | |
| | BHA-3 | 3230 | 19.34 | 384 | 376 | 744 |
| BATCH3 | B&B | 105756 | 8.6 | | | |
| | BHA-2 | 105702 | 1.5 | | | |
| BATCH4 | B&B | 60224 | 944.8 | 488 | 1553 | 1823 |
| | | same | 1919.7 | same | same | same |
| | | same | 2944.5 | same | same | same |
| | BHA-3 | 60284 | 103.0 | 2240 | 1776 | 3976 |
| | | 60306.2 | 163.2 | 2240 | same | same |
| | | 60311.7 | 1276.8 | 2240 | same | same |
| | BHA-1 | 60459.2 | 21.28 | 54 | | |
| | | 60468.3 | 63.36 | 54 | | |
| uni-mod | B&B | 60533.5 | 1.7 | | | |
| | | 60533.5 | 1.86 | 56 | 176 | 174 |
| | BHA-1 | 60533.5 | 0.19 | 50 | | |
| | BHA-2 | 60471.6 | 3.6 | | | |
| | BHA-3 | 60474.7 | 309.7 | 2240 | 1776 | 3976 |
| BATCH5 | B&B | 1400 | 66 | | | |
| | BHA-2 | 1400 | 30 | | | |
| BATCH6 | B&B | 4724 | 0.6 | | | |
| | BHA-2 | 4724 | 0.05 | | | |
| BATCH7 | B&B | 8122 | 188.5 | | | |
| | BHA-2 | 8122 | 1.15 | | | |
| BATCH8 | B&B | 1205 | * | | | |
| | BHA-2 | 11996 | 13.26 | | | |
| CONT1 | B&B | 20879 | ** | | | |
| | BHA-2 | 20992 | 13.6 | | | |
| CONT2 | B&B | 22800 | 824.6 | | | |
| | BHA-2 | 23346 | 51.6 | | | |

For some examples in Table III there are several lines. The successive lines relate to stopping *B&B* enumeration after 10,000, 20,000, and 30,000 nodes, respectively. A different number of nodes is reflected in the table as different solution time. It is surprising to see that the "optimal" *B&B* profit remains the same in all the three lines. The "optimal" BHA profit increases with the number of observations reflected as the solution time. Denote by BHA-1 and BHA-2 two versions of BHA using MRP heuristics. Denote by BHA-3 the BHA version using Simulated Annealing (SA) and the penalty function as heuristics, see [23].

The two lines marked 'uni-mod" correspond to an unmodified version of the batch scheduling problem where the uniform discrete-time interval length (1 for both examples) is comparable to the scheduling horizon.

The computational experiments were performed using a HP 9000/755 work-station. For some batch processes (BATCH3, BATCH4, BATCH8), BHA gives only suboptimal solutions. For both the continuous processes CONT1 and CONT2, BHA gives better results.

The proper schedule can lead to significant improvement in processing efficiency. On the other hand, obtaining the optimal schedule can require a large amount of computational time. Thus one needs a "good" engineering judgment to find a compromise between the cost and the benefit of schedule improvement. The Bayesian Heuristic Approach, combined with the Non-Uniform Discrete-time Model is a convenient tool for achieving that compromise.

The Bayesian Heuristic Approach is stochastic by definition. Therefore the application of this approach to optimize batch processes with random parameters is straightforward. It is well known that an a priori distribution is a convenient way to represent uncertainties in the model , see [5]. This way BHA includes uncertain parameters, too. The present formulation may be easily expanded to modeling continuous processes, a temporary unavailability of equipment, and changeovers.

# 6. Software for Global Optimization

## 6.1. Background

The global optimization software was developed considering the results of international "competition" of different algorithms of global optimization (see [6]). Some experience in real life optimization problems was also used selecting the set of optimization algorithms. The set of algorithms of global optimization includes

- four versions of the Bayesian search,
- a version of clustering,
- a version of uniform deterministic grid,
- a version of pure Monte Carlo search.

Usually it is reasonable to start optimization by a global method and to finish it by some local method. An exception is two global algorithms: the Torn version of clustering [35], and the Zilinskas version of the Bayesian technique [35]. Both these algorithms contain some simple algorithms of local search. This means that the local search is not necessary for those two methods, but it may be useful.

There are three local optimization methods:

- a method of variable metrics type with Lagrangian multipliers and penalty functions for constrained optimization of smooth functions (see [33]),
- a method of simplex type of Nelder and Mead with penalty functions for constrained optimization of non-differentiable functions.
- a method of stochastic approximation type for "noisy" functions (see [21]).

Each subroutine represents a global or a local method. The choice of method has to follow the idea that the computational complexity of the method should roughly correspond to that of the objective function:

- For computationally "expensive" functions the Bayesian methods could be recommended. Those methods need a large amount of auxiliary calculations to make each observation more efficient.
- For "cheap" functions the simple grid methods, like Monte Carlo or a uniform deterministic grid (see [34]), can be better. Here observations are not so efficient, but auxiliary calculations are negligible. This explains a relative efficiency of simple methods when optimizing simple functions.

- The clustering techniques (see [35]) may be the best choice, if we expect the number of local minima to be small.
- A relatively simple Bayesian technique is available [35] for global optimization of one-dimensional functions;
- There are optimization problems where objective functions can be roughly represented as a sum of components depending on different variables. Here the Bayesian method of line search along the coordinates usually shows very good results. This method globally optimizes one variable at a time by one-dimensional Bayesian search. The difference of this method from other methods of global optimization is that it depends on the starting point. Thus a deviation from the global minimum can be made as small as desired by applying a multi-start procedure with different uniformly distributed starting points.

All the global methods optimize in a rectangular region. Therefore we represent the linear and non-linear inequality constraints as some penalty functions. The same applies to the local method of stochastic approximation type. In local methods of simplex and variable metrics type the linear and the non-linear constraints can be defined directly. This may be done by constraint subroutines, supplied by the user in addition to the objective function.

The global optimization software is in four versions:

- portable Fortran Library,
- interactive software for Microsoft C&Fortran compiler and DOS operating system,
- interactive software for Turbo C compiler and DOS operating system,
- interactive software for C++ compiler and UNIX operating system and X-Window system.

## 6.2. Software Availability

The software is published in [23]. The interactive UNIX C++ software (LINUX 1.2.8. version) and the library of portable Fortran subroutines (LINUX 1.2.8. and DOS versions), are in the enclosed disk of the book [23] and at the address:

ftp://pit.ktu.lt/pub/optimum/gmc.tgz gmfl.tgz gmf.arj

The UNIX C++ interactive version is in the archive file 'gmc.tgz', the library of Fortran subroutines are in 'gmfl.tgz'or 'gmf.arj'.

Contact e-mail addresses:

*jonas@optimum.mii.lt jonas.mockus@ktl.mii.lt*

# 7. Dynamic Visualization in Modeling and Optimization of Ill Defined Problems

## 7.1. Introduction

The visualization is considered as a decision optimization tool in problems where the model and/or the objectives are not well defined, see [8, 23]. The problem of dynamic representation of observations in the form of averages over regions in space and time is regarded and is exemplified by epidemiological data. The objective is to detect spatial-temporal patterns such that can suggest the most efficient ways of prevention and control. We describe the software and hardware. A videotape which displays the results is available (e-mail: audris@bell-labs.com).

An application of some elements of visualization in those problems was considered by [20]. A number of visualization techniques for well-defined optimization problems are considered in [14].

In real-life applications we often encounter ill-defined problems. In this paper a problem is regarded as ill-defined if it cannot be accurately described in terms of finding the optimum of a function (as opposed to the conventional meaning of an ill-defined problem which originates in the area of integral equations). In such cases we expect to improve the mathematical model and/or the objective after inspecting the results of modeling and optimization.

In ill-defined problems the visualization is an essential part of efficient problem solving. It seems that as the model and the objective specification becomes less precise, the visualization becomes more important. Thus a dynamic visualization is important as a modeling and optimization tool in the problems where a complete formalization of the model and/or of the objectives is difficult or impossible.

In this paper a quantity is considered which varies as a smooth function in space and time but is measured as average values over a region in space-time. As an example, the incidence rate of the disease mumps is described. A natural objective is to minimize the impact of the disease using the minimum amount of resources but, unfortunately, it is not well-defined.

A common theoretical problem is to define the best visualization techniques to facilitate a better formulation of the problem and of the objective. It is important to design specific software and hardware systems to process the data efficiently and to display the results.

## 7.2. Non-interactive Dynamic Graphics

**Graphical Techniques.** Graphical methods are widely used for the presentation of results. Here visualization methods are considered as a solution tool rather than as a way of presenting results. The main advantage of visualization methods is a possibility to communicate great amounts of information ("one picture is worth a thousand words") in a short time period.

In optimization problems the objective function is a scalar or (sometimes) a vector representation of the "quality" of a very complicated model. One scalar value is, often, an incomplete specification of a model in vector-optimization problems. We, usually, are not certain how to relate different components of a vector-valued objective function or which one of many different Pareto-optimal decisions to take in vector-optimization problems. Communicating more information about the model in the process of optimization can be effectively handled using visualization methods. As the numeric optimization process is dynamic by nature so must be the visualization methods used in optimization problems.

## 7.3. Non-interactive Dynamic Maps in the Epidemiological Model

**Outline.** Analysis of geographic data is important in many applications. Often, geographic data is reported as counts or averages over a region in space for an interval in time. The intensity function underlying the data using dynamic graphics is explored. The aim of display of the intensity function is to detect spatial-temporal patterns.

This problem is exemplified with data on the mumps disease collected monthly in the United States from 1968 until 1988. For a more detailed description of this example see [7, 23].

**Objectives.** The objective is to estimate and display a smooth function with three arguments $(x, y, t)$ representing two spatial coordinates $(x, y)$ and time coordinate $t$. The available data consists of average values of the function over regions in space-time. A dynamic graphical display (animation) of the function consists of a sequence of images (frames) shown one after the other in a rapid sequence. Each image (frame) represents color map of the values of the function in the $(x, y)$ plane. The sequence number of a frame represents the time dimension $t$.

**Data.** The mumps disease is of current public health interest in the United States in part because of a large outbreak which occurred in 1986-1987, primarily among unvaccinated adolescents and young adults in states without requirements for mumps vaccination.

The weekly provisional information on the occurrence of mumps (and some other diseases) is collected by the National Notifiable Diseases Surveillance System (NNDSS). Details concerning NNDSS can be found in [3]. The raw data consists of the number of cases of mumps reported from each state for each month for the period 1968-1988. Data is not available for some states and periods. One reason for this is that mumps has become reportable at differing times in the various states. Another reason is that small numbers of cases are less likely to be reported. We presume there are other reasons, too. The data-set spans 48 states times 21 years times 12 months per year which yields 12,096 possible state-month combinations. There are 1787 missing observations, approximately 15% of those possible.

The raw counts were converted to incidence rates for each state (cases per 100,000 population) by dividing by the estimated population in units of 100,000 people (we linearly interpolated 1970 and 1980 decennial census estimates of state population).

**Video Display.**

**Raw Data Animation.** The entire dataset consists of 252 months. After several experiments we decided that displaying the data at the rate of two-thirds of one second per month was a reasonable compromise between the time required to look at the entire data set and the apparent speed with which changes take place. In NTSC video (NTSC is the television signal used in the United States and Japan), 30 video frames (images) are shown in a second, so a month would be shown in 20 frames.

The speed of interactive animation would depend on the computer speed and the complexity of the animation algorithm, consequently the number of frames per month would vary accordingly. If the recording were done so that all identical frames were recorded and then the switch were made to the next month's data, the viewer would be distracted by the jumpiness of the resulting images.

Consequently, the choice is a linear interpolation between consecutive months. Precisely, the correctly colored maps for two consecutive months are calculated and then the intermediate maps are calculated by linear interpolation in the color scale. This produces a substantially smoother appearance.

**Smooth Animation.** It seems natural to assume that the incidence rate for a disease is actually given by a smooth function over the entire United States. The incidence rates computed for each state are the integral of this unknown function over the respective state (divided by the area of the state). The problem then is to estimate the smooth function given its integral over the different states. A discussion on smoothing methods appropriate in this case can be found in [17, 7]. The locations were sampled randomly in each state, assigning the state-month value to the sampled locations, and using a kernel smoothing technique to generate a map for each month.

As with the "raw" data, we interpolate linearly the intermediate frames between the monthly smoothed maps. Thus we smooth in space and in time using different techniques. A single frame corresponding to December 1986 is displayed in Figure 2.2.

**Effect of Dynamic Maps.** The mumps disease has highest incidence before the vaccination started at the end of 1960's. In the 1970's vaccination almost completely wiped out the disease, leaving only a few cases per state per month. Mumps incidence rates in the US vary according to season. The peak occurs in early spring, while lowest incidence

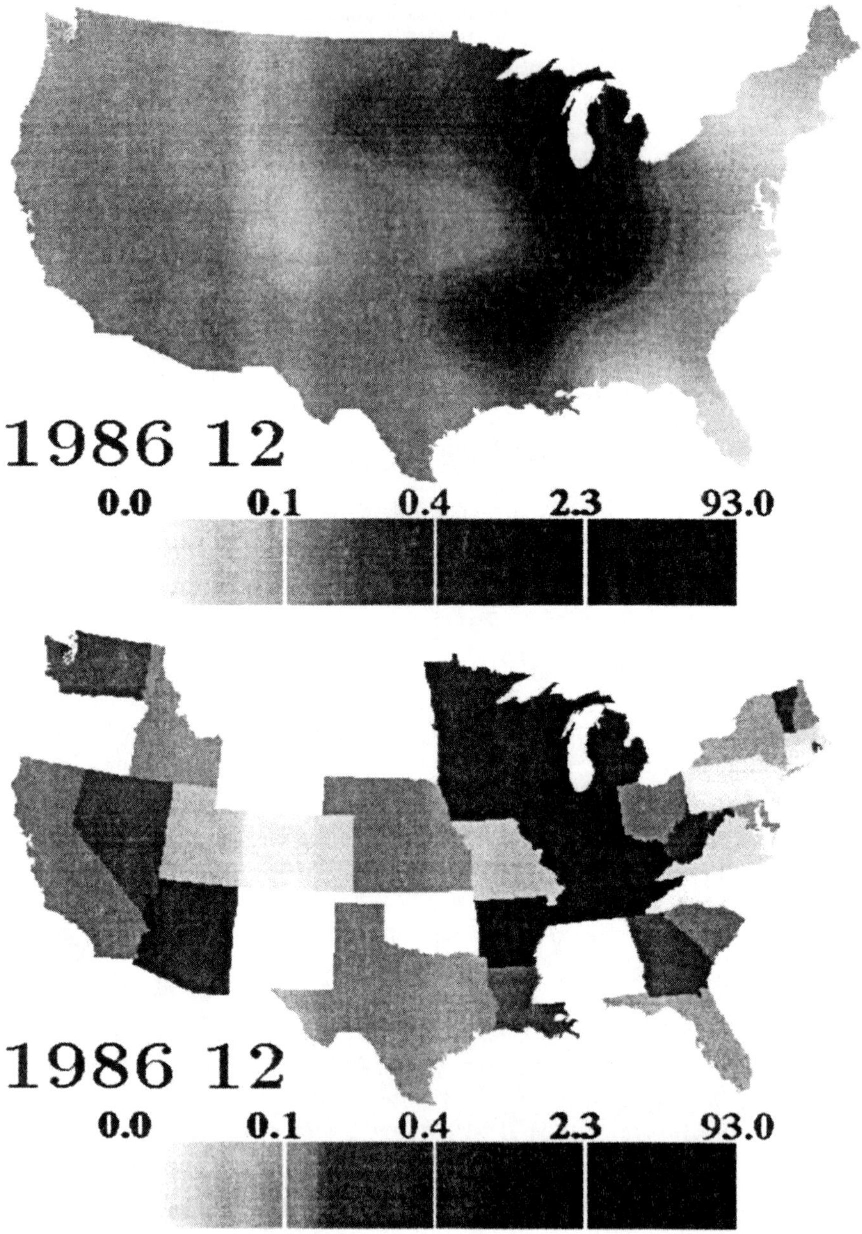


*Figure 2.2.* Incidence Rates in December 1986. Smoothed Rates (top), Raw Rates (bottom)

rates can be observed in autumn. School year could be an important factor since most of the cases are school age children.

The periodic effects are most obvious in the early years of the data set. Later, when the widespread use of the mumps vaccine reduced the typical monthly incidence rate below .1 cases per 100,000 people, the effect is not obvious in the raw version of the videotape. However, the smooth version still exhibits clear incidence differences between seasons.

The spatial spread of mumps can be seen in the raw data only after repeated viewing. It is most noticeable in the winter of 1987-1988 in the states surrounding Illinois. However, in the smoothed data the geographic spread of the disease is readily apparent. One can notice geographic areas that first reach high incidence rates in winter, and areas that are the last to be free from mumps in summer. Additional striking effect is relatively low (as compared to neighboring states) incidence rates in Louisiana in 1968-1972. This could reflect the fact that each state might use different disease reporting mechanism. The outbreak during the late winter of 1986-87 when the disease spreads from Illinois to Arkansas and Tennessee and in the subsequent winter when the disease spreads to all the neighboring states is more apparent in the smooth version of the videotape.

The vaccination programs were stopped in some states in the early 1980's and strong outbreaks of the disease occurred in 1986-1987 and in 1989, primarily among unvaccinated adolescents and young adults in the states without requirements for mumps vaccination. The explanation is well supported by the graph (see Figure 2.3) of the logarithm of incidence rates in California and Wisconsin. We can see a seasonal periodicity (high in spring and low in autumn) and an outbreak in Wisconsin in the second half of the eighties.

Clearly a non-interactive dynamic graphics system is useful for indirect optimization by filtering the general patterns. We need an interactive dynamic graphics system for the direct optimization, see [8].

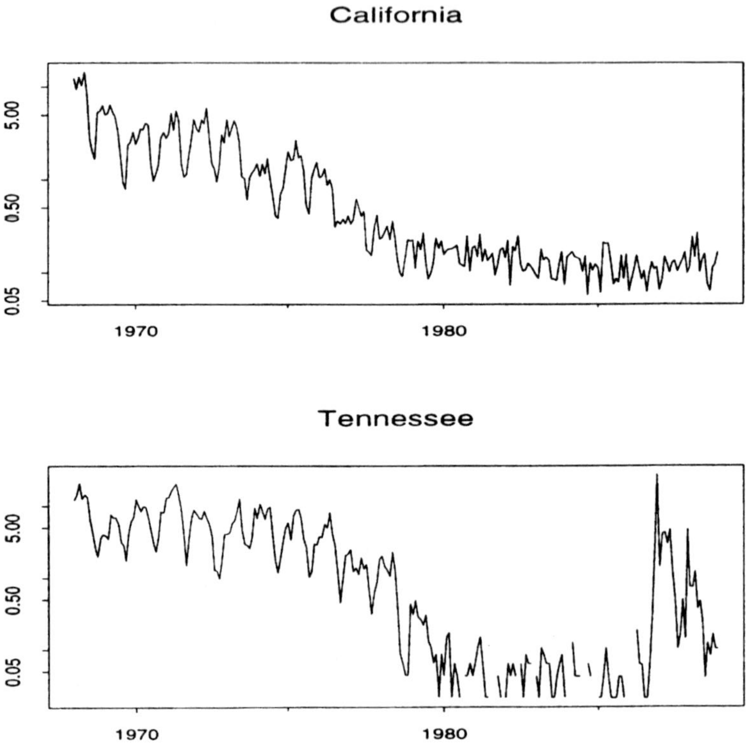


*Figure 2.3.* Log of the mumps monthly incidence rates versus months from Jan. 1968 to December 1988

# References


[1] I.P. Androulakis and V. Venkatasubramanian. A genetic algorithmic framework for process design and optimization. *Computers in Chemical Engineering*, 15:217–228, 1991.

[2] Kenneth R. Baker. *Introduction to Sequencing and Scheduling.* John Wiley & Sons, New York, 1974.

[3] T.L. Chorba, R.L. Berkelman, S.K. Safford, N.P. Gibbs, and H.F. Hull. Mandatory reporting of infectious diseases by clinicians. *Journal of the American Medical Association*, 262:3018–3019, 1989.

[4] H. Das, P.T. Cummings, and M.D. LeVan. Scheduling of serial multiproduct batch processes via simulated annealing. *Computers and Chemical Engineering*, 14:1351–1362, 1990.

[5] M. DeGroot. *Optimal Statistical Decisions.* McGraw-Hill, New York, 1970.

[6] L.C.W. Dixon and G.P. Szego. *Towards global optimisation 2.* North Holland, Amsterdam, 1978.

[7] W.F. Eddy and A. Mockus. An example of the estimation and display of a smoothly varying function of time and space - the incidence of the disease mumps. *Journal of the American Society for Information Science*, 45(9):686–693, 1994.

[8] William Eddy and Audris Mockus. Dynamic visualization in modeling and optimization of ill defined problems, case studies and generalizations. In C. A. Floudas and Panos M. Pardalos, editors, *State of the Art in Global Optimization: Computational Methods and Applications*, volume 7 of *Nonconvex Optimization and its Applications*, pages 499–520. Kluwer Academic Publishers, 1996.

[9] T.A. Feo, M.G.C. Resende, and Smith. *Operations Research*, 42:860–878.

[10] Fisher and Thompson. In *Industrial Scheduling*. Prentice-Hall, 1963.

[11] F. Glover. *Cahiers de Recherche Operationalle*, 6:131–168, 1994.

[12] D. E. Goldberg. *Genetic Algorithms in Search, Optimization, and Machine Learning*. Addison-Wesley, Reading, MA, 1989.

[13] P. Helman, B.M.E. Moret, and H.D. Shapiro. An exact characterization of greedy structures. *SIAM Journal of Discrete Mathematics*, 6:274–283, 1993.

[14] Chris Jones. *Visualization and Optimization*. Kluwer Academic Publishers, Dordrecht, The Netherlands, 1996.

[15] U.S. Karmarkar and R.S. Nambimadom. Material allocation in mrp with tardness penalties. *Journal of Global Optimization*, 9:453–482, 1996.

[16] Ker-I Ko. *Complexity Theory of Real Functions*. Birkhauser, Boston, 1991.

[17] A. Mockus. *Predicting a Space-Time Process from Aggregate Data Exemplified by the Animation of Mumps Disease*. PhD thesis, Department of Statistics, Carnegie Mellon University, 1994.

[18] A. Mockus, J. Mockus, and L. Mockus. Bayesian approach adapting stochastic and heuristic methods of global and discrete optimization. *INFORMATICA*, 5(1–2):123–166, 1994.

[19] A. Mockus, J. Mockus, and L. Mockus. Discrete optimization, information based complexity, and bayesian heuristics approach. In *Proceedings of International Symposium on Operations Research with Applications in Engineering Technology and Management (ISORA'95), August 19-22*, Beijing, 1995.

[20] A. Mockus and L. Mockus. Design of software for global optimization. *Informatica*, 1:71–88, 1990.

[21] J. Mockus. *Bayesian approach to global optimization*. Kluwer Academic Publishers, Dordrecht-London-Boston, 1989.

[22] J. Mockus. Average complexity and bayesian heuristics approach to discrete optimization. *Informatica*, 6:193–224, 1995.

[23] J. Mockus, W. Eddy, A. Mockus, L. Mockus, and G. Reklaitis. *Bayesian Heuristic Approach to Discrete and Global Optimization.* Kluwer Academic Publishers, Dordrecht, The Netherlands, 1997.

[24] J. Mockus and L. Mockus. Bayesian approach to global optimization and applications to multiobjective and constrained optimization. *Journal of Optimization Theory and Applications*, 70(1):155–171, July 1991.

[25] J. Mockus and A. Soofi. The long-run economic relationships: An optimization approach to fractional integrated and bilinear time series. *INFORMATICA*, 6:61–70, 1995.

[26] J. Mockus and A. S. Soofi. Long memory processes and exchange rate forecasting. In *Procedings of the Conference on Forecasting Financial Markets*, pages 1–15, London, U.K., 1996. Chemical Bank and Imperial College.

[27] Jonas Mockus. Application of bayesian approach to numerical methods of global and stochastic optimization. *Journal of Global Optimization*, 4(4):347–366, June 1994.

[28] L. Mockus and G. V. Reklaitis. Mathematical programming formulation for scheduling of batch operations based on nonuniform time discretization. *Computers Chem. Engng*, 20:1173–1177, 1996.

[29] L. Mockus and G. V. Reklaitis. A new global optimization algorithm for batch process scheduling. In C. A. Floudas and Panos M. Pardalos, editors, *State of the Art in Global Optimization: Computational Methods and Applications*, volume 7 of *Nonconvex Optimization and its Applications*, pages 521–538. Kluwer Academic Publishers, 1996.

[30] L. Mockus and G.V. Reklaitis. A new global optimization algorithm for batch process scheduling. In *Proceedings of State of the Art in Global Optimization: Computational Methods and Applications*, Princeton, NJ, 1995.

[31] P.M. Pardalos, L. Pitsoulis, , T. Mavridou, and M.G.C. Resende. Parallel search for combinatorial optimization: Genetic algorithms, simulated annealing and GRASP. In A. Ferreira and J. Rolim, editors, *Parallel Algorithms for Irregularly Structured Problems, Proceedings of the Second International Workshop –Irregular'95*, volume 980 of *Lecture Notes in Computer Science*, pages 317–331. Springer-Verlag, 1995.

[32] G.V. Reklaitis and L. Mockus. Mathematical programming formulation for scheduling of batch operations based on nonuniform time discretization. *Acta Chimica Slovenica*, 42(1):81–86, 1995.

[33] K. Schittkowski. Nlpql: A fortran subroutine solving constrained nonlinear programming problems. *Annals of Operations Research*, 5:485–500, 1985/86.

[34] I.M. Sobolj. On a systematic search in a hypercube. *SIAM Journal on Numerical Analysis*, 16:790–793, 1967.

[35] A. Torn and A. Zilinskas. *Global optimization.* Springer-Verlag, Berlin, 1989.

[36] P.J.M. VanLaarhoven, C.G.E. Boender, E.H.L. Aarts, and A.H.D. RinnooyKan. A bayesian approach to simulated annealing. *Probability in the Engineering and Information Sciences*, 3:453–475, 1989.

[37] M. G. Zentner, J. F. Pekny, D. L. Miller, and G. V. Reklaitis. RCSP++: A scheduling system for the chemical process industry. In *Proc. PSE'94*, pages 491–495, Kyongju, Korea, 1994.

Chapter 3

# A FINITE BRANCH-AND-BOUND METHOD FOR BILINEAR HYDRO POWER SCHEDULING

Stefan Feltenmark

*Dept. of Mathematics*
*Royal Institute of Technology*
*SE-100 44 Stockholm*
*SWEDEN*

stefanf@math.kth.se

P. O. Lindberg

*Dept. of Mathematics*
*Linköping University*
*SE-581 83 Linköping*
*SWEDEN*

polin@mai.liu.se

**Abstract** The bilinear, or head-dependent, hydro power scheduling problem, is a bilinear optimization problem of economic importance. For this problem, we develop a finitely terminating branch-and-bound algorithm, utilizing the problem structure.

**Keywords:** Bilinear optimization, branch-and-bound, finite termination, hydro power scheduling



*A. Migdalas et al. (eds.), From Local to Global Optimization,* 49–67.

# 1. Introduction

Production scheduling of hydro power systems is a problem of economic importance in many countries. In [2], the authors developed a method for production planning in a hydro power system, where the output from a power station depended explicitly on the *head* of the dam, *i. e.*, the difference between the dam water level, and the water level after the station. In its basic form, the output power from a station is a bilinear function of the head and the water flow through the turbines, the *turbine flow*.

In this paper we present a continuous branch-and-bound method for this problem. We prove that the algorithm terminates with a globally optimal solution after a finite number of iterations. Computational results for a real-world power system are given.

The paper is organized as follows. In Section 2 we describe the underlying scheduling problem and introduce some notation. In Section 3, the branch-and-bound algorithm is described in some detail. In Section 4 it is proved that the branch-and-bound algorithm terminates finitely, with a global solution to the problem. In Section 5 we give computational results when the algorithm is applied to a real-world test case. Then finally, in Section 6, we present some conclusions.

# 2. Problem statement

In this section, we state the bilinear hydro power scheduling problem. For a more thorough derivation of the problem, see [2], where also a more complex model is treated.

The power output $p$ of a hydro station depends essentially bilinearly on the discharge $x$ and on the head $h$:

$$p = \alpha x h. \tag{1}$$

The head is (in the first approximation) an affine function of the volume content $v$ of the upstream dam:

$$h = a + bv. \tag{2}$$

Using this, the water (inventory) balance for the upstream dam can, after rescaling, be written

$$h_{t+1} = h_t - x_t + w_t, \tag{3}$$

where $t = 1, \ldots, T$ indexes time, and $w$ denotes exogenous flow into the dam.

Hydro stations in a river are connected hydraulically, in that water from upstream stations travel to those downstream. Thus, let $i = 1, \ldots, I$ index the power stations, and let $\tau_i$ denote the traveling time of water from station $i$ to the next downstream station. Then, letting $J_i$ be the index set of stations immediately upstream from $i$, the water balance can be written

$$h_{i,t+1} = h_{i,t} - x_{i,t} + \sum_{j \in J_i} x_{j,t-\tau_j} + w_{i,t}, \tag{4}$$

where now $w_{i,t}$ is the water added by tributaries, rain *etc.*

The rescaling of the head, to get the water balance (3), implies that the power output now is written

$$p_{i,t} = \alpha_i x_{i,t} h_{i,t} + \beta_i x_{i,t}. \tag{5}$$

For technical reasons and by jurisdiction, there typically are upper and lower bounds on the variables: $0 = \underline{x}_i \leq x_{i,t} \leq \overline{x}_i$, and $\underline{h}_i \leq h_{i,t} \leq \overline{h}_i$.

Further, we assume that the hydro system is a profit maximizer and a price taker, *i. e.*, that power prices $\lambda_t$, typically varying hourly, are given. Finally, to prevent the dams from being emptied, there is a lower bound on final dam content, $h_{i,T+1} \geq \underline{h}_{i,T+1}$, but since we are maximizing profit, we can assume that this constraint is binding. Summing up, we have the following bilinear optimization problem, $(P)$:

$$\begin{array}{ll}
\text{minimize}_{x,h} & f(x,h) \triangleq -\sum_{i=1}^{I} \sum_{t=1}^{T} \lambda_t(\alpha_i x_{i,t} h_{i,t} + \beta_i x_{i,t}) \\
\text{subject to} & h_{i,t+1} = h_{i,t} - x_{i,t} + w_{i,t} + \sum_{j \in J_i} x_{j,t-\tau_j} \\
& \qquad\qquad\qquad\qquad i = 1, \ldots, I, \ t = 1, \ldots, T \\
& \underline{x}_i \leq x_{i,t} \leq \overline{x}_i \\
& \underline{h}_i \leq h_{i,t} \leq \overline{h}_i \\
& h_{i,T+1} = \underline{h}_{i,T+1} \\
& h_{i,1} \text{ and } x_{j,t}, \ t \geq 1 - \tau_j \text{ are given.}
\end{array}$$

Comparing with the model in [2], we see that we have excluded the spillage variables $s_{i,t}$. Spillage, *i. e.*, discharging water without turbining, usually is used only in extreme situations, to prevent the dam from overfilling and similar. Moreover, with the presence of spillage variables, the current form of the finiteness proof does not go through.

The constraint set in $(P)$ has a network flow structure, as shown in [2]. With $T$ time periods and $I$ stations, we get $2IT$ variables, $IT$ equality constraints and $IT$ bilinear terms. Hence, for our small but typical test case with three stations and to be planned hourly during a week, we get around 500 equations ($3 \times 168 = 504$), 1000 variables and 500 bilinear terms, which is quite a sizeable bilinear problem. With more stations, the problem grows proportionally.

## 3. The branch-and-bound procedure

In this section we state our branch-and-bound procedure. It is based, to a large extent, on our previous paper [2], ¿From which we will cite results without proof.

In devising a finitely terminating branch-and-bound procedure, one has to strike a balance between efficiency of the method, the complexity of the method, and the complexity of the finiteness proof. We believe that we have managed to devise an efficient, but yet not complex method. The finiteness proof, Section 4, on the other hand, might seem complex, but that also has to do with the inherent complexity of the problem. In this section, a *node* will denote a node in the branch-and-bound tree. A *rectangle*, $R_{it}$, will denote the rectangle of feasible values for $x_{it}$ and $h_{it}$, *i. e.*, discharge and head for unit $i$ at time $t$ at a given node. A *box* will be the "hyper-rectangle" of feasible values at a given node, *i. e.*, the product of all rectangles.

### 3.1. Lower bounding

In branch-and-bound one needs to obtain lower bounds on the optimal value. We do this in the classical way for bilinear programming, [1], by lower bounding the bilinear terms over the "rectangles" of simple variable bounds:

**Lemma 1** *The largest convex underestimate of $-xh$ over the rectangle $\underline{x} \leq x \leq \overline{x}$, $\underline{h} \leq h \leq \overline{h}$ is*

$$\breve{f}(x,h) = max\{\,-\underline{x}h - \overline{h}x + \underline{x}\overline{h}, - -\overline{x}h - \underline{h}x + \overline{x}\underline{h}\,\}. \qquad (6)$$

**Proof** See [2], Lemma 7. ∎

It turns out that the maximal local error is proportional to the area of the rectangle:

**Proposition 1** *The local error of the underestimate $\check{f}(x,h)$ is*

$$\delta(x,h) = -xh - \check{f}(x,h) = \min\{\,(x-\underline{x})(\overline{h}-h),\ (\overline{x}-x)(h-\underline{h})\,\}. \quad (7)$$

*It achieves its maximal value $\frac{1}{4}(\overline{x}-\underline{x})(\overline{h}-\underline{h})$ at the midpoint of the rectangle $\underline{x} \le x \le \overline{x}$, $\underline{h} \le h \le \overline{h}$.*

**Proof** See [2], Prop. 9. ∎

Due to the functional form of the lower bounds, the lower bounding problem is a linear program (LP) with a network substructure, and can be solved efficiently. Since the error of the lower bound is proportional to the sizes of the rectangles, it is important to reduce these sizes if possible. Therefore, we use range reduction to reduce the sizes of the rectangles at each node. We use both *feasibility* range reductions of the type described in [2], as well as em optimality range reductions, as pioneered by Ryoo and Sahinidis, [5].

## 3.2. Upper bounding

In order to achieve good upper bounds on the optimal value, we perform a local minimization at each node of the branch-and-bound tree, starting at the solution of the LP underestimate. Due to the network structure and other properties, this local search can be done in a very efficient way.

**Proposition 2** *Let $z^0 = (x^0, h^0)$ be a feasible solution to $(P)$, and assume that $z^c = (x^c, h^c)$ is a flow along a cycle in the network corresponding to $(P)$. Let $\varphi(\delta) = f(z^0 + \delta z^c)$. Then $\varphi$ is strictly concave.*

**Proof** Apart from the *strict* concavity, this is just a rewriting of Prop. 1 in [2], noting that there are no spillage variables $s$. The strict concavity follows from that a cycle flow must have a north-east corner point, where we get a strictly concave contribution to the objective. ∎

As a consequence we have:

**Lemma 2** *Any local optimum is an extreme point.*

**Proof** A consequence of Prop. 3 in [2]. ∎

Another consequence is that at active set reduced gradient method will move between extreme points:

**Lemma 3** *An active set reduced gradient method applied to problem (P), starting from an extreme point, will, barring cycling, move between extreme points until a local optimum is found.*

**Proof** This is Prop 4. of [2], with the addition of the cycling comment.
■

**Corollary 4** *With the addition of some anti-cycling strategy, the above method will terminate in a local minimum.*

Due to this result, a reduced gradient method will not use super-basics [4]. Moreover, because of the network structure, the bases can be represented as trees. To be able to start at a non-basic solution, e.g. the optimal solution to the lower bounding LP problem, we need and initial "purification" phase, where a non-basic solution is converted into a basic one without increasing the objective. We have implemented this modified network flow code in C, based on the code in [3].

The motivation for local minimization is two-fold. On the one hand, if we find a local solution, feasible at a given node, and with the same objective as the LP underestimate, then we have found the global optimum at the node, which hence is fathomed. On the other hand, the aim of the local search is also to find good feasible solutions to the original problem. With this purpose in mind, there is no point in searching for local optima that are feasible at the given node. Instead, having found a local optimum under the constraints at the node, we continue minimizing using the original constraints, thus trying to find a better overall local minimum. This device, moreover, happens to simplify the finiteness proof quite a lot.

## 3.3. Node selection

We always choose a node for branching that has the best lower bound.

## 3.4. Branching

There are essentially three possibilities for choice of variables to branch on: the flow, $x$, the head, $h$, or both ($x$ and $h$).

Branching on $h_{i,t}$ gives, due to range reductions, large reductions of rectangles at other times than $t$. Branching on $x$ does not give much range reductions, and should thus be inferior to branching at $h$. Branching on both $x$ and $h$ might seem more logical, but then we create four new nodes each time we branch. In the computations below, it turns out that it is in fact best to branch on $h$ only. Having decided to branch on some head variable, we have to decide on at what $h$-value to split. Since we want finite termination, we must be able to fathom a node containing the global optimum. This can only be done on the condition that the lower bound equals the upper bound. For this to be true, the error at the global optimum must be zero, whence it must be at the boundary of its box. To enforce this, one must split at a potential global optimum. We have thus chosen to always split at the best local optimum if possible, *i. e.*, if it is feasible at the node. When all head variables at the best local optimum are at their bounds, we branch on flow variables, $x$, starting with those that are close to halfway between their bounds. When we cannot split at the local optimum, we choose to split at the optimal solution of the LP underestimate.

Still we have to choose on which head-variable $h_{i,t}$ to split. We then choose the pair $(i, t)$ whose local error is largest at the branching point.

## 3.5. Algorithm

Summing up we have the following algorithm.

Branch and bound for head-dependent hydro power planning

**Step 0.** Compute lower and upper bounds, $LB_0$ and $UB_0$, to the original problem $(P) = (P_0)$. Let $(P_0)$ be the only member of the candidate list $L$. Initialize best upper and lower bounds.

**Step 1.** While $L \neq \emptyset$ do:

**Step 2.** Choose a $(P_k)$ from $L$, with least lower bound $LB_k$.

**Step 3.** Choose the splitting point $(\overline{x}, \overline{h})$ to be the best local optimum, if feasible, else the optimal solution of the LP underestimate.

**Step 4.** Choose as splitting variable $h_{i,t}$ in the rectangle with largest local error at the splitting point.

**Step 5.** Compute new lower and upper bounds for the newly generated problems, and add the problems to $L$. Update best upper and lower bounds. Also prune $L$ of problems with lower bounds greater than the best upper bound.

**Step 6.** End of while.

## 4. Finite termination

Let the *local error* for a given box (node) be the difference between the true objective and the LP underestimate at the LP solution, and similarly for a rectangle.

**Lemma 5** *There are finitely many local minima to the bilinear hydro power scheduling problem* $(P)$.

**Proof** This follows from the fact that each local optimum is an extreme point solution, Lemma 2, and hence a basic solution. ∎

The only possibility for the branch-and-bound process not to be finite, is that there is at least one infinite descending sequence of nodes. The rest of the proof aims at proving that it leads to a contradiction to assume the existence of such an infinite descending sequence.

**Lemma 6** *In an infinite descending sequence of nodes, all branchings will eventually be at the solution of the LP underestimate.*

**Proof** We branch at the LP solution unless it is possible to branch at the hitherto best feasible solution at the node. Each such feasible solution is a local optimum, of which there are only finitely many. We can branch at most $2IT$ times at a local optimum, since after that many branchings it will be at a corner of the box, and it is impossible to branch there anymore. Therefore, all branching are at an LP solution, eventually. ∎

**Lemma 7** *For an infinite descending sequence of nodes* $\{n_k\}$,

$$UB_{n_k} - LB_{n_k} \to 0, \ \ as\ k \to \infty.$$

**Proof** Consider the tail of the sequence, where all branchings are at LP solutions. There will be infinitely many branchings in rectangles of type $R^n_{it}$, for some pair $(i,t)$. Assume that $\{n_k\}$ is the corresponding subsequence.

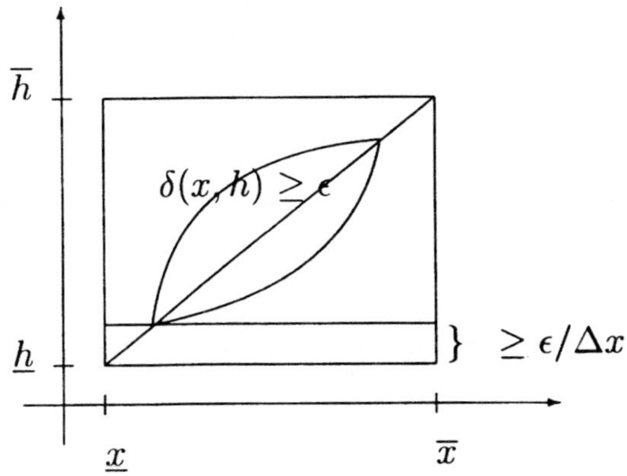


*Figure 3.1.* Level set of error function, and minimal cut-off

If $UB_{n_k} - LB_{n_k}$ does not tend to zero, then for some $\epsilon' > 0$ we must have $UB_{n_k} - LB_{n_k} \geq \epsilon'$, for all $k$. This implies that the local error in the rectangle $R_{it}^{n_k}$, where we branch, must be greater than, or equal to $\epsilon = \epsilon'/IT$. As noted above (Prop. 1), the local error in $R_{it}^{n_k}$ is (writing $\underline{x}$ for $\underline{x}_{it}^{n_k}$, etc.)

$$\delta(x,h) = min\{(x - \underline{x})(\overline{h} - h), (\overline{x} - x)(h - \underline{h})\}. \tag{8}$$

This implies that for the error to be $\geq \epsilon$, the branching point must be inside two hyperbolic curves: $(x - \underline{x})(\overline{h} - h) \geq \epsilon$ and $(\overline{x} - x)(h - \underline{h}) \geq \epsilon$, see Figure 4.

Thus, any branching point will fulfill

$$h - \underline{h} \geq \frac{\epsilon}{\overline{x} - x} \geq \frac{\epsilon}{\Delta x},$$

where $\Delta x = \Delta x_{it} = \overline{x}_{it} - \underline{x}_{it}$, and similarly

$$\overline{h} - h \geq \frac{\epsilon}{x - \underline{x}} \geq \frac{\epsilon}{\Delta x}.$$

Therefore, in each branching, the height, $\overline{h} - \underline{h}$, of the rectangle is reduced by at least $\frac{\epsilon}{\Delta x}$. Eventually, the height will be so small that $\delta(x,h) < \epsilon$ everywhere, a contradiction. ■

Let us say that a node *contains* a given point if the corresponding box does, and similarly that a sequence of nodes does if the nodes do.

**Corollary 8** *For an infinite sequence of nodes $\{n_k\}$, containing a global optimum, all local optima found by local descent, will eventually be global.*

**Proof** Since there are only finitely many local optima, the non-global local minima differ in objective value by at least some $\delta > 0$. Hence, if the statement were not true, we would not have $UB_{n_k} - LB_{n_k} \to 0$, contradicting Lemma 7. ∎

Let $f^\star$ denote the objective value of a global optimum to our problem. It is obvious that the overall lower bound will be less than, or at most equal to $f^\star$ during the branch-and-bound procedure.

**Lemma 9** *Any descending sequence of nodes $\{n_k\}$, not containing a global optimum, will terminate.*

**Proof** If the sequence were not finite, we would have $UB_{n_k} - LB_{n_k} \to 0$. Further, $UB_{n_k}$ is the objective value of some point $z_k$ in $n_k$. We can, without loss of generality, assume that $\{z_k\}$ converges to some point $\overline{z}$, that is feasible for all $n_k$. Therefore $lim_{k\to\infty} UB_{n_k} > f^\star$. But then, eventually, $LB_{n_k} > f^\star$, implying that $LB_{n_k}$ is greater than the overall lower bound, contradicting that the branching continues at $n_k$. ∎

**Lemma 10** *If the branch-and-bound procedure is infinite, the overall upper and lower bounds tend to $f^\star$.*

**Proof** Assume that the procedure is infinite. According to Lemma 9, only sequences of nodes containing a global optimum can be infinite. Since global optima also are local, there are only finitely many global optima. Thus, at least one global optimum will be contained in an infinite sequence. For that sequence $UB_{n_k} - LB_{n_k} \to 0$, and hence $UB_{n_k}$ and $LB_{n_k}$ tend to $f^\star$. Since we branch only at nodes whose lower bound equals the overall lower bound, the latter will also tend to $f^\star$. The overall upper bound is the minimum of all $UB_n$. Thus it must tend to $f^\star$, since $UB_{n_k}$ does. ∎

Before the next lemma, let us recall some facts on what happens when a non-basic variable leaves its bound. We will further assume that all variables are oriented so that this change is an increase. Suppose that the non-basic variable $y$ is increased. Then flow is sent along a cycle, and the objective varies quadratically and concavely, according to Prop. 2, say

$$\delta f = \overline{c}_y y - \frac{1}{2} q_y y^2. \tag{9}$$

Further, there is a minimal curvature $q_{min} = \min\{q_y \mid y \text{ non-basic}\} > 0$, since there are only finitely many cycles.

**Lemma 11** *A descending sequence of nodes* $\{n_k\}$ *containing a global minimum will terminate on the condition* $LB_{n_k} = UB_{n_k}$.

**Proof** Assume that the sequence in question does not terminate, *i. e.*, that it is infinite. According to Corollary 8, local search will eventually end up in a global optimum. Due to our branching rules, this global optimum will eventually be at a corner of all boxes. A prospective local search starting at the global optimum will stay there, generating reduced costs $\bar{c}_y$, non-negative for the non-basic variables and zero for the basic ones. As noted above, increasing the non-basic variable $y$ by $\delta y$ leads to a change $\delta f = \bar{c}_y \delta y - \frac{1}{2} q_y (\delta y)^2$ in the objective. Further, temporarily assuming non-degeneracy, the maximal feasible increase of $y$ is bounded from below by some $\delta_{max} > 0$. Since we start at a global optimum, the change in objective must be non-negative. Thus, we have

$$0 \leq \delta f = \bar{c}_y \delta_{max} - \frac{1}{2} q_y (\delta_{max})^2 \leq \bar{c}_y \delta_{max} - \frac{1}{2} q_{min} (\delta_{max})^2, \qquad (10)$$

implying

$$\bar{c}_y \geq \frac{q_{min} \delta_{max}}{2} > 0,$$

giving a uniform bound on $\bar{c}_y$. Observe that $\bar{c} = \nabla f - \pi^T A$, for some multipliers $\pi$, and where $A$ is the constraint matrix.

**Note:** If the non-degeneracy assumption is not fulfilled, then we can proceed in the following way. For each edge $d$ in the constraint polytope, emanating from the global optimum point, there is a corresponding basis, a multiplier vector $\pi_d$, and an entering non-basic variable $y_d$, such that $\bar{c}_{y_d}$, the $y_d$ component of $\nabla f - \pi_d^T A$, gives the rate of change of $f$ along $d$, when $y_d$ is increased. According to above, $\bar{c}_d$ is bounded from below by $\frac{q_{min} \delta_{max}}{2}$, where now $\delta_{max}$ is the minimum maximal feasible increase of non-basic variables at the global optimum. All other reduced costs are non-negative. Now choosing $\pi = \sum_d \pi_d / \sum_d 1$, the reduced cost $\bar{c} = \nabla f - \pi^T A$ for all non-basic variables will be strictly positive as will those of some basic variables.

Now choose an outgoing edge, corresponding to an increase of the non-basic variable $y$. Let $A_B$ be a corresponding basis matrix, allowing for an increase of at least $\delta_{max}$ of $y$. Let $\bar{c} = \nabla f - \pi^T A$ be the reduced cost according to above (possibly after modification for non-degeneracy). Now change $\pi^T$ to $\tilde{\pi}^T = \pi^T - \epsilon e^T A_B^{-1}$, where $e$ is the vector of all ones: $e = (1, 1, \ldots, 1)^T$, thereby changing $\bar{c}$ to $\tilde{c}$. Then $\tilde{c}_B = c_B - \tilde{\pi}^T A_B = c_B - \pi^T A_B + \epsilon e^T A_B^{-1} A_B = \bar{c} + \epsilon e^T$, which is component-wise positive. For a small enough $\epsilon$, the reduced costs of the non-basic variables will remain positive. Thus, we can assume that $\tilde{c}$ is component-wise greater than some real number $\tilde{c}_{min} > 0$.

Let the feasible set be

$$F = \{(x, h) \mid A \begin{pmatrix} x \\ h \end{pmatrix} = b,\ \underline{x} \le x \le \overline{x},\ \underline{h} \le h \le \overline{h}\},$$

Consider the variation of $\tilde{f}(x, h) = f(x, h) - \tilde{\pi}^T (A \begin{pmatrix} x \\ h \end{pmatrix} - b)$ over the whole box. Now, $\tilde{f}$ equals $f$ on $F$. On the rectangle $R_{it}$, we get

$$\begin{aligned} \delta \tilde{f}_{it} &= \tilde{c}_{h_{it}} \delta h_{it} + \tilde{c}_{x_{it}} \delta x_{it} - \frac{1}{2} \lambda_t \alpha_i \delta x_{it} \delta h_{it} = \\ &= \tilde{c}_{h_{it}} \delta h_{it} + \tilde{c}_{x_{it}} \delta x_{it} (1 \pm \frac{\lambda_t \alpha_i \delta h_{it}}{2 \tilde{c}_{x_{it}}}), \end{aligned}$$

where the + corresponds to that the global optimum is at a SE or NW corner of the rectangle (giving a single sign change in $\delta x_{i,t}$ or $\delta h_{i,t}$), and th $-$ the other case. Here $\delta h_{it}$ must be less than $\Delta h_{it}$, which tends to zero as we continue branching. On the other hand, $\tilde{c}_{x_{it}}$ and $\tilde{c}_{h_{it}}$ are $\ge \tilde{c}_{min}$, implying

$$\delta \tilde{f}_{it} \ge \tilde{c}_{min} \delta h_{it} + \tilde{c}_{x_{it}} \delta x_{it} (1 \pm \frac{\lambda_t \alpha_i \Delta h_{it}}{2 \tilde{c}_{min}}). \tag{11}$$

Eventually, in the branching process, we will have $\Delta h_{it} < \frac{\tilde{c}_{min}}{2 \lambda_t \alpha_i}$, which implies that $\delta \tilde{f}_{it} \ge 0$. Thus, eventually, all $\delta \tilde{f}_{it}$, and hence $\delta \tilde{f}$, will be non-negative.

We arrived at $\tilde{f}$ by adding to $f$ a linear term that vanishes on $F$. Hence, conv$\tilde{f}$, the convex hull of $\tilde{f}$, is arrived at by adding the same linear term to conv$f$. Thus, the LP relaxation in which one minimizes conv$f$ over $F$, can as well be seen as minimizing conv$\tilde{f}$ over $F$. But since $\delta \tilde{f} \ge 0$, when the rectangles are small enough, the LP relaxation will have its minimum at the global optimum. Hence, the corresponding node will be truncated on the condition $LB_{n_k} = UB_{n_k}$. ∎

As a corollary we have:

**Theorem 12** *The branch-and-bound procedure 3.5 terminates after a finite number of steps.*

## 5. Computational results

In this section we report on some computational experience with an implementation of Algorithm 3.5.

### 5.1. Test Cases

The test system consists of a river with three hydro stations in Uruguay. The corresponding dams are respectively large, small and medium sized, going downstream. The times to empty these dams by discharge are respectively 156 days (5 months), 40 h (1.7 days) and 240 h (10 days).

At maximal head and discharge the stations produce 230, 100 and 700 MW respectively.

We compute plans for this system in hourly time steps for a week (168 hrs), using prices that vary over the day and the week. One set of prices, set 1, is just the sum of two sinusoidals, peaking at noon. The other set, set 2, is piecewise constant for a couple of hours. The corresponding planning problem, is a network problem with about 500 nodes and 1500 arcs, implying that there are 500 bilinear terms. The underestimating LP has approximately 1000 rows and 2000 columns.

We subject this hydro system to 6 test settings. Since typically the first dam will be used to store water between weeks, and its content varies very little in a week, we assume its incoming head to be rather low, and demand the outgoing head to be 95%, 100% or 105% of that, cases L(ess), E(qual), and M(ore) respectively. For the incoming head we have a D(ry), and a W(et) case, having respectively one and three weeks of discharge as incoming content. In total, we then have 12 cases; two sets of prices in two settings of water supply and three different terminal conditions. In all cases, the external water flow has been the same, generated randomly, assuming uniform distributions.

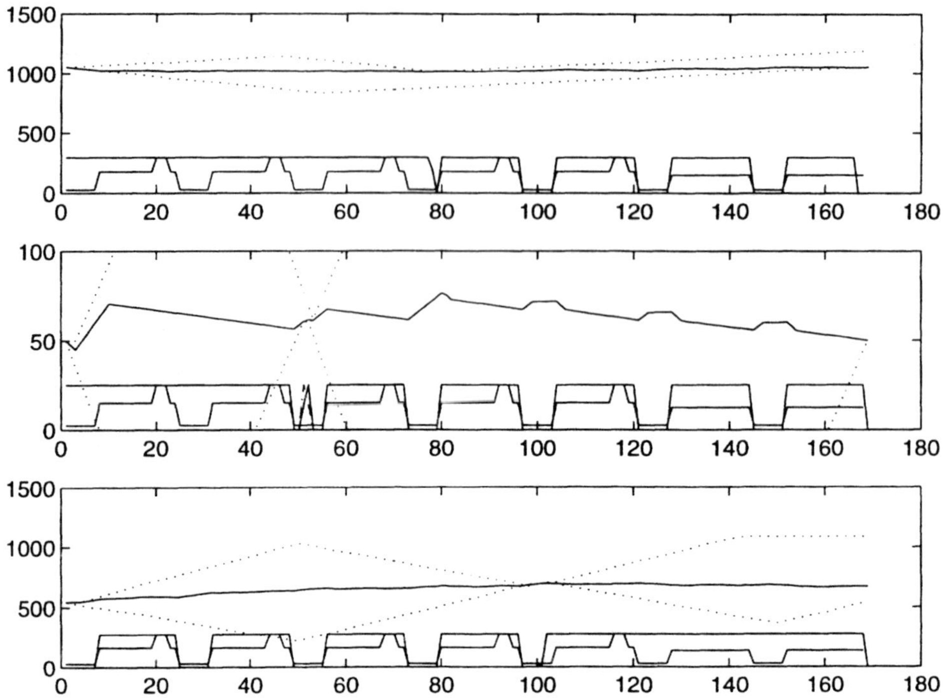


*Figure 3.2.* Optimal solution of Case WE2.

## 5.2. Performance of the algorithm

When convex hull underestimators are used, the convexification error is zero in the boundary of $R_{it}$, that is, when either the stock or the discharge attains a bound. Under normal circumstances the stock does not meet with either of its bounds, but, for the data we have, the discharge tend to be either on its upper bound or lower bound in most periods. Note that, using the convex hull under-estimators, we do not obtain spanning tree solutions from the lower bounding problems, although they are extreme solutions of the LP. We observe that the difficulty of the problem tend to grow with the number of terms in the objective that has a positive convexifiction error at the solution to the lower bounding problem. This number increases with the number of changes ¿From minimum discharge to maximum discharge or vice versa. Thus, problems with large variations in power price and little incoming content in the dams (the Dry cases) turn out to be significantly harder than problems with much incoming dam content (Wet test cases). We could solve cases

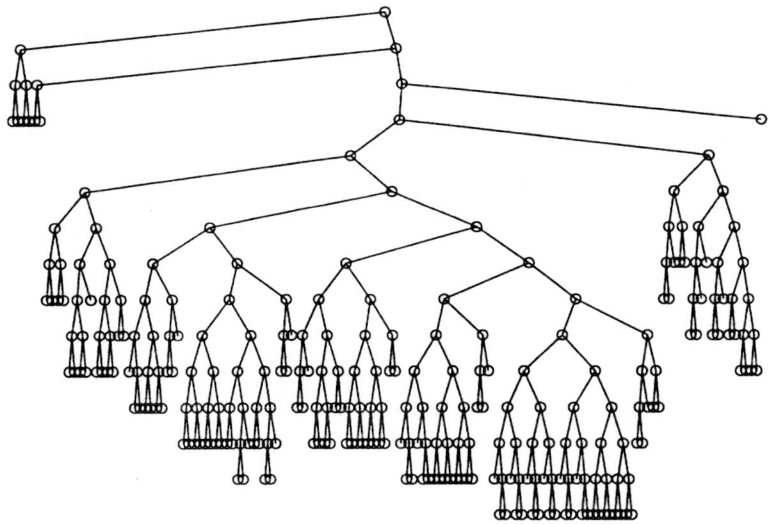

*Figure 3.3.* Branch-and-bound tree for Case WE2, 275 nodes and a depth of 15.

WE1, and WE2 to global optimum, with very few nodes in the branch and bound tree, see, Figure 3.3, while the Dry counterparts only had asymptotic convergence of the convexity gap. In practice, one would be content with a sufficient decrease of the convexity gap below some tolerance. In Figure 3.4, the 10-logarithm of the gap for the DE1 case is plotted. Another important observation is that a global optimal solution is found very early in the search tree, and often at the root node.

## 5.3. Choice of partitioning

We compared the natural partitions of the feasible set mentioned above: branching on $x$, on $h$, or on both. It turns out that branching on the stock variables, $h$, only, is superior to the other alternatives. Of course, branching on both $x$ and $h$ cannot give a worse sharpening of the lower bound in a particular node, but it generates twice as many subproblems. In figure 3.5, we see how the lower bound for the different partitions grow towards the upper bound for the test case DM1, described in Section 5.1. This picture is typical for the problems we encountered.

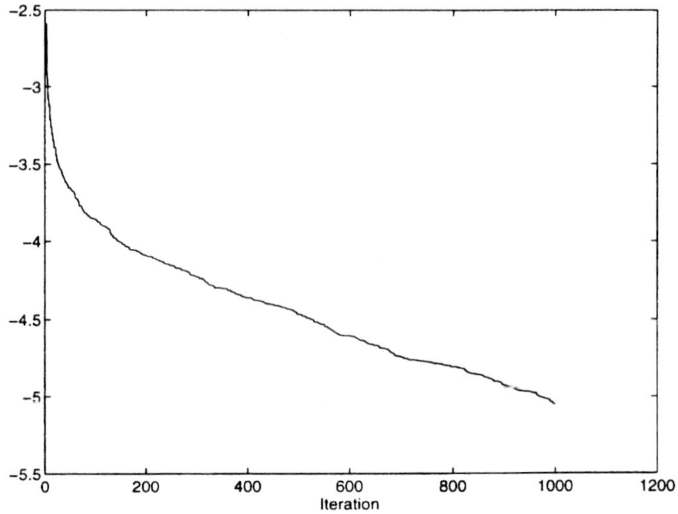


*Figure 3.4.* Logarithm of gap between upper bound and lower bound for DE1 case.

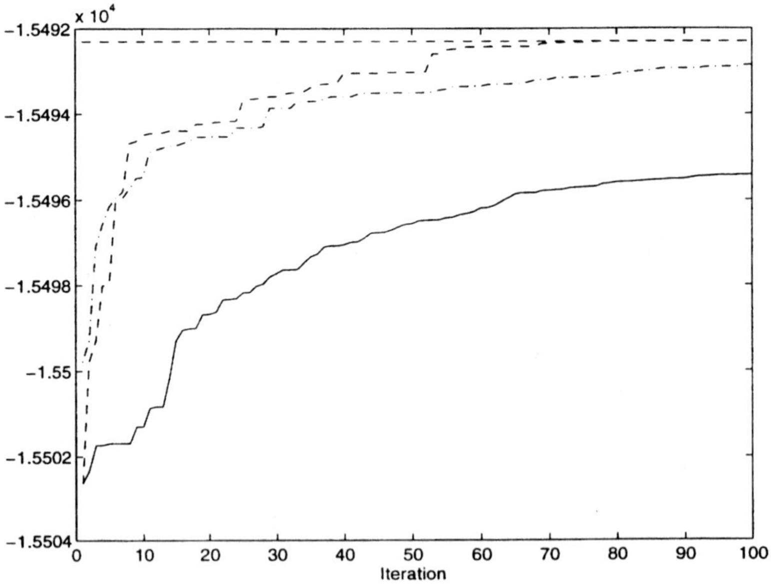


*Figure 3.5.* A comparison of lower bounds corresponding to different branching rules, case WE1. Branching on $h$ (dashed), branching on both (dash-dotted), branching on $x$ (solid).

## 6. Conclusions

We have described a branch-and-bound algorithm for global optimization of a bilinear hydro power scheduling problem. We proved that the algorithm terminates after a finite number of iterations with an optimal solution. We also presented computational results for a real hydro power system with three stations and a one week planning horizon, with a time step of one hour. For some test cases, the search trees were small enough for us to compute global solutions, while for other cases computational limitations prevented us from termination.

# References


[1] Al-Khayyal, F. A. , Falk, J. E.: Jointly constrained biconvex programming. Mathematics of Operations Research, **8** (2), 273–286, 1983.

[2] Feltenmark S., Lindberg P O: Network methods for head-dependent hydro-power scheduling. Network Optimization, by P. M. Pardalos, D. W. Hearn and W. W. Hager (eds.). Lecture Notes in Economics and Mathematical Systems, Springer-Verlag, 1997.

[3] Feltenmark S.: CNET - a simplex network flow code written in C. Optimization & Systems Theory, Department of Mathematics, 1995.

[4] Murtagh B., Saunders M.: Large scaled linearly constrained optimization. Mathematical Programming, **14**, 41–72, 1978.

[5] Ryoo H. S. and Sahinidis N. V.: A branch-and-reduce approach to global optimization. J. of Global Optim. 8 (1996), nr. 2, 107–138.

Chapter 4

# TWO EXAMPLES OF GLOBAL OPTIMIZATION BY MEANS OF LOCAL TECHNIQUES

Antanas Žilinskas[1]

*Vytautas Magnus University*
*Institute of Mathematics and Informatics,*
*Akademijos str. 4, LT-2600 Vilnius, Lithuania*

[1]The research was partly supported by Deutsche Forschungs Gemainschaft, Danish Natural Science Research Council and Lithuanian Fund of Studies and Research

**Abstract**

Two examples of global optimization of multimodal functions are considered. The first problem is maximization of throughput of a network working under finite population slotted ALOHA protocol. It is assumed that the traffic load is given and each of $n$ nodes may have its own transmission probability. The analytical solution is obtained including the results of Abramson [2] as special case. The second example is the multidimensional scaling (MDS), a well known technique for visualization and analysis of multidimensional data. The most important part of implementation of MDS is minimization of $STRESS$ function which expresses a measure of difference between set of patterns and set of images. Analysis of local optimality conditions, presented in this paper, implies the reformulation of the problem which enables the local minimization of $STRESS$ with quadratic rate of convergence, improving the efficiency of the corresponding MDS algorithm. Two presented problems demonstrate the examples of efficient application of local techniques based on thoroughful analytical investigation of a problem versus its solving by means of a general global optimization algorithm.



*A. Migdalas et al. (eds.), From Local to Global Optimization,* 69–84.

**Keywords:** maximization of throughput, slotted ALOHA, multidimensional scaling, conjugate gradients, quadratic convergence

# 1. Introduction

In applied optimization the global optimum is normally required. Sometimes global and local solutions coincide (i.e. in convex problems) and may be found easily if compared with a multimodal case, whose solutions obtained by means of a local algorithm with arbitrary starting points are normally not acceptable. Therefore global optimization algorithms are developed for various classes of multimodal problems. For functions with special structure a rich theory is developed generalizing the classical ideas of optimization theory [6]. On the other hand, random search methods, including evolutionary strategies, have been practically proved rather efficient in many problems where the objective function is obtained by means of simulation techniques [13], [15]. A rational search theory based on statistical models is developed for the case of functions whose analytical expressions are not known and only very general postulates on rationality of search may be assumed [15]. The advantage of the latter approach is the very broad field of applications of the corresponding algorithms. However, its disadvantage is the time, needed for the auxiliary computations, which very often exceeds the practically acceptable limits. Summarising we may conclude that global optimization theory offers a variety of methods and algorithms to solve practical multimodal problems. However, the usual techniques of local optimization theory should also not be forgotten, especially when an analytical expression of the objective function is available. Sometimes, specific properties of an objective function and constraints may be discovered leading to efficient solution of the problem. Although there are no recommendations for recognition of a favorable case, it pays to try: the gains in case of success are very promising.

The first problem considered below is maximizing of throughput of a computer network. Direct solution of the problem by means of a general global algorithm is very complicated because of the high dimensionality. Some reduction of the search was achieved exploiting the properties of the objective function [10]. However, thoroughful analysis provides the analytical solution of the problem. Moreover, the condition of rationality of centralised versus decentralised traffic in the network is justified.

The second problem is the well known problem of multidimensional scaling (MDS). The rationality of designing of a global algorithm using the descent methods as auxiliary routines seems rather obvious because the analytical expressions of the objective function and its derivatives are available. An evolutionary algorithm, including a local descent, was proved efficient for MDS [9]. Therefore a fast algorithm of local minimization of the so called *STRESS* function is important. Some properties of the latter are proved enabling the reformulation of the problem to guarantee a quadratic convergence of a conjugate gradient algorithm. A fast local algorithm makes the evolutionary global search for this problem more efficient.

## 2. Maximization of Throughput of Slotted ALOHA

### 2.1. Technical Formulation of the Problem

Each of $n$ nodes (terminals, stations) of a network under the ALOHA protocol may transmit a standard packet of data into the same channel, which normally means broadcasting via radio. Without loss of generality it may be assumed that data packets have a length of one time unit, i.e. the time scale is determined by the packet length. If the time intervals of broadcasting intersect (at least for two nodes) then the packets are destroyed and could not be correctly received by the destination nodes. The throughput is defined as the average number of successfully received packets per time unit. The throughput $d_{\mathrm{A}}$ of the ALOHA–protocol is well known to be $d_{\mathrm{A}} = \lambda e^{-2\lambda}$ [1], [14]. To obtain this result it is assumed that data packets arrive according to a homogeneous Poisson process with constant intensity $\lambda > 0$. The maximum throughput of ALOHA is $1/2e \approx 0.184$, which is achieved at $\lambda = 1/2$ (one packet is arriving for every two time units on the average). This is a relatively small value, and Roberts [12] published a protocol for doubling the capacity. His method is known as *slotted* ALOHA and works as follows. Time is divided into slots of just one packet length. A node is not allowed to broadcast whenever it wishes, but instead has to wait for the beginning of the next slot. Slotted ALOHA has been analyzed in an approximate model to have throughput $d_{\mathrm{sA}} = \lambda e^{-\lambda}$ which is maximum for $\lambda = 1$ with value $1/e$. Analogously to the continuous model $\lambda$ is the expected number of packets transmitted per slot. If two or more packets are transmitted in the same slot collisions occur and all packets involved are destroyed by superposition. The corresponding model assumes a large number of users $K$, each independently transmitting with equal probability $\lambda/K$. Throughput is obtained by considering the limit with $K \to \infty$ [12], [13].

Abramson [2] has investigated a particular model of this type with fixed traffic load $\lambda = 1$. He considered two groups of $n_1$ and $n_2$ users, respectively, each with different access probabilities $g_1$ and $g_2$. He concludes that the asymmetric case ($g_1$ large and $g_2$ small such that $n_1 g_1 + n_2 g_2 = \lambda = 1$) achieves large overall throughput.

We extend the results of Abramson by maximizing throughput with respect to a fixed traffic load over all access probabilities of $n$ individual users. As a special case we observe the claim of Abramson concerning asymmetricity, but only for small values of $\lambda$. For $\lambda$ approximately larger than $e = 2.718...$ again a symmetric distribution of traffic load turns out to be most favorable. For any fixed traffic load $\lambda$ we obtain access probabilities which maximize throughput. Moreover, it is proved that many possible local maxima are dominated by only two and in spite of the multimodality of the objective function, an essential maximum may be obtained without the utilization of the numerical comparison between many potential solutions used in [10].

Let us assume that each of $n$ nodes transmits in a slot independently of each other with probability $p_i$, $0 \leq p_i \leq 1$, $i = 1, \ldots, n$. The expected traffic load, i.e. the average number of packets transmitted in a slot, is $\lambda = \sum_{i=1}^{n} p_i$. The probability that a packet will be successfully transmitted in a particular slot (obviously coinciding with the throughput) is given by

$$d(p_1, \ldots, p_n) = \sum_{i=1}^{n} p_i \prod_{j \neq i} (1 - p_j). \tag{1}$$

## 2.2. Optimization Problem

Maximizing throughput with respect to a fixed traffic load $0 < \lambda \leq n$ may be formulated as to

$$\text{maximize } d(p_1, \ldots, p_n) \quad \text{such that} \quad 0 \leq p_i \leq 1, \ \sum_{i=1}^{n} p_i = \lambda > 0.$$

Let $p = (p_1, \ldots, p_n)$. In case $\lambda \leq 1$ the solution $p^* = (p_1^*, \ldots, p_n^*)$ is obvious since

$$d(p) \leq \sum_{i=1}^{n} p_i = \lambda,$$

and equality $d(p^*) = \lambda$ holds whenever $p_k^* = \lambda$ for some $k$, and $p_i^* = 0$ for $i \neq k$.

We now consider the more complicated case $\lambda > 1$, and first state some preliminary results concerning boundary points of the constraining set

$$C = \left\{ p = (p_1, \ldots, p_n) \mid 0 \leq p_i \leq 1, \sum_{i=1}^{n} = \lambda \right\}.$$

If $p_i = 1$ for at least two different components of $p$ we have $d(p) = 0$, which excludes $p$ as a maximum point. If just one component of $p$ equals 1, without loss of generality we may assume $p_n = 1$, then

$$d(p_1, \ldots, p_n) = \prod_{j=1}^{n-1} (1 - p_j).$$

## 2.3. Analysis of Local Maxima

It is easily concluded that the maximum of $d(p_1, \ldots, p_{n-1}, 1) = \prod_{j=1}^{n-1}(1-p_j)$ over $C$ is attained at $p_i^* = \frac{\lambda-1}{n-1}$, $i = 1, \ldots, n-1$, and by symmetry of $d$, for any vector with permuted components [10]. Therefore, $p^* = (p_1^*, \ldots, p_n^*)$, with $p_k^* = 1$ for just one component and $p_i^* = \frac{\lambda-1}{n-1}$ for $i \neq k$, is a candidate for a global maximum point over the boundary of $C$ with value $d(p^*) = (\frac{n-\lambda}{n-1})^{n-1}$.

The point $(0, ..., 0, \frac{\lambda-1}{k-1}, ..., \frac{\lambda-1}{k-1}, 1)$ (and symmetric points) with the throughput value $f(k) = (\frac{k-\lambda}{k-1})^{k-1}$ should be also considered as a candidate as local maximum point in $k-1$-dimensional hyperplane bounding the feasible set. But the function $f(\cdot)$ is monotonically increasing as it is shown in [11]. Therefore the local maximum $d(p^*)$ dominates the local maxima attained on the corners of lower dimensional faces of the feasible set.

We now investigate interior points of $C$ by a Lagrangian setup. We search for stationary points by solving the system of equations

$$\nabla d(p) \; + \gamma \, \nabla g(p) = 0,$$

where $g(p) = \sum_{i=1}^{n} p_i - \lambda = 0$ describes the restrictions. Carrying out differentiation yields the following system

$$\begin{aligned} \prod_{i \neq j}(1-p_i) - \sum_{k \neq j} p_k \prod_{i \neq j, i \neq k} (1-p_i) + \gamma &= 0, \quad j = 1, \ldots, n, \quad (2) \\ g(p) &= 0. \end{aligned}$$

After multiplying the $j$-th equation by $(1-p_j)$ we get

$$\prod_{i=1}^{n}(1-p_i) - \sum_{k \neq j} p_k \prod_{i \neq k}(1-p_i) + \gamma(1-p_j) = 0.$$

This leads to

$$\prod_{i \neq j}(1-p_i) + \gamma(1-p_j) = d(p) \quad \text{for all } j = 1, \ldots, n, \qquad (3)$$

and the differences of the $j$–th and $k$–th equation give

$$\Big( \prod_{i \neq j, i \neq k} (1-p_i) - \gamma \Big) (p_j - p_k) = 0 \quad \text{for all } j, k = 1, \ldots, n.$$

If $p_j \neq p_k$ for some $j \neq k$ it follows that $\prod_{i \neq j, i \neq k}(1-p_i) = \gamma$, and from $p_k \neq p_\ell$ for some $\ell \neq k$ it follows that $\prod_{i \neq k, i \neq \ell}(1-p_i) = \gamma$, which yields $p_j = p_\ell$. This shows that stationary points $p$ have at most two different components. Thus, each stationary point $p$ may be represented as $p = (a, \ldots, a, b, \ldots, b)$ with $k$ entries $a$ and $n-k$ of them equal to $b$, where $ka + (n-k)b = \lambda$, $k \in \{0, \ldots, n\}$.

$k = 0$ and $k = n$ means $a = b = \frac{\lambda}{n}$. By (2), with $\gamma = (\lambda - 1)(1 - \frac{\lambda}{n})^{n-2}$, it is easily verified that the corresponding point $p = (\frac{\lambda}{n}, \ldots, \frac{\lambda}{n})$ is stationary.

Now let $1 \leq k \leq n-1$, $0 \leq a, b < 1$, $a \neq b$, and consider the system (2) for corresponding stationary points $p$:

$$\begin{aligned} (1-a)^{k-1}(1-b)^{n-k} + \gamma(1-a) &= d(p), \\ (1-a)^{k}(1-b)^{n-k-1} + \gamma(1-b) &= d(p). \end{aligned}$$

By elementary transformations it follows that

$$\gamma = (1-a)^{k-1}(1-b)^{n-k-1}.$$

Substituting $\gamma$ in (2) and observing that

$$d(p) = (1-a)^{k-1}(1-b)^{n-k-1}(\lambda - nab) = \gamma(\lambda - nab)$$

we obtain the following system for stationary points with two different components:

$$\begin{aligned} nab - a - b &= \lambda - 2, \\ ka + (n-k)b &= \lambda, \\ 0 \le a, b < 1 \quad &, \quad a \ne b. \end{aligned} \tag{4}$$

The analysis of solutions of (4) in [11] shows that the best local maximum over the interior of $C$ is attained at the point $p^o = (\frac{\lambda}{n}, \ldots, \frac{\lambda}{n})$ with the value $\lambda \cdot (1 - \frac{\lambda}{n})^{n-1}$.

Comparison of $d(p^*)$ with $d(p^o)$ yields

$$\begin{aligned} d(p^o) > d(p^*),\ \lambda > (\frac{n}{n-1})^{n-1}, \\ d(p^o) \le d(p^*),\ \lambda \le (\frac{n}{n-1})^{n-1}. \end{aligned} \tag{5}$$

The local maxima over the interior of the $k-1$-dimensional face of the border of $C$ have similar structure as $p^o$, $\varphi(k) = d(p^o_k) = d(0, \ldots, 0, \frac{\lambda}{k}, \ldots, \frac{\lambda}{k}) = \lambda \cdot (1 - \frac{\lambda}{k})^{k-1}$. In general $\varphi$ is not monotonic function of $k$. However, the function $\varphi(k))$ is increasing function of $k$ for $\lambda > 2$ [11] and in this case the point $p^o$ dominates the maximum points (of similar structure) over interior of the faces of lower dimensionality. But the case $\lambda \le 2$ is not interesting because the point $p^o$ is dominated by the point $p^*$ according to (5).

Summarising we may conclude, that the maximum of $d(p)$ is attained at the point $p^o = (\frac{\lambda}{n}, \ldots, \frac{\lambda}{n})$ if $\lambda > (\frac{n}{n-1})^{n-1}$ and at the point $p^* = (\frac{\lambda-1}{n-1}, \ldots, \frac{\lambda-1}{n-1}, 1)$ if contrary.

# 3. Fast Local Descent Method for MDS

## 3.1. Formulation of Problem of MDS

Multidimensional scaling (MDS) is a technique widely used in different applications where the various patterns, represented by means of multidimensional vectors, should be analysed, classified or recognized [4], [7]. The practical aspects of the problem may be demonstrated by an example of visualization of the global optimization process, where analysis of two dimensional image of trial points, generated by an optimization algorithm in the multidimensional space, may be useful for interactive control of the optimization process [16].

The theoretical and algorithmic aspects of MDS are considered, e.g. in [4], [7], [8], [9]. Let us give a short formulation of the problem. The pairwise dissimilarities between $n$ objects are given by the matrix $(\delta_{ij}), i, j = 1, \ldots, n$, and it is supposed that $\delta_{ij} = \delta_{ji}$. The points $x_i \in R^m, i = 1, \ldots, n$ should be found which interpoint Euclidean distances fit given dissimilarities. The embedding Euclidean space $R^m$ normally is two dimensional ($m = 2$), although the other dimensionalities may also be of some interest. To find the points $x_i$ the *STRESS* function $f(X)$ should be minimized, where

$$f(X) = \sum_{i<j}^{n} w_{ij}(d_{ij}(X) - \delta_{ij})^2, \tag{6}$$

$X = (x_{11}, \ldots, x_{n1}, x_{12}, \ldots, x_{nm})$ and $d_{ij}(X)$ denotes the Euclidean distance between the points $x_i$ and $x_j$. It is supposed that the weights are positive: $w_{ij} > 0, i, j = 1, \ldots, n$. On one hand, MDS is related to global optimization as a difficult problem and, on the other hand, as a tool for visualization of the optimization process. It is shown in [9] that an evolutionary strategy, including a local descent, is efficient to minimize *STRESS* function. An efficient local algorithm is also an important part of some other global optimization algorithms. However, the local minimization of $f(X)$ is also not easy: normally the problems are of large dimensionality, $N = n \times m$, and the objective function is not everywhere differentiable. The known methods theoretically have a rate of convergence no better than superlinear [7],[8]. The quasi Newton methods are practically efficient for MDS of modest dimensionality and can not be implemented (at least for PC) in the case of dimensionality necessary for some interesting applications [9]. The conjugate gradient algorithms were also successfully tested for MDS problems. They have an important advantage: they may be easily implemented for the problems with several hundreds of patterns. The theoretical explanation of the efficiency of the conjugate gradient algorithm for MDS is given in this paper by means of regularization of problem (6) and by proving that in such a case the conjugate gradient algorithm converges with a quadratic rate.

## 3.2. Conditions of Local Minimum

It is well known [5] that $f(X)$ is differentiable at local minimum point, i.e. at local minimum point $X$ the following inequalities hold: $d_{ij}(X) >$

$0, i, j = 1, \ldots, n$ and:

$$\frac{\partial f(X)}{\partial x_{kh}} = 2\sum_{i \neq k} w_{ki}(1 - \frac{\delta_{ki}}{d_{ki}(X)})(x_{kh} - x_{ih}) = 0. \tag{7}$$

The positive definiteness of the Hessian at the minimum point is important for the investigation of convergence of local minimization algorithms. The second derivatives of $f(X)$ are equal to

$$\begin{aligned}
\frac{\partial^2 f(X)}{\partial x_{kh}^2} &= 2\sum_{j \neq k} w_{kj}\left(1 - \frac{\delta_{kj}((x_{kh} - x_{jh})^2 - d_{kj}^2(X))}{d_{kj}^3(X)}\right), \\
\frac{\partial^2 f(X)}{\partial x_{kh}\partial x_{sh}} &= -2w_{ks} + 2w_{ks}\frac{\delta_{ks}(-(x_{kh} - x_{sh})^2 + d_{ks}^2(X))}{d_{ks}^3(X)}, \\
\frac{\partial^2 f(X)}{\partial x_{kh}\partial x_{sr}} &= -2w_{ks}\frac{\delta_{ks}(x_{kh} - x_{sh})(x_{kr} - x_{sr})}{d_{ks}^3(X)}.
\end{aligned} \tag{8}$$

The investigation of the matrix with elements given by (8) is complicated. However, the needed conclusions on definiteness of $\nabla^2 f(X)$ may be made indirectly, analysing the directional derivatives of $f(X)$ at a local minimum point. Let us denote $\varphi(t) = f(X + t \cdot S)$, where $S = (s_{11}, \ldots, s_{nm}), s_i \in R^m, i = 1, \ldots, n$. The following expressions are valid

$$\varphi\prime(0) = 2\sum_{i<j} w_{ij}(1 - \frac{\delta_{ij}}{d_{ij}(X)})\sum_{h=1}^{m}(x_{ih} - x_{jh})(s_{ih} - s_{jh}), \tag{9}$$

$$\varphi\prime\prime(0) = 2\textstyle\sum_{i<j} w_{ij} \times$$

$$\times\left((1 - \frac{\delta_{ij}}{d_{ij}(X)})d_{ij}^2(S) + \frac{\delta_{ij}}{d_{ij}^3(X)}\left(\sum_{h=1}^{m}(s_{ih} - s_{jh})(x_{ih} - x_{jh})\right)^2\right) \tag{10}$$

If $X$ is a minimum point then the equality $\varphi\prime(0) = 0$ holds for every direction $S$. In the case $S = X$ the latter equality may be rewritten in the form

$$\varphi\prime(0) = \sum_{i<j} w_{ij}(d_{ij}^2(X) - d_{ij}(X) \cdot \delta_{ij}) = 0. \tag{11}$$

It is well known that for a twice differentiable function the second derivative at a local minimum point is non negative, but not necessary positive. From (10) it is obvious, that $\varphi''(0) = 0$ for every $S$ satisfying the equalities $s_{ih} = s_{jh}, i, j = 1, \ldots, n$. If the points $x_i$ are moving the same direction, then $d_{ij}, i, j = 1, \ldots, n$ remain constant, implying that (6) also remains constant. Therefore, this condition for $\varphi''(0) = 0$ corresponds to the invariance of (6) with respect to the translations of $R^m$.

Because of (10) and (11) the equality $\varphi''(0) = 0$ holds also for $S$ satisfying the equalities $d_{ij}(X) = d_{ij}(S)$ and

$$\sum_{h=1}^{m} (s_{ih} - s_{jh})(x_{ih} - x_{jh}) = 0.$$

It is shown in the Appendix, that the latter conditions correspond to the invariance of (6) with respect to rotations.

At a local minimum point the equality $\varphi''(0) = S \cdot \nabla^2 f(X) \cdot S$ holds for any direction $S$. Therefore the existence of $S$ with $\varphi''(0) = 0$ implies that the Hessian of $f(X)$ at local minimum point is not *positively* defined. To get a minimization problem with positively defined Hessian at the minimum point, the problem should be regularized excluding invariance with respect of translations and rotations. However, it remains not clear whether there do not exist other directions with $\varphi''(0) = 0$.

Let us consider the condition of positive definiteness of Hessian of function (6) at a local minimum point. Such a condition defines the subset of $R^N$, where the invariance of embedding space $R^m$ with respect of translations and rotations would be automatically excluded. To stress the dependence of $\varphi''(0)$ on $S$ the notation $\varphi''(0) = \varphi''(0, S)$ will be used below. The expression (10) may be rewritten as

$$\varphi''(0, S) = 2 \sum_{i<j} w_{ij} \Big( d_{ij}^2(S) - \frac{\delta_{ij}}{d_{ij}^3(X)} \Big( d_{ij}^2(S) \cdot d_{ij}^2(X) - \Big(\sum_{h=1}^{m} (s_{ih} - s_{jh})(x_{ih} - x_{jh})\Big)^2 \Big) \Big). \quad (12)$$

At a local minimum point of $f(X)$ the inequality $\varphi''(0, S) \geq 0$ holds for any $S$. Therefore the direction $S$ satisfying $\varphi''(0, S) = 0$ is a minimizer of $\varphi''(0, S)$ and, because of (12), it maximizes

$$d_{ij}^2(S) \cdot d_{ij}^2(X) - \Big( \sum_{h=1}^{m} (s_{ih} - s_{jh})(x_{ih} - x_{jh}) \Big)^2 \quad (13)$$

with respect to fixed $d_{ij}^2(S)$. It follows immediately from (13) that a minimizer of $\varphi''(0, S)$ satisfies the equality

$$\sum_{h=1}^{m}(s_{ih} - s_{jh})(x_{ih} - x_{jh}) = 0. \tag{14}$$

The comparison of (14) with (18) proves that equality $\varphi''(0, S) = 0$ is valid **only** for the directions $S$ corresponding to the orthogonal transformations of $R^m$.

## 3.3. Algorithm of Conjugate Gradients

It may be proved that a trajectory of local descent started at the point of differentiability of $f(X)$ escapes of the points of nondifferentiability [17]. Therefore a descent algorithm (based on derivatives) with a high convergence rate may be applied to minimize $f(X)$. For the test problems of modest dimensionality a quasi Newton method performed very well [9]. However, the quasi Newton methods normally are not considered as the prospective candidates because of the large dimensionality of the practical MDS problems. With respect to the convergence rate, quasi Newton and conjugate gradient methods are similar, but the latter are more suited to solve problems of large dimensionality. The Polak and Ribiere version of the conjugate gradient algorithm [3] is as follows:

$$X^{k+1} = X^k + t_k \cdot S^k, t_k = \text{argmin}_{t>0} f(X^k + t \cdot S^k), \tag{15}$$

$$S^k = G^k + \beta_k \cdot S^{k-1}, \beta_k = \frac{(G^k, G^k - G^{k-1})}{||G^{k-1}||^2}, G^k = -\nabla f(X^k), \beta_0 = 0.$$

It is well known [3] that the algorithm (15) converges to a local minimum with a quadratic rate if the following conditions are satisfied:

1 the Hessian of $f(X)$ is positive definite at the local minimum point,

2 the Hessian of $f(X)$ satisfies the Lipshits condition in the vicinity of local minimum point.

It is shown above, that the first of these conditions is not satisfied for the function $f(X)$ because of its invariance with respect to translations and rotations. This difficulty we will tackle by means of regularization. To show that the Hessian of *STRESS* satisfies the Lipschitz condition in the vicinity of local minimum point it is sufficient to show that the second derivatives of the function satisfy Lipschitz condition under the same assumption. At the local minimum point $X$ the inequality $d_{ij}(X) > 0$ holds implying that all the functions (9) are Lipschitzian.

## 3.4. Regularization

Obviously, distances in a Euclidean space are invariant with respect to translations and rotations of corresponding points. This means that any local, and also global minimum of (6) is attained at certain orbits under these groups. It is well known that sets of points with constant function values give rise to difficulties for almost all global and local optimization algorithms. To overcome these difficulties in the two-dimensional case ($m = 2$), the introduction of constraints

$$\sum_{i=1}^{n} x_{i1} = \sum_{i=1}^{n} x_{i2} = \sum_{i=1}^{n} x_{i1}x_{i2} = 0 \tag{16}$$

is heuristically proposed in [9]. The equalities above fix the origin and the orientation of the scale symmetrically with respect to all variables. By such an approach the unconstrained minimization of (6) is replaced by a constrained problem with (generally speaking, nonlinear) equality constraints. The invariance with respect of translations and rotations may be excluded by similar constraints also in the case ($m > 2$). However, we do not know any results that guarantee a quadratic convergence rate towards a local minimum points of any local minimization algorithm for such a constrained problem. Moreover, it is difficult to evaluate the trade off between difficulties caused by the invariance of the objective function with respect to translations and rotations and difficulties caused by the nonlinear constraints.

For the most important two dimensional case ($m = 2$) the equalities (16) may be analytically solved: three *dependent* variables may be expressed as functions of $2n - 3$ *independent* variables, e.g.

$$x_{n1} = -\sum_{i=1}^{n-1} x_{i1}, \quad x_{n2} = -\sum_{i=1}^{n-1} x_{i2},$$

$$x_{n-1,2} = \frac{-\sum_{i=1}^{n-2} x_{i1}x_{i2} - \sum_{i=1}^{n-1} x_{i1} \sum_{i=1}^{n-2} x_{i2}}{x_{n-1,1} + \sum_{i=1}^{n-1} x_{i1}}. \tag{17}$$

The minimization of (6) with constraints (16) reduces to unconstrained minimization of

$$F(x_{11}, \ldots, x_{n-1,1}, x_{12}, \ldots, x_{n-2,2}) = f(X),$$

where $x_{n1}, x_{n2}, x_{n-1,2}$ are substituted in $X$ with their expressions (17).

The derivatives of $F(\cdot)$ can be easily calculated combining (7) and derivatives of dependent variables with respect to independent variables. The zero value in denominator of (17) during the local descent can not be approached because it would cause the growth of $|x_{n-1,2}|$ to infinity and, consequently, similar growth of value of the objective function.

Summarising we may conclude, that the analysis of local optimality conditions implied the reformulation of the problem. The properties of the newly formulated problem guarantee quadratic convergence of the conjugate gradient algorithm.

## 3.5. Appendix

Let the point $X \in R^n$ move in the direction $S \in R^n$ which preserves the distances between the points $x_i, x_j \in R^m$, i.e. the direction $S$ corresponds to the orthogonal transformation of $R^m$. Then the following equalities hold (for $t = 0$):

$$\frac{\partial}{\partial t}\sum_{h=1}^{m}(x_{ih} - x_{jh} + t \cdot s_{ih} - t \cdot s_{jh})^2 = \sum_{h=1}^{m}(x_{ih} - x_{jh})(s_{ih} - s_{jh}) = 0. \quad (18)$$

The equalities (18) are obviously satisfied for $s_{ih} = s_{jh}, i, j = 1, \ldots, n, h = 1, \ldots, m$, corresponding to the translations of $R^n$. For the rotations of $R^m$ the folloeing equalities hold ($t = 0$):

$$\frac{\partial}{\partial t}\sum_{h=1}^{m}(x_{ih} + t \cdot s_{ih})^2 = 0, i, j = 1, \ldots, n,$$

which, together with (18), imply

$$\sum_{h=1}^{m}(x_{ih} \cdot s_{jh} + x_{jh} \cdot s_{ih}), i, j = 1, \ldots, n.$$

The explicit formula for directions $s_i$ corresponding to the rotations are given below. Let $u_k$, $k = 1, 2, 3$ be the orthonormal vectors in $R^m$, $m \geq 3$; $u_3 = 0$ in the case $m = 2$. Consider the rotation of the space around $u_3$.

The vectors $s_i$ representing the directions of movement of points $x_i$ may be expressed by the formula

$$s_i = u_1 \cdot t_{i2} - u_2 \cdot t_{i1},$$

where $t_{ik} = (x_i \odot u_k)$, and $\odot$ denotes the scalar product. The following equalities follow immediately

$$\sum_{h=1}^{m}(s_{ih} - s_{jh})(x_{ih} - x_{jh}) = \Big(u_1 \cdot (t_{i2} - t_{j2}) - u_2 \cdot (t_{i1} - t_{j1})\Big) \odot$$
$$\Big(u_1 \cdot (t_{i1} - t_{j1}) + u_2 \cdot (t_{i2} - t_{j2}) + u_3 \cdot (t_{i3} - t_{j3})\Big) =$$
$$(t_{i2} - t_{j2})(t_{i1} - t_{j1}) - (t_{i1} - t_{j1})(t_{i2} - t_{j2}) = 0.$$

It is obvious, that in the two dimensional case ($m = 2$) the projections $t_{i1}, t_{i2}$ are equal to the corresponding coordinates of point $x_i$ and $d_{ij}(X) = d_{ij}(S)$. Since $u_3$ may be chosen arbitrarily, e.g. corresponding to each coordinate axis of $R^m$, then the latter equalities hold also for $m > 2$.

# References

[1] Abramson, N. (1970), The ALOHA system — another alternative for computer communication, *AFIPS, Spring Joint Computer Conference*, **37**, pp. 281–285.

[2] Abramson, N. (1985), Development of the ALOHANET, *IEEE Transactions on Information Theory*, **IT 31**, pp. 119–123.

[3] Bazaraa, M., Sherali, H. and Shetty, C. (1993). *Nonlinear Programming*, Wiley, NY.

[4] Bong, I. and Lingoes, J. (1987), *Multidimensional Similarity Structure Analysis*, Springer, NY.

[5] de Leeuw, J. (1984), Differentiability of Kruskal's stress at a local minimum, *Psychometrika*, **49**, pp.111-113.

[6] Horst R. and Tuy H. (1990), *Global Optimization – Deterministic Approaches*, Springer, Berlin.

[7] Mathar, R. (1989), Algorithms in multidimensional scaling. In O.Opitz, (Ed.), *Conceptual and Numerical Analysis of Data*, Springer, Berlin. pp. 159-177.

[8] Mathar, R. (1995), Multidimensionale Skalierung, Mathematische Grundlagen und Algorithmische Konzepte, *Preprint*, RWTH Aachen. 44 pp.

[9] Mathar, R. and Žilinskas, A. (1993), On global optimization in two-dimensional scaling, *Acta Applicandae Mathematicae*, **33**, pp. 109-118.

[10] Mathar, R. and Žilinskas, A.(1993), Maximizing throughput of finite population slotted ALOHA, *Informatica*, **4**, No. 1-2, pp.163-171.

[11] Mathar, R. and Žilinskas, A.(1994), On channel access probabilities which maximize throughput of slotted ALOHA, *Informatica*, **5**, No.3-4, pp. 364-372.

[12] Roberts, L.G. (1972), Extensions of packet communication technology to a hand held personal terminal, *AFIPS, Spring Joint Computer Conference*, **40** , pp. 295–298.

[13] Schwefel, H.-P. (1995), *Evolution and Optimum Seeking*, Wiley, NY.

[14] Tanenbaum, A.S. (1988), *Computer Networks*, Prentice Hall, Englewood Cliffs, NJ.

[15] Tőrn, A. and Žilinskas, A. (1986), *Global Optimization*,Springer, Berlin.

[16] Žilinskas, A. (1993), On visualization of optimization process, In J.Guddat et al. (Eds.), *Parametric Optimization and Related Topics 3*, Peter Lang Verlag, Frankfurt am Main. pp.549-556.

[17] Žilinskas, A. (1996), A quadratically converging algorithm of multidimensional scaling, *Informatica*, **7**, No.2, pp.268-274.

Chapter 5

# MODELLING TECHNIQUES FOR NONLINEAR POOLING PROBLEMS

Mikael Rönnqvist

*Division of Optimization*
*Linköping Institute of Technology*
*Linköping, Sweden.*

miron@math.liu.se

**Abstract** The pooling problem is frequently found in refinery modeling. It is in some sense similar in structure to the well known blending problem but is a nonlinear problem. The standard approach in the industry to solve pooling models is to use a recursive linear programming approach. In this paper, we discuss a modeling technique which makes it possible to describe a nonlinear mixing of several, for example, crude oils in one single model which is not the case for standard approaches. This is based on establishing approximate cumulative functions of evaporation curves. Results from a case study show that the proposed model accurately describes the real application and remains efficiently solvable with standard software.

**keywords:** polling problems, blending problems, nonlinear optimization.

## 1. Introduction

The blending problem, see e.g. Winston[9], is a well known Operations Research (OR) problem arising in many applications. It can be modeled as a Linear Programming (LP) problem and is easily solvable and a global optimal solution is guaranteed. The pooling problem which also



*A. Migdalas et al. (eds.), From Local to Global Optimization,* 85–103.

is a well known OR problem is at first glance similar to the blending problem. However, there is a principal difference between a blending and a pooling problem which makes the pooling problem nonlinear. This is illustrated in Figure 5.1.

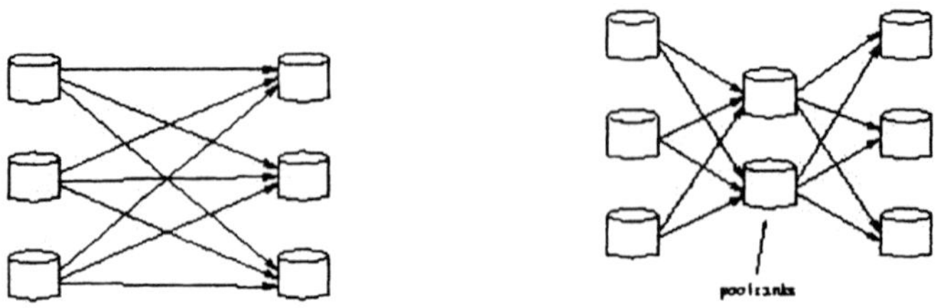

*Figure 5.1.* Illustration of a blending and a pooling problem.

In the blending problem (left in Figure), output products are blended directly from raw units. In the pooling problem (right) there is an intermediate step where the input streams are mixed together in pooling tanks and then output streams are mixed to obtain final products. The pooling problem can therefore be viewed as being two linked blending problems. Even though it may seem as if the pooling problem can be formulated as an LP problem, it is a nonlinear problem. The complication that arises is that the quality of the composition in the pooling tanks become unknown variables. Pooling problems arise in several applications. However, the industry where it is most frequently used is in the refinery industry.

Solution methods used in practice for pooling problems are often based on recursive LP models, see e.g. Fieldhouse[3] and Gill[6]. In these models, a first estimate of the quality of the pooling tanks initializes the process and an LP problem is solved in order to find the amount of different crudes to be used in the distillation process. Given a solution, it is possible to exactly compute the qualities of the pooling tanks. Thereafter, new estimates are made and another LP is formulated and solved. This process is repeated until it has converged, for example, when the qualities of the pools are within some tolerance between two consecutive iterations. Modeling systems which are based on this procedure are described in Fieldhouse [3]. There is, of course, no guarantee that such a system will find a global optimal solution. However, different tests show that the global solution is often found for practical problems if the solution process converges. Other problems with convergence using LP models is discussed in more detail in Fieldhouse [4].

Nowadays there are several nonlinear solvers available which makes it possible to solve large and detailed nonlinear models. One of the main problem in pooling models is how to model a nonlinear mixing of crudes directly in a single model. This instead of using a recursive procedure where a new approximate model is stated in each iteration. In this paper, we discuss a modeling approach which enables an accurate description of this nonlinear mixing. It is based on establishing approximate cumulative functions for the nonlinear yield functions involved in the distillation process. These cumulative functions are less nonlinear than the original yield functions which makes them more suitable in an optimization model. This because most optimization techniques are based on some approximation strategy and less nonlinear functions become more accurate in such techniques.

The outline of the paper is as follows. In Section 3 we start by describing a very simple pooling example and formulate it in a mathematical model. This model is further developed in a number of steps to include nonlinear mixing. In Section 3 we give a general description of a distillation process frequently found in oil refineries. In section 4 we decribe how the principles introduced for the example can be applied for more general problems. In Section 5 we describe a case study at the New Zealand Refining Company where this technique has been tested. Finally, we make some concluding remarks about the model and techniques used.

## 2. A simple pooling problem

To illustrate the pooling problem we will study an example taken from Haverly [7] and further discussed in Lasdon *et al.* [8]. Floudas and Pardalos [5] also give a number of testproblems with solutions pooling problems. In our modified example, see illustration in Figure 5.2, we have three input chemicals: $A$, $B$, and $C$. These chemicals are to be blended to provide two output products $Q$ and $R$. Customers have demands between 50 and 100 units of $Q$ and between 100 and 200 units of $R$, at given prices. The only impurity the chemicals are assumed to have is sulphur. Each chemical have a fixed cost per unit, $C_i, i = A, B, C$, and each product has a fixed profit per unit, $C_i, i = Q, R$. The problem is to operate the process in order to maximize profit.

The pooling appear because chemicals $A$ and $B$ are mixed in a pooling tank before any blending with chemical $C$. The nonlinearity occur due to the fact that the sulphur concentration of the pooling tank is unknown. We will use this example as an illustration and make it more complex so

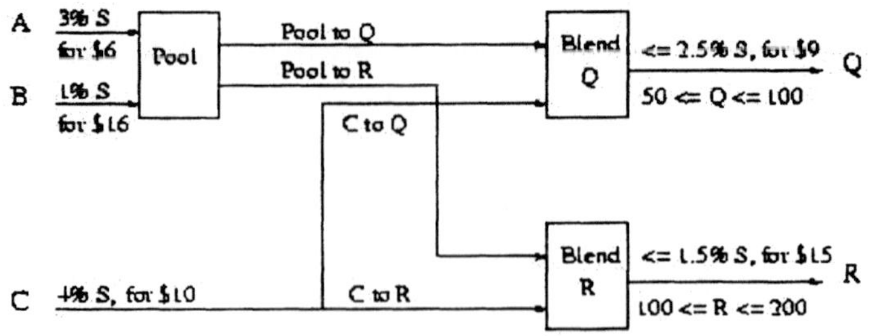


*Figure 5.2.* An example of a pooling problem.

it can be used to model a more complicated distillation process (decribed in the next section). We introduce the following variables.

| | | |
|---|---|---|
| $X_i$ | : | Amount (in units) of chemical or product $i$ used, $i = A, B, C, Q, R$. |
| $Y_j$ | : | Flow (in units) from the pooling tank to product $j$, $j = Q, R$. |
| $X_{Cj}$ | : | Flow (in units) from chemical $C$ to product $j$, $j = Q, R$. |
| $Z$ | : | Sulphur concentration (in percentage) of the pooling tank. |

We can then formulate the problem in a mathematical model given below.

$$\max \quad C_Q X_Q + C_R X_R - C_A X_A - C_B X_B - C_C X_C \tag{1}$$

subject to

$$\begin{aligned}
X_A + X_B &= Y_Q + Y_R && (2)\\
Y_Q + X_{CQ} &= X_Q && (3)\\
Y_R + X_{CR} &= X_R && (4)\\
X_{CQ} + X_{CR} &= X_C && (5)\\
Z * Y_Q + 4X_{CQ} &\leq 2.5X_Q && (6)\\
Z * Y_R + 4X_{CR} &\leq 1.5X_R && (7)\\
3X_A + X_B &= Z(X_A + X_B) && (8)\\
50 &\leq X_Q \leq 100 && (9)\\
100 &\leq X_R \leq 200 && (10)\\
1 &\leq Z \leq 3 && (11)
\end{aligned}$$

Here equation (2) gives the mass balance for the pool, whereas equations (3) and (4) give the massbalance for the two products, and equation (5) the balance for chemical $C$. The representation of the sulphur restrictions of the two output products is given in equations (6) and (7). The pool sulphur balance is stated in equation (8). We have restrictions on product demand in equations (9) and (10). Finally, we have a practical and physical restriction on the variable $Z$ in equation (11).

The original testproblem in Lasdon *et al.* [8] has two local optima and a third point that satisfies the Karush-Kuhn-Tucker conditions. We therefore know that the general pooling problem indeed is nonconvex. As in Lasdon *et al.* [8] we use the modelling language GAMS, see [2], to model and solve the problem. The best known solution to our modified problem is $X_A = 57.5, X_B = 172.5, X_C = 20.0, X_Q = 50.0, X_R = 200.0, X_{CQ} = 20.0, X_{CR} = 0.0, Y_Q = 30.0, Y_R = 200.0, Z = 1.5$. The objective function value is 145.0.

## 2.1. Temperature cut points

In industrial distillation, the actual flows from the distiller depends on so-called temperature cut points. These cut points are temperatures at which the crudes are separated into different output flows. This is illustrated in Figure 5.3. Note that the flow $X_A$ and $X_B$ actually mix before it is seperated into the two output flows $Y_Q$ and $Y_R$. The assumption we have on sulphur is easy in the sense that all chemicals mix evenly with regards to the sulphur. This is of course not generally the case and to make a first modification to take this into account we add a temperature variable $T$ to represent the cut point between the two output flows from the pooling tank, one each to the two products $Q$ and $R$. We also assume that the flow from the lower temperature interval determines the flow to product $R$. Normally the temperature would be between, say, 0 and 1000 Centigrades. In our example we have scaled the temperature so it is between 0 and 1. We then need to modify our model by introducing the following equations and restrictions.

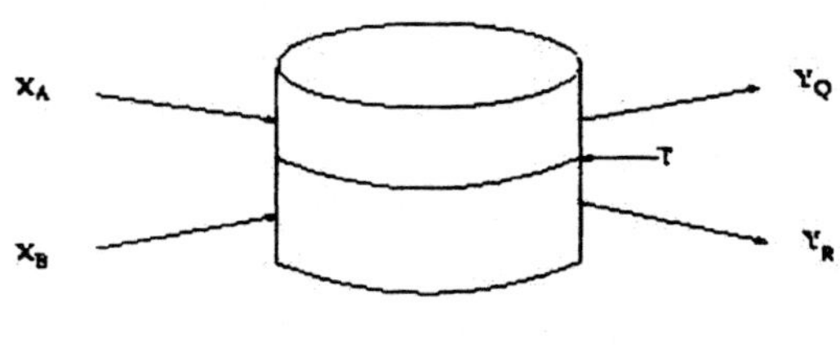


*Figure 5.3.* Illustration of a pooling tank with one temperature cut point.

$$Y_Q = (1-T)(X_A + X_B) \quad (12)$$
$$Y_R = T(X_A + X_B) \quad (13)$$
$$0 \leq T \leq 1 \quad (14)$$

As we still assume that chemicals $A$ and $B$ mix evenly, this modification will not change the optimal solution to the problem. However, as we impose more equations it becomes more difficult for the solver in GAMS to find the optimal solution. In fact, starting with default values, GAMS cannot find any feasible solution. Once GAMS is initialized with any initial solution for the variable $T$ above 0, say 0.5, GAMS finds the same solution as before (with objective value of 145.0). In this solution we have $T = 0.87$.

## 2.2. Nonlinear mixing

We will now consider the case when the chemicals do not mix evenly. To model a nonlinear mixing in our example we need to include equations that model this property. To make it simple we state that the two chemicals have a dependency on temperature and that their "mixing" density functions are

$$f_A(t) = 2t \quad (15)$$
$$f_B(t) = 1, \quad (16)$$

where $t$ is the temperature. Given these density functions we have that chemical $B$ has a constant mixing rate throughout the temperature scale whereas chemical $A$ has a linearly increasing contribution with respect to the temperature. To find the actual flow for product $R$ we need to integrate these density functions from 0 to the cut point $T$. This is, of course, not practical to perform in an optimization model. Instead we find the cumulative functions so that we can express the flow as a simple function value. By this, we can include it directly in the model. In our example, we then get the following cumulative functions.

$$F_A(t) = \text{int}\,{}_0^t f_A(u)du = t^2 \quad (17)$$
$$F_B(t) = \text{int}\,{}_0^t f_B(u)du = t \quad (18)$$

With these functions we can reformulate the mathematical model. Instead of discussing differences with the original model we state the entire new model. In this formulation we have introduced two new variables $F_A$ and $F_B$ which determines the fraction of each chemicals that will flow in the two output flows. In this case, the fraction given determines the flow to product $R$. These two variables are also used to balance the flow and the sulphur contents.

$$\max \quad C_Q X_Q + C_R X_R - C_A X_A - C_B X_B - C_C X_C \tag{19}$$

subject to

$$
\begin{aligned}
X_A + X_B &= Y_Q + Y_R && (20)\\
Y_Q + X_{CQ} &= X_Q && (21)\\
Y_R + X_{CR} &= X_R && (22)\\
X_{CQ} + X_{CR} &= X_C && (23)\\
F_A &= T^2 && (24)\\
F_B &= T && (25)\\
3(1 - F_A)X_A + (1 - F_B) * X_B + 2X_{CQ} &\leq 2.5X_Q && (26)\\
3F_A * X_A + F_B * X_B + 4X_{CR} &\leq 1.5X_R && (27)\\
(1 - F_A)X_A + (1 - F_B)X_B &= Y_Q && (28)\\
F_A * X_A + F_B * X_B &= Y_R && (29)\\
50 &\leq X_Q \leq 100 && (30)\\
100 &\leq X_R \leq 200 && (31)\\
0 &\leq F_A \leq 1 && (32)\\
0 &\leq F_B \leq 1 && (33)
\end{aligned}
$$

Only one temperature cut point makes the formulation easier than if we have two or several cut points (as in a practical situation). Also, in our example we have only one impurity, sulphur. In a real situation we would have several "impurities" to consider. These aspects would require several cumulative functions that governs the physical mixing of the chemicals. The best known solution for this modified problem is $X_A = 64.38, X_B = 172.21, X_C = 15.41, X_Q = 50.0, X_R = 200.0, X_{CQ} = 15.41, X_{CR} = 0.0, Y_Q = 34.59, Y_R = 200.0, F_A = 0.78, F_B = 0.88, T = 0.88$. The objective function value is 186.3.

## 3. Refinery distillation

In refinery modeling, the input streams represent crude oils, pooling tanks the crude distillers, and output streams the flow from the distillers to the product tanks. There are many processes that take place when crude oil is turned into higher value products such as petrol, jet fuel, gas oil and bitumen. Crude oil is not a homogeneous substance but is made up of many different hydrocarbon compounds. These compounds can contain any number of carbon atoms between one and about two hundred. As a general rule, the more carbon atoms there are in a compound, the higher the boiling point will be. This boiling point property is utilized in the separation process known as distillation. Crude is pumped from storage tanks into a distillation column which is then heated to a very high temperature (typically one thousand degrees centigrade). A simple schematic of a distillation column is illustrated in Figure 5.4.

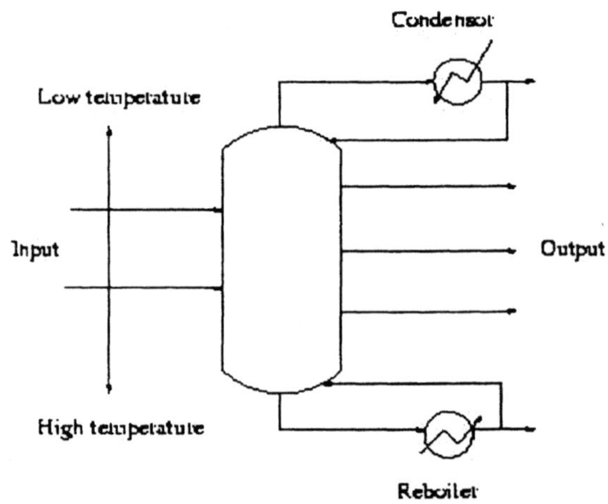


*Figure 5.4.* An illustration of a distillation column.

Further up the distillation column, the temperature will be lower and some of the crude that had previously evaporated will condense again. The material that has condensed here is called the fuel-oil fraction, and if the separation in the distillation column was perfect then all of the fuel-oil fraction would evaporate before any of the residue fraction evaporated. Still further up the distillation column the temperature will be even lower, and some more of the original crude that had a boiling point that was too low for the fuel-oil fraction (let alone the residue fraction) would condense here. This fraction is called the gas-oil fraction. The next fraction up is the kero fraction, and the lightest fraction of the crude is called the overheads fraction. A cut point between two consecutive fractions is defined as the temperature at which all of the lighter fraction will (in theory) be gaseous, and all of the heavier fraction will be liquid (or solid) at some reference conditions.

The objective in destillation is to minimize the overall cost which include costs of crude oils and production costs. Decisions that needs to be made are how much of each crude oil to use and which fraction of each that are to be processed in each of the distilling units. Decisions are also needed about temperature cut points in each distiller. This will determine the flows (and their properties) from the distillers to the tanks with the final products. There are various restrictions on the products. This include, for example, limits on density and sulphur concentration.

## 4. Modeling of nonlinear mixing

As discussed earlier in our example, there is a need to incorporate nonlinear mixing of chemicals (or crudes in refineries) in practical pooling problems. To develop a suitable modeling technique we will study mixing of different crude oils in a crude distilling unit (CDU) as we described earlier in Section 3. We will use evaporation curves and density specifications for crudes as the base for our studies. The same technique as we develop can also be used for any similar nonlinear behaviour in the mixing of crudes.

As an illustration of this we give an example of two crudes used in refinery distilling which have different density profiles. One important aspect of a crude oils is its evaporation properties. In Figure 5.5 we give an example of the evaporation-temperature plots of one of the light crude oils Maui, which is found in New Zealand. Contrasting to Maui was one of the more heavy crudes from the Middle East, Arab. It is clear that a major part of Maui evaporates at a low temperature whereas the Arab crude evaporates mainly at a higher temperature. Therefore, Arab would yield far more residue than Maui would.

### 4.1. Density restriction and evapouration curves

To model density restrictions we need to establish the average density of each of the output streams from the CDUs. Moreover, to compute

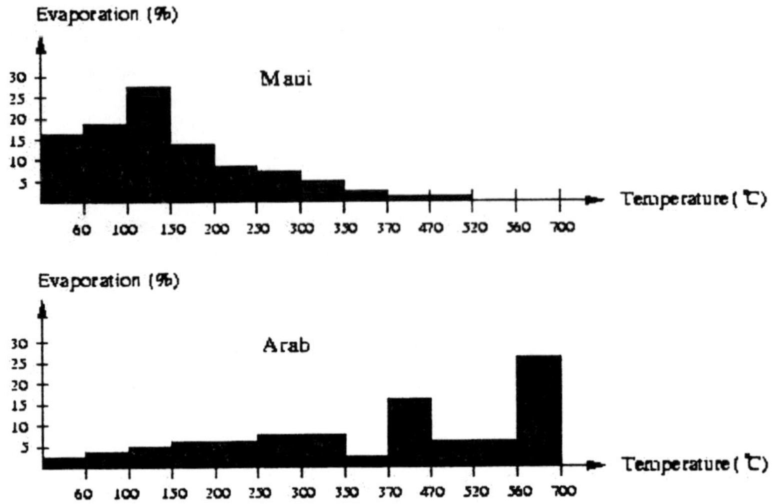


*Figure 5.5.* Examples of Evaporation-Temperature plots, here for the Maui and the Arab crude.

this we need to know the proportion of each crude that flows in each of the output streams (Note that there is not enough to know the flow of each crude into the CDUs and the flow out from the CDUs.). The quality of the output streams are, in turn, determined by the temperature cut points which in general are unknowns. Therefore, we must must be able to compute the fraction of any crude oil that evaporates between two given cut points. Basically, as discussed earlier, there is a need to integrate the distillation yield functions between any two cut points (which themselves are variables). The approach we aggain use is to establish cumulative functions i.e. the integrand of the distillation functions. Once these are determined we can analytically perform the integration between two cut points as the difference in objective function of the cumulative function given the cut points.

Evaporation curves are found by testing each crude oil separately. To do this, the volume of un-evaporated crude oil is measured at a certain temperature. The temperature is then increased and the remaining volume is re-measured. This is then repeated in a stepwise fashion until a full description of the evaporation is found. Evaporation values are therefore only given for a limited number of temperatures and these curves are just approximations of the true evaporation curves. We have already given examples of two such evaporation curves in Figure 5.5. As we can assume that evaporation percentage is directly proportional to density, it is possible to establish an average linear function for conversion purposes. This has the form $\rho = k_0 + k_1 t$, where $\rho$ and $t$ are the density and the temperature respectively, and $k_0$ and $k_1$ are constants.

## 4.2. Approximation scheme

There are two methodologies to compute these approximate cumulative functions. Firstly, we can approximate the distillation yield functions by a polynomial and then integrate it to obtain the cumulative function. Secondly, we can approximate the cumulative function directly with a polynomial. We have chosen the latter because the cumulative function is less nonlinear which makes it easier to establish high quality approximation. It is also easy to ensure that the cumulative function is exactly 1.0 (or 100%) at the maximum distillation temperature. Moreover, it is easier to include the restriction that the cumulative function must be monotonically increasing.

Given a distillation yield function $f_c(t)$ for crude $c$ we need to find

$$F_c(t) = \text{int}\ {}_0^t f_c(u)du, \tag{34}$$

where $F_c(t)$ is the cumulative function of $f_c(t)$ and $t$ is the temperature. However, $f_c(t)$ is given (see e.g. Figure 5.5), with certain values at a limited number of temperature points. As discussed earlier we approximate $F_c(t)$ with a polynomial function $a_0 + a_1 t + a_2 t^2 + \ldots + a_k t^k$ where the coefficients $(a_1, a_2, \ldots, a_k)$ are found by solving a constrained least square problem. Figure 5.6 gives the distillation yield function and the cumulative function for Maui. The dashed line illustrates the approximating polynomial for the cumulative function. (In our tests we used a maximum degree of four for this polynomial.)

Suppose $\hat{t}_i, \hat{F}_i, i = 1, \ldots, p$, are the measured cumulative evaporation values and their corresponding temperature points. The constrained least square problem can then be formulated as

$$\min \sum_{i=1}^{p} \left( \hat{F}_i - (a_1 \hat{t}_i + a_2 \hat{t}_i^2 + a_3 \hat{t}_i^3 + a_4 \hat{t}_i^4) \right)^2 \tag{35}$$

subject to

$$a_1 T_{max} + a_2 T_{max}^2 + a_3 T_{max}^3 + a_4 T_{max}^4 \quad = \quad 1.0 \tag{36}$$

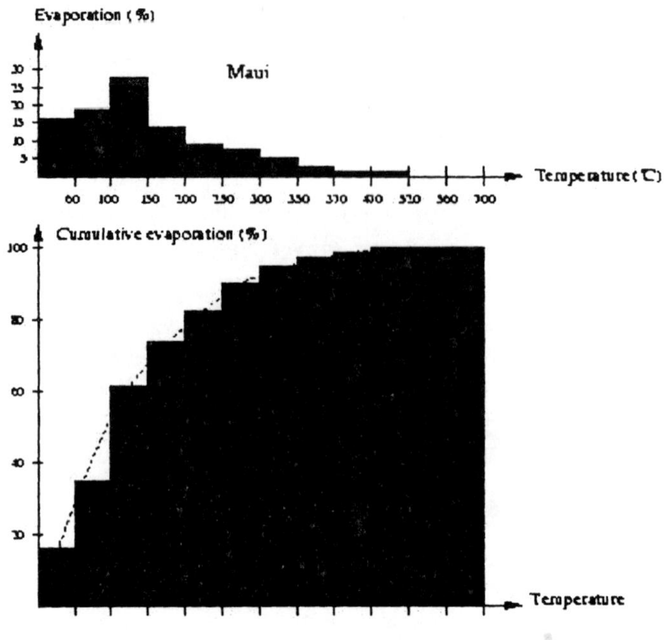


*Figure 5.6.* Approximate cumulative function for Maui.

$$a_1 + 2a_2\hat{t}_i + 3a_3\hat{t}_i^2 + 4a_4\hat{t}_i^3 \geq 0, \quad i = 1, \ldots, p. \tag{37}$$

The constraint (36) is to enforce the condition $F_c(T_{max}) = 1.0$ where $T_{max}$ is the maximum evaporation temperature. The restriction $F_c(0) = 0.0$ is already taken into account by omitting the constant term, $a_0$, in the polynomial. The remaining $p$ constraints in (37) is added to enforce the property that $F_c(t)$ is monotonically increasing. Ideally, we want $F_c'(y) = a_1 + 2a_2t + 3a_3t^2 + 4a_4t^3 \geq 0 \; \forall \; t$ but this would give an infinite number of constraints. In various tests we have found that using these $p$ constraints gives a very high quality approximation. When they were omitted we found that many crude oils obtained cumulative functions that were not monotonically increasing giving erroneous results.

Given cumulative functions, as Figure 5.7 illustrates, the percentage of crude from crude A for example, going to the first (lightest) output corresponds to the total percentage of evaporated crude at the first cut point (*cut* 1 in figure). Amount of crude A to the lightest output is then the cumulative function value, i.e. $F_A(cut\ 1)$, multiplied with the flow of crude A to this distiller. The total flow from the distiller to the lightest output is then just the sum of the corresponding parts over all crudes.

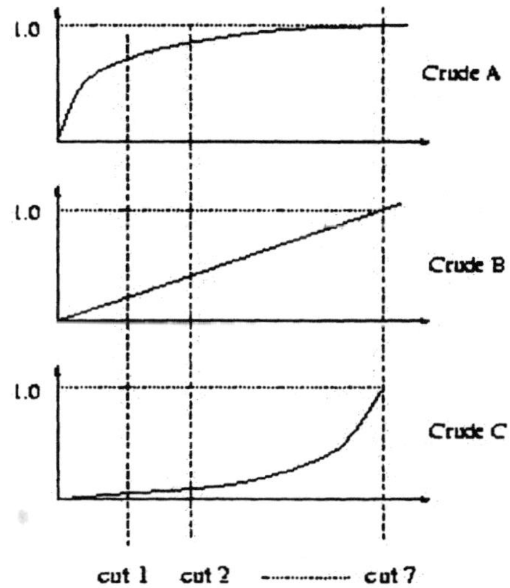


*Figure 5.7.* Illustration of how cut points are applied to cumulative distillation functions in one distiller.

## 5. A case study

The modelling technique described earlier has been tested in a project undertaken at the New Zealand Refining Company (NZRC), see Amos *et al.* [1] for a detailed description. The simplified schematic which were used at NZRC is given in Figure 5.8. The NZRC operates two CDUs and produces normally 7 output products. Standard refineries uses few crudes but at NZRC several oil companies co-operate and each company buy its own crudes which increases the overall number quite drastic. In this study we assume that the crudes mix nonlinearly with respect to density but that the sulphur mix evenly.

To formulate the model we need the following data.

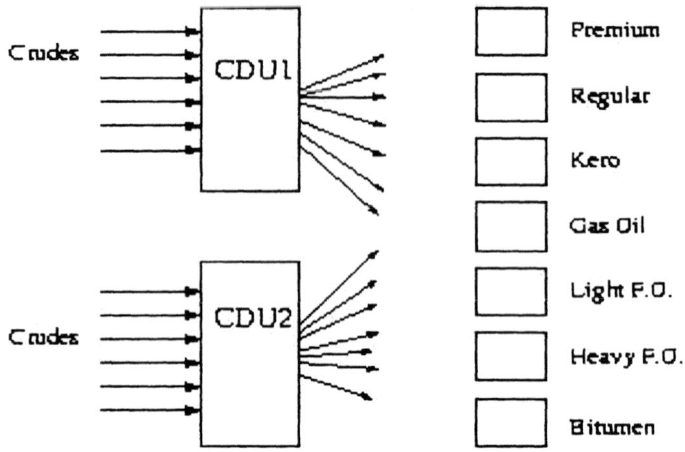


*Figure 5.8.* A simplified schematic of the pooling process.

| | | |
|---|---|---|
| $n$ | : | number of crude distillers. |
| $m$ | : | number of crude oils available. |
| $w$ | : | number of output products. |
| $C_c$ | : | cost per barrel of crude $c$. |
| $V_c$ | : | maximum available volume of crude $c$. |
| $S_c$ | : | sulphur content of crude $c$. |
| $\underline{S}_o$, $\overline{S}_o$ | : | lower and upper limit of sulphur specification for output product $o$. |
| $D_o$ | : | demand volume for output $o$. |
| $T_{max}$ | : | maximum evaporation temperature of the crude oils. |
| $\underline{D}_o$, $\overline{D}_o$ | : | lower and upper limit of density specification for output $o$. |

The model also uses the following variables.

| | | |
|---|---|---|
| $x_{cd}$ | = | volume of crude $c$ going to distiller $d$. |
| $y_{do}$ | = | volume of crude fraction going from distiller $d$ to output $o$. |
| $z_d$ | = | sulphur concentration in distiller $d$. |
| $w_{di}$ | = | temperature of cut point $i$ in distiller $d$. |
| $t_{do}$ | = | average evaporation temperature of the flow from distiller $d$ to output $o$. |
| $d_{do}$ | = | average density of the flow from distiller $d$ to output $o$. |

In addition cumulative functions for the distillation yield of the crudes are needed. Note that these are established *a priori* any model is formulated. The overall model is then as follows.

$$\min \sum_{c=1}^{m} \sum_{d=1}^{n} C_c x_{cd} \tag{38}$$

subject to

$$\sum_{d=1}^{n} x_{cd} \leq V_c, \quad c = 1, \ldots, m \tag{39}$$

$$\sum_{d=1}^{n} y_{do} = D_o, \quad o = 1, \ldots, w \tag{40}$$

$$z_d \sum_{c=1}^{m} x_{cd} - \sum_{c=1}^{m} S_c x_{cd} = 0, \quad d = 1, \ldots, n \tag{41}$$

$$\sum_{d=1}^{n} z_d y_{do} - \overline{S}_o \sum_{d=1}^{n} y_{do} \leq 0, \quad o = 1, \ldots, w \tag{42}$$

$$\sum_{d=1}^{n} z_d y_{do} - \underline{S}_o \sum_{d=1}^{n} y_{do} \geq 0, \quad o = 1, \ldots, w \tag{43}$$

$$\sum_{c=1}^{m} F_c(w_{d1}) x_{cd} - y_{d1} = 0, \quad d = 1, \ldots, n \tag{44}$$

$$\sum_{c=1}^{m} F_c(w_{do}) x_{cd} - \sum_{c=1}^{m} F_c(w_{d,o-1}) x_{cd} - y_{do} = 0, \quad d = 1, \ldots, n, o = 2, \ldots, w \tag{45}$$

$$\sum_{d=1}^{n} d_{do} y_{do} \leq \overline{D}_o, \quad o = 1, \ldots, w \tag{46}$$

$$\sum_{d=1}^{n} d_{do} y_{do} \geq \underline{D}_o, \quad o = 1, \ldots, w \tag{47}$$

$$2t_{do} - w_{do} = 0 \quad d = 1, \ldots, n, \; o = 1 \tag{48}$$

$$2t_{do} - w_{do} - w_{d,o-1} = 0 \quad d = 1, \ldots, n, \; o = 2, \ldots, w \tag{49}$$

$$d_{do} - k_1 t_{do} = k_0 \quad d = 1, \ldots, n, \; o = 1, \ldots, w \tag{50}$$

$$w_{do} - w_{d,o+1} \leq 0, \quad d = 1, \ldots, n, \; o = 1, \ldots, w \tag{51}$$

$$w_{do} \geq 0, \quad d = 1, \ldots, n, \; o = 1, \ldots, w \tag{52}$$

$$x_{cd} \geq 0, \quad c = 1, \ldots, m, \; d = 1, \ldots, n \tag{53}$$

$$y_{do} \geq 0, \quad d = 1, \ldots, n, \; o = 1, \ldots, w \tag{54}$$

Here, Equations (39) and (40) give limits on supply of crudes and demand of products, respectively. Equation (41) state the relationship for the sulphur concentration in the CDUs. Equations (42) and (43) give the sulphur restrictions. Equations (44) and (45) give the equation that controls the flows out from the CDUs to the output products based on the cut points. Density restrictions are stated through equations (46)-(50). The remaining equations are nonnegativity restrictions on variables and the fact that the temperatures for consecutive cut points must be increasing.

## 5.1. Results

The modeling language GAMS was used to formulate and solve the model. GAMS was also used to solve the constrained least square problems needed for the approximate cumulative functions. Six different crudes and seven products were used in the test. Some data about crudes and products is given in Tables 5.1 and 5.2. The final pooling problem in this test had 201 variables and 232 constraints. The overall solution time (including generation and execution time) was just 3.9 seconds. Crude volume going into each CDU is given in Table 5.3. As the table shows, not all the crude oils were selected for processing. Although Dubai oil is not the most expensive crude, the decision not to select it would have been based on the other factors, i.e. its sulphur content and density distribution. Temperature cut points and flow from the two CDUs to the output products are given in Tables 5.4 and 5.5. The maximum temperature was 700.

| Crude | Maui | Arun | Kutubu | Murban | Dubai | Arab |
|---|---|---|---|---|---|---|
| Cost/barrel ($NZ) | 29.5 | 30.3 | 29.3 | 28.3 | 26.1 | 23.5 |
| Max volume (barrels) | 600 | 600 | 600 | 600 | 600 | 600 |
| Sulphur contents (%) | 0.01 | 0.02 | 0.04 | 0.80 | 2.04 | 2.92 |

*Table 5.1.* Some data for the available crudes.

# 6. Concluding remarks

The pooling problem is an important practical problems arising in refineries. Standard solution approaches based on LP techniques is not capable of handling a nonlinear mixing of crudes. We have introduced a modelling technique where approximate cumulative functions are used

| | Premium | Regular | Kero | Gasoil |
|---|---|---|---|---|
| Output demand (barrels) | 400 | 450 | 350 | 750 |
| Min (Max) sulphur contents(%) | 0.0 (.03) | 0.0 (.2) | 0.0 (.4) | 0.2 (.6) |
| Min (Max) density (kg/l) | 0.72 (.75) | 0.72 (.75) | 0.76 (.79) | 0.82 (.85) |
| | Light FO | Heavy FO | Bitumen | |
| Output demand (barrels) | 150 | 200 | 100 | |
| Min (Max) sulphur contents(%) | 0.9 (1.8) | 0.9 (1.8) | 1.5 (2.5) | |
| Min (Max) density (kg/l) | 0.86 (.89) | 0.90 (.93) | 1.01 (1.04) | |

*Table 5.2.* Data for output products.

| CRUDE | Maui | Arun | Kutubu | Murban | Dubai | Arab |
|---|---|---|---|---|---|---|
| CDU1 | 473.1 | 600.0 | 600.0 | 12.3 | - | - |
| CDU2 | 128.9 | - | - | 203.4 | - | 384.4 |

*Table 5.3.* Results for the flows of crudes.

| Cut-point | 1 | 2 | 3 | 4 | 5 | 6 | 7 |
|---|---|---|---|---|---|---|---|
| CDU1 | 59.8 | 134.9 | 207.6 | 495.8 | 700.0 | 700.0 | 700.0 |
| CDU2 | 0.0 | 25.8 | 55.1 | 239.5 | 341.0 | 542.0 | 700.0 |

*Table 5.4.* The cut points in temperature in each CDU.

to model the nonlinear evaporation functions which makes this possible. The proposed model also includes cut points as decision variables which also is a major difference as compared to traditional approaches. The constrained least square problem to establish each cumulative function is convex which guarantees an optimal solution and they are easily solved. Furthermore, the cumulative functions are much less nonlinear than the original distillation functions. Results from a case study show that the proposed model indeed accurately describes the real problem and that it remains efficiently solvable.

| | Premium | Regular | Kero | Gasoil | Light FO | Heavy FO | Bitumen |
|---|---|---|---|---|---|---|---|
| CDU1 | 400.0 | 413.3 | 310.1 | 525.1 | 36.9 | - | - |
| CDU2 | - | 36.7 | 39.9 | 224.9 | 113.1 | 200.0 | 100.0 |

*Table 5.5.* Results for the flows from the CDUs.

# References


[1] Fransiska A, Gill G, and Rönnqvist M, Modelling the pooling problem at the New Zealand Refinery Company, Journal of the Operational Research Society, Vol. 48, No. 8, pp 767-778, 1997.

[2] Brooke A, Kendrick D, and Meeraus A (1992). GAMS: A User's Guide, Release 2.25, The Scientific Press, San Fransisco.

[3] Fieldhouse M (1993). The Pooling Problem, Chapter 13 in Optimization in Industry, (Eds) T.A. Ciriani and R.C. Leachman, John Wiley & Sons.

[4] Fieldhouse M (1995). More Convergent Recursion, Paper presented at the MUGA 16 Conference, March 19-21, Fremantle, Australia.

[5] Floudas CA and Pardalos PM (1990), Lecture Notes in Computer Science: A Collection of Test problems for Constrained Global optimization Algorithms, Chapter 6, Springer-Verlag, Berlin.

[6] Gill G (1995). Linear Programming as a Tool for Refinery Planning, Proceedings of the 31st Annual Conference of the Operational Research Society of New Zealand, Wellington, New Zealand, August 31 - September 1, 103-110.

[7] Haverly CA, Studies of the Behaviour Recursion for the Pooling problem, SIGMAP Bulletin, No 25, pp 19-28, 1978.

[8] Lasdon LL, Plummer J, and Waren AD, Nonlinear Programming, in Avriel M and Golany B (eds) Mathematical Programming for Industrial Engineers, Marcel Dekker, New york, 1996.

[9] Winston WL (1993). Operations Research: Applications and Algorithms, 3rd Edition, Duxbury Press, Belmont, USA.

# Chapter 6

# OPTIMIZATION AND HIERARCHIES FOR LUMPED DISTRIBUTED NETWORKS

Alain Bui

*LaRIA, Universit Picardie Jules Verne, 5 rue du Moulin Neuf, 80000 Amiens*

alain.bui@u-picardie.fr

**Abstract** The present paper is a generalization to lumped networks of our previous results [2, 5] on a stochastic model for performance evaluation of various distributed algorithms. In this model, a network is represented by N appropriate finite Markov chains, connected by relations $\mathcal{R}$ between entries of their transition matrices. "Solutions" $\rho$ for a functioning of such a network are represented by vectors whose components are entries of the mentioned transition matrices satisfying $\mathcal{R}$. The criterion of good functioning is based on the definition of a function $F$, named the guide function, to be optimized under constraints. The choice of $F$ depends on the context. Two approaches are proposed: the first one uses mean recurrence times of ergodic states of the Markov chains; the other one uses mathematical expectations of random numbers of passages from states to others. These two approaches can lead to the same guide function $F$, which induces a hierarchy on "solutions" - the first order hierarchy, so named because conditional expectation and mathematical expectation are (probabilistic) first order moments. In this hierarchy, the best "solutions" $\rho$ are optimal points of the optimization problem. When two "solutions" are equivalent in the first order hierarchy, we make use of variances for deciding between them. The better is the one with, globally, smaller variances. A function of variances is then defined, which generates a new hierarchy, named the second order hierarchy, because variance and conditional variance are (probabilistic) second order central moments. We show that the two mentioned approaches lead to two different ways for the second order hierarchy. As applications, we examine some particular problems: a problem of routing algorithms, the mutual exclusion problem and the dining philosophers' problem. In



*A. Migdalas et al. (eds.), From Local to Global Optimization,* 105–132.

addition to optimization, fuzzy sets are built which provide a comprehensive view of sets of "solutions".

**Keywords:** distributed computing,lumped networks, Markov chains, guide function, solution hierarchies, routing algorithms, the mutual exclusion problem, the dining philosophers' problem.

# 1. Introduction

The study of good functioning properties is an interesting problem in distributed computing. In our previous papers (cf. [2, 5]), we solved problems of rating parameters for a good functioning of Markov networks by optimizing a guide function $F$ with appropriate values of the parameters. This choice of $F$ is made either with the help of a probabilistic reasoning concerning the mean recurrence times of ergodic states, or with the help of a statistic reasoning. In the present paper, we deal with the following innovation: we extended results to lumped networks. On the other hand, in particular problems, new forms of constraints have been examined.

We quote [1, 7, 12, 13, 14] for generalities (models and applications) on Distributed Computing, and we refer to [10, 11] for fundamental results on Probability theory and Stochastic processes, and [9] for Optimization.

# 2. Preliminaries

Let $(X_t)_{t=0,1,2,\dots}$ be a homogeneous Markov chain with discrete time defined on a probability space $(\Omega, \mathcal{A}, Pr)$ and which (finite) state space is denoted by $\mathcal{E}$. Let $P$ be its transition matrix, which entries are $p(x,e)$, $(x,e) \in \mathcal{E} \times \mathcal{E}$, and let $P^t$, be the $t^{th}$ power of $P$, which entries are $p_t(x,e), (x,e) \in \mathcal{E} \times \mathcal{E}$.

In the lumping of states, when we consider a set $B$ of states ($B \subset \mathcal{E}$) and a state $x \in \mathcal{E}$, we denote by $p(x,B)$ the transition probability from $x$ to $B$ instead of $\sum_{e \in B} p(x,e)$.

Consider a finite partition $\{B_i, i \in S\}$ of the state space $\mathcal{E}$, and let $\mathcal{B}$ be the Boolen algebra generated by $\{B_i, i \in S\}$. Every $B \in \mathcal{B}$ can be expressed as an union of some $B_i \in \mathcal{B}$. Consider now a sample $(X_0(\omega), X_1(\omega), ..., X_n(\omega))$ corresponding to the point $\omega \in \Omega$. For $(B, B') \in \mathcal{B} \times \mathcal{B}$, let $n(B \times B'; \omega)$ be the number of direct transitions from $B$ to $B'$, in this sample. Recall that $n(B \times B')$ as function of $\omega$, is a random variable, while $n(B \times B'; \omega)$ is a number.

Suppose now that, with respect to the partition $\{B_i, i \in S\}$, the Markov chain is lumpable (cf. [11]), that is to say, with the grouping of states into the $B_i$, $i \in S$, the transition from set $B_i$ to set $B_j$ is also markovian, and generates a new Markov chain, named the lumped Markov chain. A necessary and sufficient condition for that (cf. [11]) is that for every pair of sets $B_i$ and $B_j$, $p(x, B_j)$ have the same value for every $x \in B_i$. These common values $m(i,j)$, $(i,j) \in S \times S$, form the transition matrix $M$ for the lumped Markov chain $(Y_t)_{t=0,1,2,...}$ which state space is $S$. Let us denote by $m_t(i,j)$ the entries of $M^t$. Suppose now that this lumped chain has only one acyclic ergodic class of states. Then $\lim_{t\to\infty} M^t$ exists and is a matrix with identical rows $\mu = (\mu(j), j \in S)$, that is to say that $\forall (i,j) \in S^2, \lim_{t\to\infty} m_t(i,j) = \mu(j)$. In the following, we agree to consider that the lumped Markov chain is in the permanent case, that is to say that its absolute stationary probability is $\mu$. We name condition $(\mathcal{H})$, the abovementioned conditions imposed to the lumped Markov chains.

Consider now a sample $(Y_0(\omega), ..., Y_n(\omega))$ corresponding to the point $\omega \in \Omega$, and let us denote by $n_{ij}(\omega)$ the number of direct transitions from $i$ to $j$ in this sample. As function of $\omega$, $n_{ij}$ is a random variable. $n_{ij}(\omega)$ is its value corresponding to $\omega \in \Omega$.

By the expression

$$n_{ij}(\omega) = \sum_{t=0}^{n-1} \mathbf{I}_{(i,j)} \left( Y_t(\omega), Y_{t+1}(\omega) \right)$$

where $\mathbf{I}_{(i,j)}(y, y') = 1$ if $(y, y') = (i,j)$ and $\mathbf{I}_{(i,j)}(y, y') = 0$ if $(y, y') \neq (i,j)$, we can easily obtain the mathematical expectation of $n_{ij}$

$$E(n_{ij}) = n.\mu(i).m(i,j).$$

We establish now some preliminary results which will be used latter:

**Proposition 3** *If, with respect to the partition $\{B_i, i \in S\}$, the Markov chain is lumpable, and if it satisfies the condition $(\mathcal{H})$, then $\forall i \in S$ and $\forall x \in \mathcal{E}$, $\lim_{t\to\infty} p_t(x, B_i)$ exists and is independant of $x$. Let $\Pi(B_i)$ be this limit. Then the mathematical expectation of the random variable $n(B_i \times B_j)$ is*

$$E[n(B_i \times B_j)] = n \cdot \Pi(B_i) \cdot p(x, B_j)$$

*for any state $x$ of $B_i$. In particular,*

$$E\left[n\left(B_i \times \mathcal{E}\right)\right] = n \cdot \Pi\left(B_i\right)$$

*Proof.* First, let us prove that

$\forall t \in \{1, 2, \ldots\}$, $\forall (i, j) \in S \times S$, $m_t(i, j) = p_t(x, B_j)$ for any state $x \in B_i$.

Indeed, for $t = 2$,

$$m_2(i, j) = \sum_{s \in S} m(i, s).m(s, j) = \sum_{s \in S} p(x, B_s).p(y, B_j)$$

for any state $x \in B_i$, and any state $y \in B_s$, because of lumpability. Hence,

$$\begin{aligned} m_2(i, j) &= \sum_{s \in S} \sum_{y \in B_s} p(x, y).p(y, B_j) \\ &= \sum_{y \in \bigcup_{s \in S} B_s} p(x, y).p(y, B_j) \\ &= \sum_{y \in \mathcal{E}} p(x, y).p(y, B_j) \\ &= p_2(x, B_j) \end{aligned}$$

Suppose now that the relation is true for $t - 1$, then

$$\begin{aligned} m_t(i, j) &= \sum_{s \in S} m(i, s).m_{t-1}(s, j) \\ &= \sum_{s \in S} p(x, B_s).p_{t-1}(y, B_j) \ (x \in B_i \text{ and } y \in B_s) \\ &= \sum_{s \in S} \sum_{y \in B_s} p(x, y).p_{t-1}(y, B_j) \\ &= \sum_{y \in \bigcup_{s \in S} B_s} p(x, y).p_{t-1}(y, B_j) \\ &= \sum_{y \in \mathcal{E}} p(x, y).p_{t-1}(y, B_j) \\ &= p_t(x, B_j), \end{aligned}$$

that is to say that this relation is valid for $t$. Hence $\lim_{t \to \infty} p_t(x, B_j)$ exists, and is independant of $x$, since $\lim_{t \to \infty} m_t(i, j) = \mu(j)$ exists, and is independant of $i$. We point out that the existence of the "individual" limit $\lim_{t\to\infty} p_t(x,y)$, $y \in B_j$ is not proved; we haved only showed that the "collective" limit $\lim_{t\to\infty} p_t(x, B_j) = \Pi(B_j)$ exists.

Then $E[n(B_i \times B_j)] = n\cdot \Pi(B_i)\cdot p(x, B_j)$, $x \in B_i$, results from $E(n_{ij}) = n \cdot \mu(i) \cdot m(i,j)$. ■

As corollary, we have the following result:

**Proposition 4** *If, with respect to partition $\{B_i, i \in S\}$, the Markov chain is lumpable, if the lumped Markov chain satisfies the condition $(\mathcal{H})$ and if $i_0 \in S$ is a recurrent state of the lumped chain, then*

$$T_{i_0} \cdot E[n(B_{i_0} \times \mathcal{E})] = n,$$

*where $T_{i_0}$ denotes the mean recurrence time of state $i_0$ in the lumped chain.*

**Proof** First, let us recall that $T_{i_0}$ is the conditional expectation of the random number $f_{i_0}$ of transitions of the lumped chain before entering $i_0$ for the first time when starting from initial state $i_0$, i.e. $E(f_{i_0}/Y_0 = i_0)$.

State $i_0$ being recurrent, its mean recurrence time in the lumped chain is $1/\mu(i_0)$, with $\mu(i_0) > 0$. Then, we have $T_{i_0} = 1/\Pi(B_{i_0})$. Consequently, following proposition 3,

$$\forall j \in S,\ \tfrac{1}{n} \cdot E[n(B_{i_0} \times B_j)] = \tfrac{1}{T_{i_0}} \cdot p(x, B_j),$$

which implies that

$$\tfrac{1}{n} \cdot \textstyle\sum_{j\in S} E[n(B_{i_0} \times B_j)] = \tfrac{1}{T_{i_0}} \cdot \sum_{j\in S} p(x, B_j)$$

viz.

$$\tfrac{1}{n} \cdot E\left[n\left(B_{i_0} \times \left(\textstyle\bigcup_{j\in S} B_j\right)\right)\right] = \tfrac{1}{T_{i_0}} \cdot p(x, \textstyle\bigcup_{j\in S} B_j)$$

viz.

$$\tfrac{1}{n} \cdot E[n(B_{i_0} \times \mathcal{E})] = \tfrac{1}{T_{i_0}} \cdot p(x, \mathcal{E}) = \tfrac{1}{T_{i_0}}$$

■

Consider now $n_{iA} = \sum_{j\in A} n_{ij}$ in the lumped chain, and let us denote by $\sigma^2(n_{iA})$ its variance. In particular, $n_{iS} = \sum_{j\in S} n_{ij}$ is nothing but $n(B_i \times \mathcal{E})$ in the initial chain. By simplification, we denote it by $n_{i\cdot}$.

**Proposition 5**

$$\sigma^2\left(\frac{n_{iA}}{\sqrt{n}}\right) = \mu(i) \cdot m(i, A)[1 - \mu(i) \cdot m(i, A)]$$

$$+2\sum_{t=0}^{n-2}\frac{n-t-1}{n}\mu(i)\sum_{j\in A}m(i,j)\cdot[m_t(j,i)-\mu(i)]\,m(i,A).$$

*In particular,*

$$\sigma^2\left(\frac{n_{i\cdot}}{\sqrt{n}}\right)=\mu(i)\,(1-\mu(i))+2\mu(i)\sum_{t=0}^{n-2}\frac{n-t-1}{n}\left[m_{t+1}(i,i)-\mu(i)\right]. \quad (1)$$

*Proof.* Since

$$n_{ij}(\omega)=\sum_{t=0}^{n-1}\mathbf{1}_{(i,j)}\left(Y_t(\omega),Y_{t+1}(\omega)\right),$$

one has

$$(n_{ij})^2=\sum_{t=0}^{n-1}\left[\mathbf{1}_{(i,j)}\left(Y_t,Y_{t+1}\right)\right]^2+2\sum_{(t,t'):t>t'}\mathbf{1}_{(i,j)}\left(Y_t,Y_{t+1}\right)\cdot\mathbf{1}_{(i,j)}\left(Y_{t'},Y_{t'+1}\right).$$

It implies that

$$\begin{aligned}E\left[(n_{ij})^2\right] &= \sum_{t=0}^{n-1}E\left[\left(\mathbf{1}_{(i,j)}\left(Y_t,Y_{t+1}\right)\right)^2\right]\\ &\quad +2\sum_{(t,t'):t>t'}E\left[\mathbf{1}_{(i,j)}\left(Y_t,Y_{t+1}\right)\cdot\mathbf{1}_{(i,j)}\left(Y_{t'},Y_{t'+1}\right)\right]\\ &= n\mu(i)m(i,j)+2\sum_{t=0}^{n-2}(n-t-1)\mu(i)m(i,j)m_t(j,i)m(i,j).\end{aligned}$$

As

$$2\sum_{t=0}^{n-2}(n-t-1)+n=n^2,$$

we have the variance

$$\begin{aligned}\sigma^2(n_{ij}) &= E\left[(n_{ij})^2\right]-\left[2\sum_{t=0}^{n-2}(n-t-1)+n\right]\left[E\left(\frac{n_{ij}}{n}\right)\right]^2\\ &= n\mu(i)m(i,j)+2\sum_{t=0}^{n-2}(n-t-1)\mu(i)m(i,j)m_t(j,i)m(i,j)\end{aligned}$$

$$
\begin{aligned}
& - \left[ 2 \sum_{t=0}^{n-2} (n-t-1) + n \right] [\mu(i) m(i,j)]^2 \\
= \; & n\mu(i)m(i,j) \left[1 - \mu(i)m(i,j)\right] \\
& +2 \sum_{t=0}^{n-2} (n-t-1) \mu(i) m(i,j) \left[m_t(j,i) - \mu(i)\right] m(i,j)
\end{aligned}
$$

And then,

$$
\sigma^2 \left( \frac{n_{ij}}{\sqrt{n}} \right) = \mu(i)m(i,j) \left[1 - \mu(i)m(i,j)\right]
$$

$$
+2 \sum_{t=0}^{n-2} \frac{n-t-1}{n} \mu(i) m(i,j) \left[m_t(j,i) - \mu(i)\right] m(i,j) \qquad (2)
$$

More generally, if $A$ is a subset of state of $\mathcal{E}$, let $n_{iA}$ may be expressed as

$$
\sum_{t=0}^{n-1} \mathbf{1}_{(i,A)}(y_t, y_{t+1}),
$$

where

$$
\mathbf{1}_{(i,A)}(y_t, y_{t+1}) = \begin{cases} 1 \text{ if } y_t = i, \;\; y_{t+1} \in A \\ 0 \text{ otherwise} \end{cases}
$$

Therefore,

$$
E\left(n_{iA}\right) = n \cdot \mu(i) \cdot m(i,A)
$$

and

$$
\begin{aligned}
E\left( (n_{iA})^2 \right) \; = \; & \sum_{t=0}^{n-1} E\left[ \left( \mathbf{1}_{(i,A)}\left(Y_t, Y_{t+1}\right) \right)^2 \right] \\
& +2 \sum_{(t,t'):t>t'} E\left[ \mathbf{1}_{(i,A)}\left(Y_t, Y_{t+1}\right) \cdot \mathbf{1}_{(i,A)}\left(Y_{t'}, Y_{t'+1}\right) \right] \\
= \; & n \cdot \mu(i) \cdot m(i,A) \\
& +2 \sum_{t=0}^{n-2} (n-t-1) \cdot \mu(i) \sum_{j \in A} m(i,j) \cdot m_t(j,i) \cdot m(i,A).
\end{aligned}
$$

And we easily obtained the expression of $\sigma^2 \left(n_{iA}/\sqrt{n}\right)$ in proposition 5. In particular, since $\sum_{j \in S} m(i,j) = 1$ and $\sum_{j \in S} m(i,j) \cdot m_t(j,i) = m_{t+1}(i,i)$, we deduce from $\sigma^2 \left(n_{iA}/\sqrt{n}\right)$ the expression of $\sigma^2 \left(n_{iS}/\sqrt{n}\right)$, viz. the expression of $\sigma^2 \left(n_{i\cdot}/\sqrt{n}\right)$.■

# 3. Lumped networks and hierarchies based on optimization

Let us recall that we used a stochastic model for networks which allows a behavioral study of distributed computing, and it usefulness is proved (cf. [2, 5, 6]). This model can be easily adapted to lumped networks. In the following, when we deal with many Markov chains, an index $k$ will be added concerning the $k^{th}$ chain, like ${}^kX_t$, ${}^k\Pi(B)$, etc., except the case of supersposition of index $k$ like ${}^kn\left({}^kB_i \times B_j\right)$, etc. In this particular case, the first index $k$ can be dropped and we denote simply $n\left({}^kB_i \times B_j\right)$ etc.

Consider $N$ processors, represented by $N$ random functions with discrete time. $\left({}^kX_t\right)_{t=0,1,2,\ldots}$, defined on a probability space $(\Omega, \mathcal{A}, Pr)$ evolving as $N$ finite homogeneous Markov chains, which state spaces are respectively ${}^k\mathcal{E}$, $k \in \{1, \ldots, N\}$. Let us denote by ${}^kP$ the transition matrix of the $k^{th}$ chain, and by $\left({}^kP\right)^t$ the $t^{th}$ power of ${}^kP$.

Suppose that, for every $k \in \{1, \ldots, N\}$, there exists a partition $\leftarrow {}^kB_i,\ i \in {}^kS \rightarrow$ of the state space ${}^k\mathcal{E}$, with respect to which, the $k^{th}$ Markov chain is lumpable. Let us denote by $({}^kY_t)_{t=0,1,2,\ldots}$ the $k^{th}$ lumped chain: its state space is ${}^kS$, its transition matrix is ${}^kM = \left({}^km(i,j)\right),\ (i,j) \in {}^kS \times {}^kS$, where ${}^km(i,j) = {}^kp\left(e, {}^kB_j\right),\ e \in {}^kB_i$. Moreover, suppose that for every $k \in \{1, \ldots, N\}$, the lumped chain $({}^kY_t)_{t=0,1,2,\ldots}$ satisfies the conditions $(\mathcal{H})$ mentionned in the preceding paragraph (one acyclic ergodic class and permanent case with the stationary absolute probability ${}^k\mu = \left({}^k\mu(i), i \in {}^kS\right)$).

These $N$ Markov chains form a network if there is a finite number of relations $\mathcal{R}_h,\ h \in H$, between the matrices ${}^kM,\ k \in \{1, \ldots, N\}$.

## 3.1. Two equivalent approaches in the first order hierarchy

Following the context, we have to define a real function $F$ of these ${}^km(i,j),\ (i,j) \in {}^kS \times {}^kS,\ k \in \{1, \ldots, N\}$, then to find solutions $\rho$, viz. the ${}^km(i,j)$'s, which satisfy the relations $\mathcal{R}_h, h \in H$, and which optimize the function $F$. So, a solution $\rho = (\rho_1, \ldots, \rho_N)$ is a $N$-uple which $k^{th}$ component is the stochastic matrix ${}^kM$. A solution $\rho'$ is said to be better than a solution $\rho$ iff $F(\rho') < F(\rho)$ (resp. $F(\rho') > F(\rho)$), when the rule is the minimization (resp. the maximization). This solution $\rho'$ is optimal (when it exits) iff it is the best, viz. iff $F(\rho') = \inf_\rho F(\rho)$ (resp.

$F(\rho') = \sup_\rho F(\rho))$. Two solutions $\rho$ and $\rho'$ are said to be equivalent iff $F(\rho') = F(\rho)$. The ideal is to find the optimal solution $\rho$ and make use of it when it exists. Moreover, total or partial "maps" of sets of solutions will be useful.

The choice of $F$ must be argufied. Two approaches can be used. The first one uses the mean recurrence times ${}^kT_i$, $i \in {}^kU$, ${}^kU$ being a set of "useful" ergodic states of the lumped chains, $k \in \{1, \ldots, N\}$. In the second approach, we consider $({}^kX_0, \ldots, {}^kX_n)$, $k \in \{1, \ldots, N\}$; $F$ can be defined with the help of $(1/n)E\left[n\left({}^kB_i \times {}^k\mathcal{E}\right)\right]$, $k \in \{1, \ldots, N\}$.

We pointed out that $\left({}^kX_0(\omega), \ldots, {}^kX_n(\omega)\right)$, $k \in \{1, \ldots, N\}$, are samples with the same size $n$; they allows the observation of the evolutions of the network until time $n$. $E\left[n\left({}^kB_i \times {}^k\mathcal{E}\right)\right]$ is the theoritical mean of the random number of visits of the $k^{th}$ chain to the set ${}^kB_i$. By proposition 3, we see that this theoritical mean divided by $n$ is independant of $n$.

These two approaches can be equivalent. Indeed, in the definition of $F$ in the first approach,

$$F\left({}^kT_i,\ i \in {}^kU,\ k \in \{1, \ldots, N\}\right),$$

it is sufficient (following proposition 4) to substitute $n/\left(E\left[n\left({}^kB_i \times {}^k\mathcal{E}\right)\right]\right)$ for ${}^kT_i$, and then we rediscover the $F$ of the second approach.

We named first order hierarchy, the hierarchy of the solutions generated by the guide function $F$ since it is defined through the mathematical expectations (first order moments): either the ${}^kT_i$'s or the $E\left[n\left({}^kB_i \times {}^kB_j\right)\right]$'s. So the two abovementioned approaches can be equivalent in the first order hierarchy.

## 3.2. Two non equivalent approaches in the second order hierachy

When two solutions $\rho$ and $\rho'$ are equivalent in the first order hierarchy, how to decide between them? Our idea is make use of variances for deciding between them, since the variance expresses the dispersion of values around the mean. This new criterion of preference generates a second order hierarchy; we name it the second hierarchy since variance and conditional variance are second order central moments.

Recall that under hypothesis $(\mathcal{H})$, the lumped chains $(Y_t)_{t=0,1,2,\dots}$ have only one acyclic ergodic set.

The first approach is based on the mean recurrence times ${}^kT_i$, of ergodic states $i$. More generally, let $i$ and $i'$ be two ergodic states of the $k^{th}$ lumped chain, and consider ${}^kT_{ii'} = E\left({}^kf_{i'}/{}^kY_0 = i\right)$, viz. the conditional mathematical expectation of the random number ${}^kf_{i'}$ of transitions of the $k^{th}$ lumped chain before entering $i'$ for the first time when starting from initial state $i$. (${}^kT_i$ is the particular case of ${}^kT_{ii'}$ when $i = i'$)

Let ${}^k\tau$ be the matrix which entries are ${}^kT_{ii'}$, let ${}^kQ$ be $lim_{t\to\infty}\left({}^kM\right)^t$, let ${}^kZ = \left(I - {}^kP + {}^kQ\right)^{-1}$, let ${}^kZ_{dg}$ result from ${}^kZ$ by setting off-diagonal matrix entries equal to 0, let ${}^k\Delta$ be the diagonal matrix with $i^{th}$ entry $1/{}^k\mu(i)$, and let $J$ be the matrix with all entries 1, then

$$ {}^k\tau = \left(I - {}^kZ + J\cdot {}^kZ_{dg}\right)\cdot {}^k\Delta \tag{3} $$

The conditional variances $\sigma^2\left({}^kf_{i'}/{}^kY_0 = i\right)$ are given by the matrix ${}^kV$ (which entries are ${}^kv_{ii'}$):

$$ {}^kV = {}^k\tau\cdot\left[\left(2\cdot {}^kZ_{dg}\cdot {}^k\Delta\right) - I\right] + 2\left[{}^kZ\cdot {}^k\tau - J\left({}^kZ\cdot {}^k\tau\right)_{dg} - {}^k\tau_{sq}\right], \tag{4} $$

where ${}^k\tau$ is given by expression (3), where ${}^k\tau_{sq}$ results from ${}^k\tau$ by squaring each entry, and where $\left({}^kZ\cdot {}^k\tau\right)_{dg}$ results form $\left({}^kZ\cdot {}^k\tau\right)$ by setting off-diagonal entries equal to zero.

Since ${}^kV$ depends on ${}^kM$, i.e. it depends on $\rho$, we denote it by ${}^kV(\rho)$. Since $\sum_{k=1}^{N}\sum_{i\in {}^kU} {}^kv_{ii}(\rho)$ is a kind of global dispersion of values around the means of ${}^kf_i$, $k \in \{1,\dots,N\}$, we use it for a criterion of comparison: let $\rho$ and $\rho'$ be equivalent in the first order hierarchy; $\rho$ is said to be better than $\rho'$ in the second order hierarchy iff

$$ \sum_{k=1}^{N}\sum_{i\in {}^kU} {}^kv_{ii}(\rho) < \sum_{k=1}^{N}\sum_{i\in {}^kU} {}^kv_{ii}(\rho'). \tag{5} $$

The second approach consists in using variances of $n\left({}^kB_i \times \mathcal{E}\right)$, (i.e. of ${}^kn_{i\cdot}$), $i \in {}^kU$, $k \in \{1, \ldots, N\}$, more precisely in using

$$W = \lim_{n\to\infty} \sum_{k=1}^{N} \sum_{i\in {}^kU} \sigma^2\left({}^kn_{i\cdot}/\sqrt{n}\right).$$

Of course, $W$ depends on $\rho$, so we denote it by $W(\rho)$.

Let us justify this choice. First $n$ being the size of the samples, the greater it is, better is the precision. Secondly, $i$ being ergodic, it is well-known (cf. [10]) that for every $j \in S$, $m_t(j,i)$ converges to $\mu(i)$ exponentially when $t \to \infty$; therefore, by its expression in proposition 5, $\sigma^2\left({}^kn_{i\cdot}/\sqrt{n}\right)$ converges to a finite limit when $n \to \infty$. $W(\rho)$ being then finite and representing a global dispersion of values around the means of the ${}^kn_{i\cdot}$'s, we can use it for the comparison in the second order hierarchy as follows: let $\rho$ and $\rho'$ be equivalent in the first order hierarchy; $\rho$ is said to be better than $\rho'$ in the second order hierarchy iff

$$W(\rho) < W(\rho'). \tag{6}$$

In practice, the choice of the ${}^kU$'s, on which depend $F$, $W$, etc., depends on the context, as show the examples hereinafter. Details concerning these problems can be found in [2, 6], though the constraints are not the same. In the present paper, we only study the technical part concerning optimization and hierarchies.

## 4. Applications to some problems of distributed algorithms

### 4.1. Routing algorithms

In this problem, for any $k \in \{1, \ldots, N\}$, the $k^{th}$ Markov chain is the following: the state space ${}^k\mathcal{E}$ is partitioned into five subsets $\{{}^kB_1, \ldots, {}^kB_5\}$ and the transition matrix ${}^kP = \left({}^kp(x,e)\right)$ is such that

$\forall x \in {}^kB_1$,

$$\begin{cases} {}^kp\left(x, {}^kB_2\right) = 1 \\ {}^kp\left(x, {}^kB_j\right) = 0, \quad j \in \{1,3,4,5\} \end{cases}$$

$\forall x \in {}^kB_2$,

$$\begin{cases} {}^kp\left(x, {}^kB_1\right) = 0 \\ {}^kp\left(x, {}^kB_2\right) = \Gamma - b_k \\ {}^kp\left(x, {}^kB_3\right) = b_k \\ {}^kp\left(x, {}^kB_4\right) = c \\ {}^kp\left(x, {}^kB_5\right) = 1 - \Gamma - c, \quad \text{where } 0 < b_k < \Gamma,\ 0 < \Gamma + c < 1. \end{cases}$$

$\forall x \in {}^kB_3$,

$$\begin{cases} {}^kp\left(x, {}^kB_2\right) = d_k \\ {}^kp\left(x, {}^kB_3\right) = 1 - d_k & \text{where } 0 < d_k < 1 \\ {}^kp\left(x, {}^kB_j\right) = 0, & j \in \{1,4,5\} \end{cases}$$

$\forall x \in {}^kB_4$,

$$\begin{cases} {}^kp\left(x, {}^kB_1\right) = 1 \\ {}^kp\left(x, {}^kB_j\right) = 0, \quad j \in \{2,3,4,5\} \end{cases}$$

$\forall x \in {}^kB_5$,

$$\begin{cases} {}^kp\left(x, {}^kB_1\right) = 1 \\ {}^kp\left(x, {}^kB_j\right) = 0, \quad j \in \{2,3,4,5\} \end{cases}$$

It is easy to see that

**Proposition 6** *With the respect to the partition* $\{{}^kB_1, \ldots, {}^kB_5\}$ *of the state space* ${}^k\mathcal{E}$, *the* $k^{th}$ *Markov chain is lumpable. The* $k^{th}$ *lumped chain is a homogeneous Markov chain with 5 states, which transition matrix* ${}^kM = \left({}^km(i,j)\right)$ *is such that*

$$\begin{array}{ll} {}^km(1,2) = 1, & \\ {}^km(2,1) = \Gamma - b_k, & {}^km(2,3) = b_k,\ {}^km(2,4) = c,\ {}^km(2,5) = 1 - \Gamma - c, \\ {}^km(3,2) = d_k, & {}^km(3,3) = 1 - d_k, \\ {}^km(4,1) = 1, & \\ {}^km(5,1) = 1. & \end{array}$$

*The other* ${}^km(i,j)'s$ *are 0.*

In the present case, for any $k \in \{1, \ldots, N\}$, the five states of the $k^{th}$ lumped chain have the following meaning: initialization state, waiting state, update state, link crash failure state and link insertion state. It is clear that every matrix ${}^kM$ is identified by the pair $(b_k, d_k)$. The guide function $F$, when defined, shall be a function of the $(b_k, d_k)$, $k \in \{1, \ldots, N\}$ and in the optimization of this function, every solution is a $2N$-uple $((b_1, d_1), \ldots, (b_N, d_N))$ denoted by $(b, d)$ by simplification.

In the context of this problem, the $N$ lumped chains are connected by the relations $\sum_{k=1}^{N} b_k - B \geq 0$ and $\sum_{k=1}^{N} d_k - D \geq 0$ where $B$ and $D$ are numbers $> 0$.

**Proposition 7** *With the partition $\{{}^kB_1, \ldots, {}^kB_5\}$, the $k^{th}$ lumped chain has only one acyclic ergodic class, and in the permanent regime, we have*

$$\begin{aligned}
E\left[n\left({}^kB_1 \times {}^k\mathcal{E}\right)\right] &= n(1-\Gamma)d_k/L_k \\
E\left[n\left({}^kB_2 \times {}^k\mathcal{E}\right)\right] &= nd_k/L_k \\
E\left[n\left({}^kB_3 \times {}^k\mathcal{E}\right)\right] &= nb_k/L_k \\
E\left[n\left({}^kB_4 \times {}^k\mathcal{E}\right)\right] &= ncd_k/L_k \\
E\left[n\left({}^kB_5 \times {}^k\mathcal{E}\right)\right] &= n(1-\Gamma-c)d_k/L_k
\end{aligned}$$

*where $L_k = b_k + (3 - 2\Gamma)d_k$.*

**Proof** Indeed, denoting by ${}^k\mu = \left({}^k\mu(1), \ldots, {}^k\mu(5)\right)$ the stationary absolute probability of the $k^{th}$ lumped chain, the resolution of the equation ${}^k\mu \cdot {}^kM = {}^k\mu$ gives ${}^k\mu(1) = (1-\Gamma)d_k/L_k$ , ${}^k\mu(2) = d_k/L_k$, ${}^k\mu(3) = b_k/L_k$, ${}^k\mu(4) = cd_k/L_k$, ${}^k\mu(5) = (1-\Gamma-c)d_k/L_k$. Then using, $E\left[n\left({}^kB_i \times {}^k\mathcal{E}\right)\right] = n \cdot {}^k\mu(i)$, we obtain the announced results. ∎

As exposed in [4], the criterion of choice is that state 3 should appear as frequent as possible and that state 2 less frequent as possible. So, for every $k \in \{1, \ldots, N\}$, ${}^kU = \{2, 3\}$. Following proposition 4 and proposition 7, we have

**Proposition 8** *The guide function $F$, to be minimized, can be expressed either by*

$$\sum_{k=1}^{N} \frac{E\left[n\left({}^kB_2 \times {}^k\mathcal{E}\right)\right]}{E\left[n\left({}^kB_3 \times {}^k\mathcal{E}\right)\right]} \quad \textit{or by} \quad \sum_{k=1}^{N} \frac{{}^kT_3}{{}^kT_2}$$

*Both of them lead to the expression*

$$F\left[(b_1, d_1), \ldots, (b_N, d_N)\right] = \sum_{k=1}^{N} \frac{d_k}{b_k}.$$

In order to solve the problem, we restrict the range of the $b_k$'s and the $d_k$'s, and then the problem becomes the following one:

*Problem "Pb1":*

$$\begin{array}{ll} \text{Minimize} & \displaystyle\sum_{k=1}^{N} \frac{d_k}{b_k} \\ \text{under the constraints} & \displaystyle\sum_{k=1}^{N} b_k \geq B, \ \sum_{k=1}^{N} d_k \geq D, \\ & \varepsilon \leq b_k \leq \Gamma - \varepsilon, \ \eta \leq d_k \leq 1 - \eta, \ k \in \{1, \ldots, N\}, \end{array}$$

where $\varepsilon$ and $\eta$ are arbitrary small positive numbers such that $N(\Gamma - \varepsilon) \geq B, \ N(1 - \eta) \geq D, \ N\eta < D$.

**Proposition 9** *(i) Optimal solutions $(b, d)$ in the first order hierarchy (viz. solutions of the problem "Pb1") are those such that*

$$b_1 = \ldots = b_k = \ldots = b_N = \Gamma - \epsilon \ \textit{and} \ \sum_{k=1}^{N} d_k = D$$

*(ii) Let $\Phi : (b, d) \longrightarrow [0, 1]$ be the function*

$$\Phi(b, d) = \frac{N(1 - \eta) - (\Gamma - \epsilon)F(b, d)}{N(1 - \eta) - D},$$

*then $\Phi^{-1}(]0, 1])$ is a fuzzy subset of solutions in the first order hierarchy.*

**Proof**

(i) A feasible solution $(b,d)$ of problem "Pb1" can be optimal if $b_k = \Gamma - \epsilon$, $k \in \{1,\ldots,N\}$. Indeed, whatever be the feasible solution $(b',d)$ such that $b'_k \neq \Gamma - \epsilon$, for at least one $k \in \{1,\ldots,N\}$,

$$\sum_{k=1}^{N} \frac{d_k}{b'_k} > \sum_{k=1}^{N} \frac{d_k}{\Gamma - \epsilon}$$

shows that $(b,d)$ is a better feasible solution. Hence, we can set $b_k = \Gamma - \epsilon$, $k \in \{1,\ldots,N\}$, and problem "Pb1" is equivalent to the following linear program "LP":

*Problem "LP":*

$$\text{minimize} \quad \sum_{k=1}^{N} d_k$$

$$\text{where} \quad \sum_{k=1}^{N} d_k \geq D, \ \eta \leq d_k \leq 1 - \eta, \ k \in \{1,\ldots,N\}.$$

Since $N\eta < D$, a feasible solution of "LP" is optimal iff $\sum_{k=1}^{N} d_k = D$.

Indeed, if $(b,d')$ is such that $\sum_{k=1}^{N} d'_k > D$ then there exists $k_0$ such that $\eta < d'_{k_0}$, and a better feasible solution can be obtained by slightly diminishing $d'_{k_0}$ while keeping the other $b_k$'s, $d_k$'s unchanged. Therefore, an optimal solution of problem "Pb1" is any $(b,d)$ such that $b_k = \Gamma - \epsilon$ , $\epsilon \leq b_k \leq \Gamma - \epsilon$, $k \in \{1,\ldots,N\}$ and $\sum_{k=1}^{N} d_k = D$.

(ii) The maximum of $F$ being $N(1-\eta)/(\Gamma - \epsilon)$, it is easy to see that $\Phi = (\max F - F)/(\max F - \min F)$ is the announced expression, which allows a partial "map" of the set of solutions in the first order hierarchy.

■

**Remark 1** *Since there exists several optimal solutions in the first order hierarchy, the problem of deciding between these equivalent solutions arises. We can make use of the second order hierarchy. In the approach 3.2, by a programming software we had the analytic expression of $V(b,d)$ (cf. [5]), and consequently, the one of $V(b,d)$ when*

*$b_1 = \ldots = b_N = \Gamma - \epsilon$ . It's a function $\tilde{V}(d)$ of the form $\sum_{k=1}^{N} \left({}^kC_1/d_k{}^2 + {}^kC_2/d_k + {}^kC_3 + {}^kC_4 d_k + {}^kC_5 d_k{}^2\right)$ to be minimized under the constraint $\sum_{k=1}^{N} d_k = D$.*

*(More precisely, ${}^kC_1 = (\Gamma - \epsilon)(2 - \Gamma + \epsilon)$, ${}^kC_2 = (\Gamma - \epsilon)(4\Gamma - 5)$, ${}^kC_3 = 4\Gamma(1 - \Gamma)$, ${}^kC_4 = \left[2(3 - 2\Gamma)^2 + (\Gamma - \epsilon)(3 - 4\Gamma)\right]/(\Gamma - \epsilon)^2$, ${}^kC_5 = -(3 - 2\Gamma)^2/(\Gamma - \epsilon)^2$ ).*

*Actually, we are able to compare, in the second order hierarchy, two given equivalent solutions in the first order hierarchy, but the optimal solution in the second order hierarchy remains to be investigated. On the other hand, the approach 3.2 allows only a comparison in the second order hierarchy by numerical values of two given equivalent solutions in the first order hierarchy. Investigation problem of the optimal solution in the second order hierarchy remains then open.*

**Remark 2** *If the condition $\sum_{k=1}^{N} b_k \geq B$ is changed to $\sum_{k=1}^{N} b_k = B$, then the problem is equivalent to a concave minimization problem under linear constraints. Indeed, setting*

$$g(d) = \min \leftarrow \sum_{k=1}^{N} \frac{d_k}{b_k} | \sum_{k=1}^{N} b_k = B, \epsilon \leq b_k \leq \Gamma - \epsilon, k \in \{1, \ldots, N\} \rightarrow \quad (7)$$

*Problem "Pb1" reduces to*

$$\min \leftarrow g(d) | \sum_{k=1}^{N} d_k \geq D, \eta \leq d_k \leq 1 - \eta, k \in \{1, \ldots, N\} \rightarrow . \quad (8)$$

*Note that for each fixed d, (8) is a convex program. Since for each fixed b, the function $d \longrightarrow \sum_{k=1}^{N}(d_k/b_k)$ is linear, g is the pointwise minimum of a family of linear functions, hence g is concave. There are several available algorithms for solving the concave minimization problem (8) (see [9]). In view of the special structure of the constraints, one suitable method is by outer approximation.*

*On the hand, in the case where $\sum_{k=1}^{N} d_k \geq D$ is also changed to $\sum_{k=1}^{N} d_k = D$, we proposed a solution in [5].*

## 4.2. The mutual exclusion problem

It's the case of a network where for any $k \in \{1,\dots,N\}$ the $k^{th}$ chain is the following: its state space ${}^k\mathcal{E}$ is partitionned into four subsets $\{{}^kB_1,\dots,{}^kB_4\}$ and the transition matrix ${}^kP = \left({}^kp(x,e)\right)$ is such that

$\forall x \in {}^kB_1$,

$$\begin{cases} {}^kp\left(x,{}^kB_1\right) = \alpha_k \\ {}^kp\left(x,{}^kB_2\right) = \beta_k \\ {}^kp\left(x,{}^kB_3\right) = 1-\alpha_k-\beta_k \\ {}^kp\left(x,{}^kB_4\right) = 0 \end{cases}$$

where $\alpha_k > 0, \beta_k > 0$ and $\alpha_k + \beta_k < 1$.

$\forall x \in {}^kB_2$,

$$\begin{cases} {}^kp\left(x,{}^kB_1\right) = 1-\gamma \\ {}^kp\left(x,{}^kB_2\right) = \gamma \\ {}^kp\left(x,{}^kB_3\right) = 0 \\ {}^kp\left(x,{}^kB_4\right) = 0 \end{cases}$$

where $0 < \gamma < 1$.

$\forall x \in {}^kB_3$,

$$\begin{cases} {}^kp\left(x,{}^kB_1\right) = 0 \\ {}^kp\left(x,{}^kB_2\right) = 0 \\ {}^kp\left(x,{}^kB_3\right) = \delta \\ {}^kp\left(x,{}^kB_4\right) = 1-\delta \end{cases}$$

where $0 < \delta < 1$.

$\forall x \in {}^kB_4$,

$$\begin{cases} {}^kp\left(x,{}^kB_1\right) = 1-\theta \\ {}^kp\left(x,{}^kB_2\right) = 0 \\ {}^kp\left(x,{}^kB_3\right) = 0 \\ {}^kp\left(x,{}^kB_4\right) = \theta \end{cases}$$

where $0 < \theta < 1$

**Proposition 10** *With respect to the partition $\{{}^kB_1,\dots,{}^kB_4\}$ of the state space ${}^k\mathcal{E}$, the $k^{th}$ Markov chain is lumpable. The lumped chain* *is a homogeneous Markov chain with four states, which transition matrix* ${}^kM = \left({}^km(i,j)\right)$ *is:*

$$\begin{array}{lll} {}^km(1,1) = \alpha_k, & {}^km(1,2) = \beta_k, & {}^km(1,3) = 1 - \alpha_k - \beta_k, \\ {}^km(2,1) = 1 - \gamma, & {}^km(2,2) = \gamma, & \\ {}^km(3,3) = \delta, & {}^km(3,4) = 1 - \delta, & \\ {}^km(4,1) = 1 - \theta, & {}^km(4,4) = \theta, & \end{array}$$

*the other* ${}^km(i,j)$*'s are 0.*

In the present case, for any $k \in \{1, \dots, N\}$, the four states of the $k^{th}$ lumped chain have the following meaning : request state, refusal state, acceptation state and execution state.

As an extension of [6], we consider that the chains are connected by relations

$$\left\{ \begin{array}{l} \sum_{k=1}^N \alpha_k = a \\ \sum_{k=1}^N (1 - \alpha_k - \beta_k) = b \end{array} \right.$$

($a > 0$ and $b > 0$ being two constraints such that $a + b < N$) viz.

$$\left\{ \begin{array}{l} \sum_{k=1}^N \alpha_k - a = 0 \\ N - a - b - \sum_{k=1}^N \beta_k = 0 \end{array} \right. \tag{9}$$

where

$$\alpha_k > 0, \beta_k > 0, \ \alpha_k + \beta_k < 1, \ k \in \{1, \dots, N\}$$

**Proposition 11** *With respect to the partition* $\{{}^kB_1, \dots, {}^kB_4\}$*, the* $k^{th}$ *lumped chain has only an acyclic ergodic class, and in the permanent regime, we have:*

$$\begin{array}{l} E\left[n\left({}^kB_1 \times {}^k\mathcal{E}\right)\right] = n(1-\gamma)(1-\delta)(1-\theta)/L'_k \\ E\left[n\left({}^kB_2 \times {}^k\mathcal{E}\right)\right] = n(1-\delta)(1-\theta)\beta_k/L'_k \\ E\left[n\left({}^kB_3 \times {}^k\mathcal{E}\right)\right] = n(1-\gamma)(1-\theta)(1-\alpha_k-\beta_k)/L'_k \\ E\left[n\left({}^kB_4 \times {}^k\mathcal{E}\right)\right] = n(1-\gamma)(1-\delta)(1-\alpha_k-\beta_k)/L'_k \end{array}$$

*where* $L'_k = (1 - \gamma + \beta_k)(1 - \delta)(1 - \theta) + (1 - \alpha_k - \beta_k)(1 - \gamma)(2 - \delta - \theta)$.

*Proof.* Indeed, the resolution of the equation ${}^k\mu \cdot {}^kM = {}^k\mu$ gives

$$\begin{array}{ll} {}^k\mu(1) = (1-\gamma)(1-\delta)(1-\theta)/L'_k, & {}^k\mu(2) = (1-\delta)(1-\theta)\beta_k/L'_k \\ {}^k\mu(3) = (1-\gamma)(1-\theta)(1-\alpha_k-\beta_k)/L'_k, & {}^k\mu(4) = (1-\gamma)(1-\delta)\times \\ & \times(1-\alpha_k-\beta_k)/L'_k \end{array}$$

Then, using $E\left[n\left({}^kB_i \times {}^k\mathcal{E}\right)\right] = n \cdot {}^k\mu(i)$, we obtain the announced results.■

As exposed in [6], the criterion of choice is the maximization of the frequence of the refusal state and the minimisation of the frequence of the acceptation state. So, for every $k \in \{1, \ldots, N\}$, ${}^kU = \{2, 3\}$.

As a consequence of proposition 4 and proposition 11, we have

**Proposition 12** *The guide function $F$, to be minimized, can be expressed either by*

$$\sum_{k=1}^{N} \frac{E\left[n\left({}^kB_3 \times {}^k\mathcal{E}\right)\right]}{E\left[n\left({}^kB_2 \times {}^k\mathcal{E}\right)\right]} \text{ or by } \sum_{k=1}^{N} \frac{{}^kT_2}{{}^kT_3}.$$

*Both of them lead to the expression*

$$F\left[(\alpha_1, \beta_1), \ldots (\alpha_N, \beta_N)\right] = \frac{1-\gamma}{1-\delta} \sum_{k=1}^{N} \frac{1-\alpha_k-\beta_k}{\beta_k}.$$

Then we have the following proposition

**Proposition 13** *(i) Equity is an optimal solution in the first order hierarchy.*

*(ii) Let $\eta > 0$ and $\epsilon > 0$ be two numbers such that*

$$\eta \leq a/N, \ \epsilon \leq (N-a-b)/N, \ \eta + \epsilon > (N-b-1)/(N-1),$$

*let $\chi$ be the set of the $(\alpha, \beta)$'s such that $\alpha_k \geq \eta$, $\beta_k \geq \epsilon$ , $k \in \{1, \ldots, N\}$ and let $\Phi : \chi \longmapsto [0,1]$ be the function*

$$\Phi(\alpha, \beta) = \frac{\frac{(N-1)(1-\eta)}{\epsilon} + \frac{N-a-(N-1)(1-\eta)}{N-a-b-(N-1)\epsilon} - \frac{1-\delta}{1-\gamma} F(\alpha, \beta)}{\frac{(N-1)(1-\eta)}{\epsilon} + \frac{N-a-(N-1)(1-\eta)}{N-a-b-(N-1)\epsilon} - \frac{Nb}{N-a-b}}$$

*Then the set $\Phi^{-1}\left(]0,1]\right)$ is a fuzzy subset of solutions in the first order hierarchy.*

*Proof.*

(i) It is easy to see that the problem of optimization of $F$ under constraints (9), by the Lagrange multipliers method, leads to $\alpha_k =$

$a/N$, $\beta_k = (N-a-b)/N$, $k \in \{1,\ldots,N\}$. This $(\alpha,\beta)$ minimizes $F$, as proved (generalizing [6]) as follows:

Consider the variables $A_k, \beta_k$ where $A_k = \sqrt{1-\alpha_k}$, $k \in \{1,\ldots,N\}$. With these variables, the guide function $F$ is transformed into the function $\zeta$ such that

$$\zeta\left[(A_1,\beta_1),\ldots,(A_N,\beta_N)\right] = \frac{1-\gamma}{1-\delta}\sum_{k=1}^{N}\frac{A_k^2}{\beta_k} - N\frac{1-\gamma}{1-\delta}$$

to be minimized under the constraints:

$$0 < A_k < 1,\ k \in \{1,\ldots,N\},$$
$$\sum_{k=1}^{N} A_k^2 = N-a \text{ and } \sum_{k=1}^{N}\beta_k = N-a-b. \tag{10}$$

In fact, that amounts to minimize the function $\xi$ such that

$$\xi\left[(A_1,\beta_1),\ldots,(A_N,\beta_N)\right] = \sum_{k=1}^{N}\xi_k(A_k,\beta_k) = \sum_{k=1}^{N}\frac{A_k^2}{\beta_k}$$

under the above constraints. $\xi_k$ is a convex function, because of its hessian

$$\nabla^2\xi_k = \begin{pmatrix} 2/\beta_k & -2A_k/\beta_k^2 \\ -2A_k/\beta_k^2 & -2A_k/\beta_k^3 \end{pmatrix}$$

which corresponds to a positive semi-definite quadratic form, viz.

$$(x_1\ \ x_2)\nabla^2\xi_k\begin{pmatrix} x_1 \\ x_2 \end{pmatrix} = \frac{2}{\beta_k}\left(x_1 - \frac{A_k}{\beta_k}x_2\right)^2 \geq 0.$$

$\xi$ is then a convex function, since it is the sum of the convex functions $\xi_k$'s. Consequently, the solution $(\alpha,\beta)$ such that $\alpha_k = a/N$, $\beta_k = (N-a-b)/N$, $k \in \{1,\ldots,N\}$, viz. equity, which corresponds to the solution $(A,\beta)$ such that $A_k = \sqrt{(N-1)/N}$, $\beta_k = (N-a-b)/N$, $k \in \{1,\ldots,N\}$, for $\xi$, makes $F$ minimum.

(ii) Now, in order to construct a fuzzy subset of solutions of the problem of minimization of $F$ under constraints (9), let us consider variables $A_k$'s and $\beta_k$'s: since $1-\alpha_k-\beta_k > 0$ and $A_k^2 = 1-\alpha_k$, we have

$$\forall k \in \{1,\ldots,N\}\ :\ \epsilon \leq \beta_k < A_k^2 \leq 1-\eta. \tag{11}$$

Concerning the $A_k^2$ 's at the border of $\chi$, if $(N-1)$ of them are equal to $1-\eta$, then by (10), the $N^{th}$ is equal to $N-a-(N-1)(1-\eta)$. This $N^{th}$ coordinate satisfies (11), since $N-a-(N-1)(1-\eta) \leq N-N\eta-(N-1)(1-\eta) = 1-\eta$ by the choice of $\eta$.

Concerning the $\beta_k$'s at the border of $\chi$, if $(N-1)$ of them are equal to $\epsilon$, then the $N^{th}$ is equal to $N-a-b-(N-1)\epsilon$. This $N^{th}$ coordinate satisfies (11), since $N-a-b-(N-1)\epsilon \geq N\epsilon-(N-1)\epsilon = \epsilon$, by the choice of $\epsilon$.

${A_N}^2$ and $\beta_N$ satisfy (11) because $\eta$ and $\varepsilon$ are chosen such that $\eta+\epsilon > (N-b-1)/(N-1)$.

Now, $\xi$ being convex, its maximum is to be found in the border of $\chi$, that is to say that there are only two possibilities of such solutions (modulo a permutation of indices): either

$$\begin{cases} A_1^2 = \ldots = A_{N-1}^2 = 1-\eta, & A_N^2 = N-a-(N-1)(1-\eta) \\ \beta_1 = \ldots = \beta_{N-1} = \varepsilon, & \beta_N = N-a-b-(N-1)\epsilon \end{cases}$$

or

$$\begin{cases} A_1^2 = \ldots = A_{N-2}^2 = 1-\eta, A_{N-1}^2 = N-a-(N-1)(1-\eta), \\ \qquad\qquad\qquad\qquad\qquad\qquad A_N^2 = 1-\eta \\ \beta_1 = \ldots = \beta_{N-2} = \varepsilon,\ \beta_{N-1} = \varepsilon,\ \beta_N = N-a-b-(N-1)\epsilon \end{cases}$$

The first solution gives the following value of $\xi$:

$$\xi' = (N-1)\frac{1-\eta}{\varepsilon} + \frac{N-a-(N-1)(1-\eta)}{N-a-b-(N-1)\epsilon}$$

The second solution gives the following value of $\xi$:

$$\xi'' = (N-2)\frac{1-\eta}{\varepsilon} + \frac{N-a-(N-1)(1-\eta)}{\varepsilon} + \frac{1-\eta}{N-a-b-(N-1)\epsilon}$$

Now,

$$\begin{aligned} \xi' - \xi'' &= \frac{1-\eta}{\varepsilon} - \frac{N-a-(N-1)(1-\eta)}{\varepsilon} + \\ &\quad \frac{N-a-(N-1)(1-\eta)-(1-\eta)}{N-a-b-(N-1)\epsilon} \\ &= \frac{1}{\varepsilon}[(1-\eta) - N + a + (N-1)(1-\eta)] + \\ &\quad \frac{N-a-N(1-\eta)}{N-a-b-(N-1)\epsilon} \\ &= (N(1-\eta)-N+a)\left[\frac{1}{\varepsilon} - \frac{1}{N-a-b-(N-1)\epsilon}\right] \geq 0 \end{aligned}$$

It proves that $\xi'$ is $\max \xi$, while $\min \xi = \frac{Nb}{N-a-b}$ (since it corresponds to solution $\alpha_k = \frac{a}{N}, \beta_k = \frac{N-a-b}{N}$, $k \in \{1, \ldots, N\}$). Taking

$$\Phi(\alpha, \beta) = \frac{\max \xi - \xi(\alpha, \beta)}{\max \xi - \min \xi}$$

we have the announced result of the proposition.

■

**Remark 3** *In the present case, the useful purpose remains in deciding between two solutions $(\alpha, \beta)$ and $(\alpha', \beta')$ such that $\Phi(\alpha, \beta) = \Phi(\alpha', \beta') \in ]0, 1[$. The second order hierarchy leads to the comparison*

*(i) in the first approach, by use of the function $V$:*

$$V(\alpha, \beta) = \sum_{k=1}^{N} {}^kV(\alpha, \beta)$$

*where ${}^kV(\alpha, \beta)$ is given by (4),*

*(ii) in the second approach, by use of the function $W$:*
$W(\alpha, \beta) =$

$$\sum_{k=1}^{N} \Big( {}^k\mu(2) \left(1 - {}^k\mu(2)\right) +$$
$$2 \cdot {}^k\mu(2) \lim_{n \to \infty} \sum_{t=0}^{n-2} \frac{n-t-1}{n} \left[ {}^k\, m_{t+1}(2,2) - {}^k\mu(2) \right] \Big) +$$
$$\sum_{k=1}^{N} \Big( {}^k\mu(3) \left(1 - {}^k\mu(3)\right) +$$
$$2 \cdot {}^k\mu(3) \lim_{n \to \infty} \sum_{t=0}^{n-2} \frac{n-t-1}{n} \left[ {}^k m_{t+1}(3,3) - {}^k\mu(3) \right] \Big)$$

*where ${}^k\mu(2)$, ${}^k\mu(3)$ and $L'_k$ are given in proposition 11.*

Note that in the present case, we don't obtain from (4) a simple analytic expression of ${}^kV(\alpha, \beta)$. By way of compensation, a programming software allows a comparison by numerical values. It is the same with the second approach. On the other hand, we have only an approximate value of $W(\alpha, \beta)$ even if $n$ is large.

## 4.3. The dining philosophers'problem

In this problem, for any $k \in \{1, \ldots, N\}$, the $k^{th}$ Markov chain is the following: the state space ${}^k\mathcal{E}$ is partitionned into four subsets $\{{}^kB_1, {}^kB_2, {}^kB_3, {}^kB_4\}$ and the transition matrix ${}^kP = \left({}^kp(x,e)\right)$ is such that $\forall x \in {}^kB_1$,

$$\begin{cases} {}^kp\left(x, {}^kB_1\right) = 1 - \alpha_k \\ {}^kp\left(x, {}^kB_2\right) = \alpha_k \\ {}^kp\left(x, {}^kB_3\right) = 0 \\ {}^kp\left(x, {}^kB_4\right) = 0 \end{cases}$$

where $0 < \alpha_k < 1$.

$\forall x \in {}^kB_2$,

$$\begin{cases} {}^kp\left(x, {}^kB_1\right) = 0 \\ {}^kp\left(x, {}^kB_2\right) = 1 - \beta_k \\ {}^kp\left(x, {}^kB_3\right) = \beta_k \\ {}^kp\left(x, {}^kB_4\right) = 0 \end{cases}$$

where $0 < \beta_k < 1$.

$\forall x \in {}^kB_3$,

$$\begin{cases} {}^kp\left(x, {}^kB_1\right) = 0 \\ {}^kp\left(x, {}^kB_2\right) = 0 \\ {}^kp\left(x, {}^kB_3\right) = 1 - \theta_k \\ {}^kp\left(x, {}^kB_4\right) = \theta_k \end{cases}$$

where $0 < \theta_k < 1$.

$\forall x \in {}^kB_4$,

$$\begin{cases} {}^kp\left(x, {}^kB_1\right) = 1 \\ {}^kp\left(x, {}^kB_2\right) = 0 \\ {}^kp\left(x, {}^kB_3\right) = 0 \\ {}^kp\left(x, {}^kB_4\right) = 0 \end{cases}$$

**Proposition 14** *With respect to the partition $\{{}^kB_1, \ldots, {}^kB_4\}$ of the state space ${}^k\mathcal{E}$, the $k^{th}$ Markov chain is lumpable. The lumped chain* *is a homogeneous Markov chain with four states, which transition matrix* ${}^kM = \left({}^km(i,j)\right)$ *is:*

$$\begin{array}{ll} {}^km(1,1) = 1-\alpha_k, & {}^km(1,2) = \alpha_k, \\ {}^km(2,2) = 1-\beta_k, & {}^km(2,3) = \beta_k, \\ {}^km(3,3) = 1-\theta_k, & {}^km(3,4) = \theta_k, \\ {}^km(4,1) = 1, & \end{array}$$

*the other* ${}^km(i,j)$*'s are 0.*

In the present case, it is supposed that for any $k \in \{1,\ldots,N\}$, $\theta_k$ is a given fixed number. Every matrix ${}^kM$ is then identified by the pair $(\alpha_k, \beta_k)$ and the guide function $F$ shall be a function of $(\alpha,\beta) = ((\alpha_1,\beta_1),\ldots,(\alpha_N,\beta_N))$. As an extension of the problem, described in [6], we consider that the lumped chains are connected by the relations

$$\left\{ \begin{array}{l} \alpha_k + \beta_{k+1} = c_k, \ k \in \{1,\ldots,N-1\} \\ \alpha_N + \beta_1 = c_N, \\ \text{where } 0 < c_k < 2, \ k \in \{1,\ldots,N\} \end{array} \right.$$

**Proposition 15** *With respect to the partition* $\{{}^kB_1,\ldots,{}^kB_4\}$, *the* $k^{th}$ *lumped chain has only an acyclic ergodic class,*

$$\begin{array}{l} E\left[n\left({}^kB_1 \times {}^k\mathcal{E}\right)\right] = n\beta_k\theta_k/L''_k \\ E\left[n\left({}^kB_2 \times {}^k\mathcal{E}\right)\right] = n\alpha_k\theta_k/L''_k \\ E\left[n\left({}^kB_3 \times {}^k\mathcal{E}\right)\right] = n\alpha_k\beta_k/L''_k \\ E\left[n\left({}^kB_4 \times {}^k\mathcal{E}\right)\right] = n\alpha_k\beta_k\theta_k/L''_k \end{array}$$

*where* $L''_k = \alpha_k\beta_k + \beta_k\theta_k + \theta_k\alpha_k + \alpha_k\beta_k\theta_k$.

**Proof** Indeed, the resolution of the equation ${}^k\mu \cdot {}^kM = {}^k\mu$ gives

$$\begin{array}{ll} {}^k\mu(1) = \beta_k\theta_k/L''_k, & {}^k\mu(2) = \alpha_k\theta_k/L''_k \\ {}^k\mu(3) = \alpha_k\beta_k/L''_k, & {}^k\mu(4) = \alpha_k\beta_k\theta_k/L''_k \end{array}$$

Then using $E\left[n\left({}^kB_i \times {}^k\mathcal{E}\right)\right] = n \cdot {}^k\mu(i)$, we obtain the announced result.
■

As exposed in [6], the criterion of choice is to cause state 3 to endure (state 3 being the state "with two forks"). So, for every $k \in \{1,\ldots,N\}$, ${}^kU = \{3\}$, and we have to minimize the recurrence times ${}^kT_3$'s or to minimize the $1/E\left[n({}^kB_3 \times {}^k\mathcal{E})\right]$.

**Proposition 16** *The guide function $F$, to be minimized, can be expressed either by*

$$\sum_{k=1}^{N} \frac{1}{E\left[n\left({}^kB_3 \times {}^k\mathcal{E}\right)\right]} \text{ or by } \sum_{k=1}^{N} {}^kT_3.$$

*Both of them lead to the expression*

$$F\left[(\alpha_1, \beta_1), \ldots (\alpha_N, \beta_N)\right] = \sum_{k=1}^{N} \left[(1 + \theta_k) + \theta_k\left(\frac{1}{\alpha_k} + \frac{1}{\theta_k}\right)\right].$$

In fact, the useful part of this expression is $\sum_{k=1}^{N} \theta_k(\frac{1}{\alpha_k} + \frac{1}{\theta_k})$, and when necessary $\sum_{k=1}^{N}(1 + \theta_k)$ can be dropped. Then we have the following proposition

**Proposition 17** *(i) The solution $(\alpha, \beta)$ such that*

$$\alpha_k = \frac{c_k}{1+\sqrt{\theta_{k+1}/\theta_k}}, \qquad \beta_{k+1} = \frac{c_k\sqrt{\theta_{k+1}/\theta_k}}{1+\sqrt{\theta_{k+1}/\theta_k}},$$

$k \in \{1, \ldots, N-1\}$, *and*

$$\alpha_N = \frac{c_N}{1+\sqrt{\theta_1/\theta_N}}, \qquad \beta_1 = \frac{c_N\sqrt{\theta_1/\theta_N}}{1+\sqrt{\theta_1/\theta_N}}$$

*is the optimal one.*

*(ii) Let $\eta > 0$ and $\epsilon > 0$ and let $\chi$ be the set of $(\alpha, \beta)$'s such that $\alpha_k \geq \eta$, $\beta_k \geq \epsilon$, $k \in \{1, \ldots, N\}$ and let $\Phi : \chi \to [0, 1]$ be the function*

$$\Phi(\alpha, \beta) = \frac{\sum_{k=1}^{N} \theta_k \left[1 - \frac{\eta\epsilon}{\eta+\epsilon}\frac{\alpha_k+\beta_k}{\alpha_k\beta_k}\right]}{\sum_{k=1}^{N} \theta_k - \frac{\eta\epsilon}{\eta+\epsilon}\sum_{k(mod\ N)} \frac{\theta_k+\theta_{k+1}}{c_k}\sqrt{\frac{\theta_k}{\theta_{k+1}}}\left(1 + \sqrt{\frac{\theta_{k+1}}{\theta_k}}\right)^2}$$

*Then the set $\Phi^{-1}(]0, 1])$ is a fuzzy subset of solutions in the first order hierarchy.*

*Proof.* It is easy to see that $F$ is a convex function of several variables $((\alpha_1, \beta_1), \ldots, (\alpha_N, \beta_N))$ since it is a sum of convex functions of a single variable.

(i) The Lagrange multipliers method gives

$$\frac{\theta_k}{\alpha_k^2} = \frac{\theta_{k+1}}{\beta_{k+1}^2}, \quad k \in \{1, \ldots, N-1\}$$

and

$$\frac{\theta_N}{\alpha_N^2} = \frac{\theta_1}{\beta_1^2}.$$

Then

$$\beta_{k+1} = \sqrt{\frac{\theta_{k+1}}{\theta_k}} \cdot \alpha_k, \; k \in \{1, \ldots, N-1\}$$

and

$$\beta_1 = \sqrt{\frac{\theta_1}{\theta_N}} \cdot \alpha_N.$$

Since $\alpha_k + \beta_{k+1} = c_k, \; k \in \{1, \ldots, N-1\}$ and $\alpha_N + \beta_1 = c_N$, we have the announced optimal solution.

(ii) The expression of $\Phi(\alpha, \beta)$ comes from

$$\frac{\left(\frac{1}{\eta} + \frac{1}{\epsilon}\right) \sum_{k=1}^N \theta_k - \sum_{k=1}^N \theta_k \left(\frac{1}{\alpha_k} + \frac{1}{\beta_k}\right)}{\left(\frac{1}{\eta} + \frac{1}{\epsilon}\right) \sum_{k=1}^N \theta_k - \sum_{k (\mathrm{mod}\ N)} \theta_k \left(\frac{1+\sqrt{\frac{\theta_{k+1}}{\theta_k}}}{c_k} + \frac{1+\sqrt{\frac{\theta_k}{\theta_{k-1}}}}{c_{k-1}\sqrt{\frac{\theta_k}{\theta_{k-1}}}}\right)}$$

■

**Remark 4** *Remark 4.3 on the preceding example remains valid for the present problem, concerning second order hierarchy, except $W(\alpha, \beta)$ which becomes*

$$\sum_{k=1}^{N} \left( {}^k\mu(3)\left(1 - {}^k\mu(3)\right) + \right.$$

$$\left. 2 \cdot {}^k\mu(3) \lim_{n \to \infty} \sum_{t=0}^{n-2} \frac{n-t-1}{n} \left[ {}^k m_{t+1}(3,3) - {}^k\mu(3) \right] \right)$$

## Acknowledgment

I am grateful to professor Hoang Tuy for several remarks which have helped to improve the paper.

# References


[1] D. Brand and P. Zafiropulo, On communicating finite-state machines. *Journal of the ACM*, 1983, 30, (2), pp. 323-342.

[2] A. Bui, Performance evaluation of fault-tolerant routing algorithms: an optimization problem. *Optimization*, 1994, 29, pp. 267–280.

[3] A. Bui, A bounding problem concerning distributed routing algorithms. *Rev. R. de Math. Pures et Appl.*, 1994, 39, (1), pp. 1–15.

[4] A. Bui, Analyse du fonctionnement d'un type d'algorithmes distribués. *Rev. R. de Math. Pures et Appl.*, 1995, 40, (7–8), pp. 583–593.

[5] A. Bui, A problem of hierarchy for some distributed algorithms. *Mathematica*, (to appear), 1997.

[6] M. Bui, Tuning Distributed Control Algorithms for Optimal Functioning. *J. of Global Optimization*, 1992, 2, pp. 177–199.

[7] T.L. Casavant and J.G. Kuhl, A communicating finite automata approach to modeling distributed computation and its application to distributed decision making. *IEEE Trans. on Computers*, 1990, 39, (5), pp.628-639.

[9] R. Horst and H. Tuy, *Global Optimization.* Springer, 3rd ed., 1996.

[10] M. Iosifescu, *Finite Markov Processes and their Applications.* Wiley, 1980.

[11] J. G. Kemeny and J. L. Snell, *Finite Markov chains.* Springer, 1983.

[12] B. Rozoy, On distributed languages and models for distributed computation. *Res. Rept. No. 563, University Paris 11,* 1990.

[13] A. Segall, Distributed Network Protocols. *IEEE Trans. on Information Theory*, 1983, 29, (1), pp. 23–25.

[14] W.D. Tajibnapis, A Correctness Proof of a Topology Information Maintenance Protocol for a Distributed Computer Network. *Communications of the ACM*, 1977, 20, (7), pp. 477–485.

# Chapter 7

# ON TUY'S 1964 CONE SPLITTING ALGORITHM FOR CONCAVE MINIMIZATION

Christophe Meyer

*Ecole Polytechnique de Montréal, Département de Mathématiques et de Génie Industriel, C.P. 6079, succ. Centre-ville, Montréal (Québec) H3C 3A7, Canada.*

**Abstract** Since the work of Zwart, it is known that cycling may occur in the cone splitting algorithm proposed by Tuy in 1964 to minimize a concave function over a polytope. In this paper, we show that despite this fact, Tuy's algorithm is convergent in the sense that it always finds an optimal solution. This result also holds for a variant of Tuy's algorithm proposed by Gallo, in which a cone is split into a smaller subset of subcones (in term of inclusion). As shown by an example, this variant may also cycle. The transformation of these two algorithms into finite step procedures is discussed.

**Keywords:** concave minimization, cone splitting algorithm, covering.

## 1. Introduction

We consider the following concave minimization problem

$$(CP) \qquad \min\{f(x) \mid x \in P\}$$

where $f$ is a concave function defined on $\mathbb{R}^n$ and $P$ is a full dimensional polytope of $\mathbb{R}^n$.

In his landmark paper, Tuy [17] proposed in 1964 several ideas and algorithms for solving this problem, that still influence today's methods (for



*A. Migdalas et al. (eds.), From Local to Global Optimization,* 133–153.

surveys on concave minimization, see for example Horst and Tuy [10], Benson [2]). One of these algorithms consists to cover the polytope by polyhedral cones. A one-dimensional nonlinear problem and a linear program are then solved for each cone. Depending on the optimal value of the linear program, the cone is shown to not contain a better solution that the best one at hand (i.e., can be fathomed), or is replaced by a set of subcones that cover it. The algorithm terminates when all cone have been fathomed. Tuy [17] asserted that "it can be proved that the process must terminate in a finite number of steps", but Zwart [25] provided a counter-example on which the algorithm cycles. Gallo [3] (see also Gallo and Ülkücü [4]) then proposed a modification which consists to split a cone into a smaller subset of subcones than Tuy does. This modified algorithm, if run on Zwart's problem, terminates after a finite number of iterations but no proof of convergence is given. Also no example is known on which Gallo's modification is shown to cycle (in [24], Vaish observes that on Zwart's problem, some cones are generated several times; they do not however form a cycle). In order to avoid the possibility of cycling, covering was replaced by partition (see Zwart [26], Bali [1]), and the efforts were invested in the development of such cones partitioning algorithms. The first convergent algorithm of this family is due to Thoai and Tuy [16] and uses bisection to subdivide the cones. Several convergent algorithms have followed, differing essentially by the subdivision strategy [6][20][22]. The best of them uses Normal subdivision, a concept introduced by Tuy [19], which consists to use almost only $\omega$-subdivisions, and bisections occasionally in such a way to ensure the convergence ($\omega$-subdivision consists to subdivide a cone with respect to a point, $\omega$, which is a byproduct of the fathoming procedure; in contrast, bisection is a purely geometrical subdivision process which does not take into account any information about the objective function and thus performs less satisfactorily). Recently Jaumard and Meyer [14] have proved the convergence of the cones partitioning algorithm with a pure $\omega$-subdivision strategy, which shows that bisections are not necessary at all. Parallely, branch-and-bound variants have been developed (see, e.g., [16][5]) and the cones partitioning algorithms have been generalized to handle more general problem such that concave minimization over a convex set, optimization problems with reverse convex constraints, D.C. optimization (see, e.g., [7][9][11][21][23]).

An other direction in which Tuy's 1964 algorithm has evolved is the polyhedral annexation method (see Tuy [18]). This approach generates a sequence of expanding polytopes on which $f$ is greater than or equal to the value of the current best known solution. At each iteration, linear programs whose objectif functions correspond to the facets of the current approximating polytope $Q$ are solved. Either it is proved that $Q$ contains $P$, in which case the method finishes, or a point is found in $P \backslash Q$. In this latter case, this point is used to expand the polytope $Q$. The facets of the new approximating polytope are determined by enumerating the extreme points of an associated polytope in the dual space.

Numerical experiments [8] indicate that the cones partitioning algorithm and the polyhedral annexation method have comparable performance and that they are much better than the outer-approximation approach.

To the knowledge of the author, cones covering was never used again in conical algorithms. Covering are however used in simplicial algorithms, in which simplices replace cones: see Horst, Thoai and De Vries [12][13]. The aim of this paper is to try to rehabilitate cones covering algorithms. To do this, we show that Tuy's original algorithm and Gallo's variant always find an optimal solution of the concave minimization problem after a finite number of iterations. An upper bound of this number is given in terms of the dimension $n$ and of the number of constraints $m$ defining the polytope. However, as shown on an example, these two algorithms can repeat infinitely the same sequence of cones if not adequately modified (this was already known for Tuy's algorithm). Two such modifications are briefly discussed.

The paper is organized as follows. In Section 2, the basic operations of the cone splitting algorithm are recalled, i.e., the construction of an initial cover, the fathoming test and the subdivision procedure. In Section 3, we present Tuy's 1964 algorithm and Gallo's variant. We also rapidly discuss how to transform them into finite algorithms. In Section 4, we illustrate Gallo's variant on an example and show that it may also cycle. Conclusions are drawn in the last section.

## 2. Basic operations

In this section, we recall the basic operations needed to define the algorithms, namely the construction of an initial conical cover (Section 2.1), the fathoming test (Section 2.2) and the subdivision procedure (Section 2.3).

## 2.1. Initial cover

Assume that a nondegenerate vertex of $P$ is available. By performing a change of variables if necessary, we may assume that this vertex is the origin $O$. Let $K^{01}$ be the polyhedral cone of origin $O$ defined by the $n$ adjacent vertices to $O$. This cone forms our initial cover of $P$. (In Tuy [17], it is assumed in addition that $O$ is a local minimum of problem $(CP)$ but this is not necessary to the convergence of the algorithm.)

If no nondegenerate vertex of $P$ is available, we can reduce the problem to the previous case as follows. By translating $P$ if necessary, we may assume that $P$ strictly contains the origin $O$. Let $U^0 = conv\{u^{01}, u^{02}, \ldots, u^{0,n+1}\}$ be a simplex containing strictly $O$ (for example, we may take $u^{0j} = e^j$ where $e^j (j = 1, 2, \ldots, n)$ are the $n$ unitary vectors and $u^{0,n+1} = -e$ where $e$ is the vector whose components are all equal to 1). Let $K^{0i}$ denote the cone of origin $O$ and directions $u^{0j}$, $j \in \{1, 2, \ldots, n+1\} \backslash \{i\}$ for $i = 1, 2, \ldots, n+1$. Then $\bigcup_{i=1}^{n+1} K^{0i}$ defines a partition of $\mathbb{R}^n$ in the sense of Horst and Tuy [10]. The point $O$ is a nondegenerate vertex of the polytopes $P^i = P \cap K^{0i}, i = 1, \ldots, n+1$ and by solving the problem $(CP)$ for each of these polytopes, we clearly solve the initial problem.

In this paper, we only consider polyhedral cone of origin $O$ with exactly $n$ independent edges. Moreover, each edges will be defined by an extreme point of $P$. If $u^j, j = 1, \ldots, n$ are the directions of a cone, this cone will be noted $cone\{u^1, u^2, \ldots, u^n\}$ (since the cone is vertexed at $O$, $u^j$ can be viewed indifferently as a point or as a vector of $\mathbb{R}^n$, for $j = 1, \ldots, n$).

## 2.2. Fathoming of a cone

In order to define the fathoming procedure of Tuy [17], we first recall the definition of $\gamma$-extensions. Basically the $\gamma$-extension along an halfline of origin $O$ (with $\gamma \leq f(O)$) is the farthest point $y$ on the halfline with value $f(y) \geq \gamma$. More precisely, if $u$ denotes the direction of the halfline, the $\gamma$-extension is the point $y = \overline{\theta} u$ with

$$\overline{\theta} = \max\{\theta \mid f(\theta u) \geq \gamma;\ \theta \leq \theta_0\}$$

where $\theta_0$ is a large fixed number.

Computing a $\gamma$-extension amounts to solve a one-dimensional unimodal program. See, e.g., Horst and Thoai [8] for a simple procedure to solve this problem.

Now consider a cone $K = cone\{u^1, \ldots, u^n\}$ and let $y^j = \theta_j u^j, j = 1, \ldots, n$ be the $\overline{f}$-extensions along the directions $u^1 \ldots, u^n$ respectively, where $\overline{f}$ denotes the value of the best known solution. Let $S = conv\{O, y^1, \ldots, y^n\}$ be the simplex defined by the origin $O$ and the $\overline{f}$-extensions. By concavity of $f$ and definition of the $\overline{f}$-extensions, we have

$$\min_{x \in S} f(x) = \min\{f(O), f(y^1), \ldots, f(y^n)\} \geq \overline{f}.$$

Assume that $K \cap P$ is included in $S$. Then

$$\min_{x \in K \cap P} f(x) \geq \min_{x \in S} f(x) \geq \overline{f}$$

which shows that the cone $K$ contains no feasible point with value better than $\overline{f}$.
A sufficient condition for the inclusion $K \cap P \subseteq S$ to hold is that no point of $P$ lies on the other side, with respect to $O$, of the hyperplane passing through the points $y^j, j = 1, \ldots, n$. If $\alpha x = 1$ denotes the equation of this hyperplane, this condition can be checked by solving the linear program

$$LP(Y) \quad \max \quad \alpha x \qquad \text{s.t.} \quad Ax \leq b$$

where $Y = (y^1, \ldots, y^n)$.
If the optimal value $\rho$ satisfies $\rho \leq 1$, then the inclusion $K \cap P \subseteq S$ holds and the cone can be fathomed. An alternate equivalent formulation of this linear program that we will use is obtained by setting $x = \sum_{j=1}^{n} \lambda_j u^j$:

$$LP'(Y) \quad \max \quad \sum_{j=1}^{n} \frac{\lambda_j}{\theta_j} \qquad \text{s.t.} \quad A \sum_{j=1}^{n} \lambda_j u^j \leq b.$$

Also we denote by $\omega$ an optimal solution of problem $LP(Y)$. Such a point will be used in the next section to subdivide the cone $K$.

## 2.3. Subdivision of a cone

Let $K = cone\{u^1, \ldots, u^n\}$ be a cone to be subdivided and let $\omega = \sum_{j=1}^{n} \lambda_j u^j$ be an optimal solution of the linear program $LP$ associated to $K$ (see Section 2.2). For each $j$, define $K^j$ as the cone obtained from $K$ by replacing the $j^{th}$ edge of $K$ by the halfline of origin $O$ passing through $\omega$. Let $J_{\neq} = \{j \mid \lambda_j \neq 0\}$ and $J_{>} = \{j \mid \lambda_j > 0\}$. We have the following result.

**Proposition 18**

$$K \subseteq \bigcup_{j \in J_>} K^j \subseteq \bigcup_{j \in J_{\neq}} K^j.$$

**Proof** From the inclusion $J_> \subseteq J_{\neq}$, we have immediately $\bigcup_{j \in J_>} K^j \subseteq \bigcup_{j \in J_{\neq}} K^j$.

Let us show that $K \subseteq \bigcup_{j \in J_>} K^j$. Let $x$ be a point of $K$. Since $K$ is nondegenerate, there exists a vector $\nu \geq 0$ such that $x = \sum_{j=1}^{n} \nu_j u^j$. Now let $\ell$ such that $\dfrac{\nu_\ell}{\lambda_\ell} = \min_{j \in J_>} \dfrac{\nu_j}{\lambda_j}$. We have

$$\begin{aligned} x &= \sum_{j=1, j\neq \ell}^{n} \nu_j u^j + \frac{\nu_\ell}{\lambda_\ell}\left(\omega - \sum_{j=1, j\neq \ell} \lambda_j u^j\right) \\ &= \sum_{j=1, j\neq \ell}^{n} \left(\nu_j - \frac{\nu_\ell}{\lambda_\ell}\lambda_j\right) u^j + \frac{\nu_\ell}{\lambda_\ell}\omega. \end{aligned}$$

By definition of $\ell$, $\nu_j - \dfrac{\nu_\ell}{\lambda_\ell}\lambda_j \geq 0$ for $j = 1, \ldots, n, j \neq \ell$ and $\dfrac{\nu_\ell}{\lambda_\ell} \geq 0$. Hence $x$ belongs to $K^\ell$, which proves the inclusion $K \subseteq \bigcup_{j \in J_>} K^j$. ■ By Proposition 18, the set of cones $\{K^j\}_{j \in J_{\neq}}$ and $\{K^j\}_{j \in J_>}$ both define a cover of the cone $K$. The first was proposed by Tuy [17] while the second is due to Gallo [3]. This subdivision process is called $\omega$-subdivision. The cones $K^j$ are said *subcones of* $K$.

## 3. Algorithm

The cone splitting algorithm we present in this Section is in its simplest form. We do not consider here sophistications such as, for example, restart after a better solution has been found and/or additions of cuts (see Horst and Tuy [10] for these kinds of variants). Tuy's 1964 initial algorithm and Gallo's variant differ by the realization of Step 5.

**Step 1 (initialization):** construct an initial conical cover $\mathcal{C}^0$ of $P$ as indicated in Section 2.1. Initialize the incumbent value $\overline{f}^0$ and solution $\overline{x}^0$ with the best point among $O$ and the intersection points of the edges of $K \in \mathcal{C}^0$ with the boundary $\delta P$ of $P$. Set $q$ to 0.

**Step 2 (fathoming):** for all cone $K^{qk}$ in $\mathcal{C}^q$, compute the $\overline{f}^q$-extensions $y^{qk1}, \ldots, y^{qkn}$ and solve the problem $LP(Y^{qk})$, obtaining the optimal solution $\omega^{qk}$ and value $\rho^{qk}$ (see Section 2.2). Remove from $\mathcal{C}^q$ all cones $K^{qk}$ for which $\rho^{qk} \leq 1$.

**Step 3 (optimality test):** if $\mathcal{C}^q = \emptyset$, stop: $\overline{x}^q$ is an optimal solution of problem $(CP)$ with value $\overline{f}^q$.

**Step 4 (update of the incumbent):** if for some $\omega^{qk}$, $f(\omega^{qk}) < \overline{f}^q$, set $\overline{x}^{q+1}$ to $\omega^{qk}$ and $\overline{f}^{q+1}$ to $f(\omega^{qk})$. Otherwise set $\overline{x}^{q+1}$ to $\overline{x}^q$ and $\overline{f}^{q+1}$ to $\overline{f}^q$.

**Step 5 (subdivision):** subdivide all cones $K^{qk}$ in $\mathcal{C}^q$ via the point $\omega^{qk}$ as indicated in Section 2.3. Let $\mathcal{C}^{q+1}$ be the set of subcones of all cones of $\mathcal{C}^q$. Increment $q$ and return to Step 2.

Actually, in Step 2, we have to compute all $\overline{f}^q$-extensions only if $\overline{f}^q$ has been improved at the previous iteration. If $\overline{f}^q = \overline{f}^{q-1}$, it is only necessary to compute the $\overline{f}^q$-extensions along the new edges $\omega^{q-1,k}$ for $k$ such that $K^{q-1,k} \in \mathcal{C}^{q-1}$.

If in Step 5, we subdivide the cones as proposed by Tuy, we obtain Tuy's 1964 algorithm (TUY64 for short), otherwise we obtain Gallo's variant. We will refer to this variant as to the Tuy-Gallo algorithm and note it T-G75.

The following result shows that algorithm T-G75 finally finds an optimal solution of problem $(CP)$.

**Theorem 1** *After at most $N(m,n) = \binom{\binom{m}{n}-1}{n}$ iterations of algorithm T-G75, $\overline{x}^q$ is an optimal solution of problem $(CP)$.*

**Proof** Let $\tilde{x}$ be any point of $P$. We will show that after a certain number of iterations of algorithm T-G75, the incumbent value $\overline{f}^q$ satisfies $\overline{f}^q \leq f(\tilde{x})$. In order to do so, we construct a sequence of cones $\{K^{qk_q}\}, q = 0, 1, \ldots$ containing $\tilde{x}$ as follows. Assume that $K^{qk_q} = cone\{x^{q1}, \ldots, x^{qn}\}$ contains $\tilde{x}$, and let $y^{qj} = \theta_j^q x^{qj}, j = 1, \ldots, n$ be the $\overline{f}^q$-extensions (we use $x^{qj}$ and $y^{qj}$ rather than $x^{qk_qj}$ and $y^{qk_qj}$ in order to simplify the notation). Since $K^{qk_q}$ is nondegenerate, there exists an unique $\tilde{\mu}^q \geq 0$ such that

$$\tilde{x} = \sum_{j=1}^{n} \tilde{\mu}_j^q x^{qj}. \tag{1}$$

If $\rho^{qk_q} \leq 1$ in Step 2, $f(\tilde{x}) \geq \min_{x \in K^{qk_q} \cap P} f(x) \geq \overline{f}^q$ (see Section 2.2), hence we are done. Otherwise, we consider in Step 5 the subcone $K^{q+1,k_{q+1}}$ of $K^{qk_q}$ obtained by replacing the point $x^{q\ell_q}$ by $\omega^{qk_q} = \sum_{j=1}^{n} \lambda_j^{qk_q} x^{qj}$ with $\ell_q$ satisfying

$$\frac{\tilde{\mu}_{\ell_q}^q}{\lambda_{\ell_q}^{qk_q}} = \min_{j|\lambda_j^{qk_q}>0} \left\{ \frac{\tilde{\mu}_j^q}{\lambda_j^{qk_q}} \right\}.$$

Note that this choice of $\ell_q$ ensures that $\tilde{x}$ belongs to $K^{q+1,k_{q+1}}$ (see the proof of Proposition 18). By definition of the subdivision, $K^{q+1,k_{q+1}} = cone\{x^{q+1,1}, \ldots, x^{q+1,n}\}$ with $x^{q+1,j} = x^{qj}$ for $j = 1, \ldots, n$, $j \neq \ell_q$ and $x^{q+1,\ell_q} = \omega^{qk_q}$. Since $x^{q+1,\ell_q} = \omega^{qk_q} = \sum_{j=1}^{n} \lambda_j^{qk_q} x^{qj}$ and by (1), we deduce that

$$\tilde{x} = \sum_{j=1}^{n} \tilde{\mu}_j^{q+1} x^{q+1,j}$$

with

$$\tilde{\mu}_j^{q+1} = \begin{cases} \tilde{\mu}_j^q - \dfrac{\tilde{\mu}_{\ell_q}^q}{\lambda_{\ell_q}^{qk_q}} \lambda_j^{qk_q} & \text{if } j \neq \ell_q \\ \dfrac{\tilde{\mu}_{\ell_q}^q}{\lambda_{\ell_q}^{qk_q}} & \text{if } j = \ell_q. \end{cases}$$

Since $\overline{f}^{q+1} \leq \overline{f}^q$, we have $\theta_j^{q+1} \geq \theta_j^q$ for $j \neq \ell_q$, hence

$$\sum_{j=1, j \neq \ell_q}^{n} \frac{\tilde{\mu}_j^{q+1}}{\theta_j^{q+1}} \leq \sum_{j=1, j \neq \ell_q}^{n} \frac{\tilde{\mu}_j^{q+1}}{\theta_j^q} = \sum_{j=1, j \neq \ell_q}^{n} \frac{1}{\theta_j^q} \left( \tilde{\mu}_j^q - \frac{\tilde{\mu}_{\ell_q}^q}{\lambda_{\ell_q}^{qk_q}} \lambda_j^{qk_q} \right)$$

$$= \sum_{j=1}^{n} \frac{1}{\theta_j^q} \left( \tilde{\mu}_j^q - \frac{\tilde{\mu}_{\ell_q}^q}{\lambda_{\ell_q}^{qk_q}} \lambda_j^{qk_q} \right) = \sum_{j=1}^{n} \frac{\tilde{\mu}_j^q}{\theta_j^q} - \frac{\tilde{\mu}_{\ell_q}^q}{\lambda_{\ell_q}^{qk_q}} \rho^{qk_q} \leq \left(1 - \frac{\tilde{\mu}_{\ell_q}^q}{\lambda_{\ell_q}^{qk_q}}\right) \sum_{j=1}^{n} \frac{\tilde{\mu}_j^q}{\theta_j^q},$$

the last inequality holding because $\rho^{qk_q} \geq \sum_{j=1}^{n} \frac{\tilde{\mu}_j^q}{\theta_j^q}$ as $\tilde{\mu}^q$ is a feasible solution to $LP'(Y^q)$. It follows that

$$\sum_{j=1}^{n} \frac{\tilde{\mu}_j^{q+1}}{\theta_j^{q+1}} \leq \left(1 - \frac{\tilde{\mu}_{\ell_q}^q}{\lambda_{\ell_q}^{qk_q}}\right) \sum_{j=1}^{n} \frac{\tilde{\mu}_j^q}{\theta_j^q} + \frac{\tilde{\mu}_{\ell_q}^q}{\lambda_{\ell_q}^{qk_q}} \frac{1}{\theta_{\ell_q}^{q+1}}. \tag{2}$$

If $\sum_{j=1}^{n} \frac{\tilde{\mu}_j^q}{\theta_j^q} \leq 1$, $\tilde{x} \in S^q = conv\{O, y^{q1}, \ldots, y^{qn}\}$ hence $f(\tilde{x}) \geq \overline{f}^q$ (see Section 2.2) and we are done. Assume therefore that $\sum_{j=1}^{n} \frac{\tilde{\mu}_j^q}{\theta_j^q} > 1$. Since $\overline{f}^{q+1} \leq f(\omega^{qk_q})$ (Step 4 of the algorithm) we have $\theta_{\ell_q}^{q+1} \geq 1$. Hence

$$\frac{1}{\theta_{\ell_q}^{q+1}} \leq 1 < \sum_{j=1}^{n} \frac{\tilde{\mu}_j^q}{\theta_j^q}.$$

It follows from (2) that

$$\sum_{j=1}^{n} \frac{\tilde{\mu}_j^{q+1}}{\theta_j^{q+1}} < \sum_{j=1}^{n} \frac{\tilde{\mu}_j^q}{\theta_j^q}$$

which shows that a same cone cannot occur more than one time in the sequence. Let $N_K$ denote the total number of distinct cones generated by the algorithm. Then after at most $N_K$ iterations we must have $\sum_{j=1}^{n} \frac{\tilde{\mu}_j^q}{\theta_j^q} \leq 1$, and hence $\overline{f}^q \leq f(\tilde{x})$. Since a cone is defined by $n$ extreme points, $N_K$ can be bounded from above by $\binom{N_V - 1}{n}$ where $N_V$ is the number of extreme points of $P$. This number is in turn bounded by $\binom{m}{n}$. Hence $N_K \leq N(m, n) = \binom{\binom{m}{n} - 1}{n}$.

Now take for $\tilde{x}$ an optimal solution $x^*$ of problem $(CP)$. Then $\overline{f}^{N_K} \leq f(x^*)$, which proves the optimality of $\overline{x}^{N_K}$. ■

**Corollary 1** *After at most* $N(m,n) = \binom{\binom{m}{n}-1}{n}$ *iterations of algorithm TUY64,* $\overline{x}^q$ *is an optimal solution of problem* $(CP)$.

**Proof** Obvious by noting that at each iteration, the set of cones generated by algorithm TUY64 contains the set of cones generated by algorithm T-G75. ∎

Zwart [25] has shown on an example that algorithm TUY64 can be infinite. In the next Section, we provide a similar example for algorithm T-G75. The remaining of this section is devoted to a brief discussion on how we can stop these algorithms while guaranteeing that an optimal solution of problem $(CP)$ has been found.

Theorem 1 and Corollary 1 give a first way to transform these algorithms into finite ones, namely to stop them after $N(m,n)$ iterations. Although the bound $N(m,n)$ could probably be improved, this method seems not to be practical, due to the huge number of iterations needed.

A second solution is to maintain a list of the cones generated by the algorithm. At each time a cone is generated, we check if it already belongs to the list. If yes, we do not reprocess the cone again; if no, we add it in the list and continue the algorithm. Since a cone is defined by $n$ extreme points of $P$ and since an extreme point may belong to several cones, it may be helpful to also keep a list of the extreme points. The disadvantage of this method is that the two lists may require a lot of memory (we have to store all cones rather than only those of the current iteration as it was the case for the first approach). Also searching in the lists if a cone or an extreme point is present may be time-consuming.

## 4. Examples

The first example of this Section shows that algorithm T-G75 can actually be infinite. The second one shows that a cycle may exist even if we use the more sophisticated fathoming procedure of Bali-Zwart [1][26] which consists to include the constraint $x \in K$ in the linear programs.

**Example 1** *Consider the following problem*

$$\min \quad f(x) = -x_1^2 - (x_2+2)^2 - x_3^2 - \left(x_4 + \frac{7}{2}\right)^2$$

$$
\text{s.t.} \quad \begin{cases}
3x_1 + 7x_2 + 6x_3 + 2x_4 & \leq 0 & (2) \\
2x_1 - 2x_2 - x_3 + 3x_4 & \leq 0 & (3) \\
3x_1 + 7x_3 - 4x_4 & \leq 22 & (4) \\
x_1 & \geq 0 & (5) \\
x_2 & \geq -2 & (6) \\
x_3 & \geq 0 & (7) \\
x_4 & \geq -\frac{7}{2}. & (8)
\end{cases}
$$

*Let $P$ be the polytope defined by the constraints (2)-(8). An extreme point of $P$ is $O = (0,0,0,0)$ with value $f(O) = -\frac{65}{4}$. The 4 extreme points adjacent to $O$ are*

$$
\begin{array}{lllll}
x^1 & = ( & 0 & -2 & 0 & -\frac{4}{3}) \\
x^2 & = ( & 0 & 1 & 0 & -\frac{7}{2}) \\
x^3 & = ( & \frac{110}{31} & -\frac{22}{31} & 0 & -\frac{88}{31}) \\
x^4 & = ( & 0 & -2 & \frac{5}{2} & -\frac{1}{2}).
\end{array}
$$

*They define the initial cone $K^{(0)} = K^{01}$. The $\gamma$-extensions of these points with $\gamma = -\frac{65}{4}$ are $y^1 = 3x^1$, $y^2 = \frac{82}{53}x^2$, $y^3 = \frac{248}{231}x^3$ and $y^4 = \frac{23}{21}x^4$. The equation of the hyperplane passing through the points $y^j, j = 1,2,3,4$ is*
$\frac{1}{41}\left(\frac{65}{16}x_1 - \frac{3}{2}x_2 + \frac{280}{23}x_3 - 8x_4\right) = 0$. *Hence the linear program to solve is* $\max_{x \in P} \frac{1}{41}\left(\frac{65}{16}x_1 - \frac{3}{2}x_2 + \frac{280}{23}x_3 - 8x_4\right)$. *An optimal solution is $\omega^{01} = x^5 = (0, \ -2, \ \frac{8}{7}, \ -\frac{7}{2})^t = \frac{297}{350}x^1 + \frac{107}{175}x^2 + \frac{16}{35}x^4$ with a value $\rho^{01} = \frac{1033}{943}$. Since $\rho^{01} > 1$, we construct the cones $K^{(1)} = K^{11}$, $K^{(2)} = K^{12}$ and $K^{(3)} = K^{13}$ obtained by replacing respectively $x^1$, $x^2$ and $x^4$ by $x^5$. Continuing the algorithm yields the results shown in Tables 7.1 and II.*

*Table 7.1 shows the extreme points generated by the algorithm. For each of them, we give the constraints of $P$ satisfied at equality, the value of $f$ and the $\gamma$-extension for $\gamma = -\frac{65}{4}$.*

*Table II gives the list of cones that were generated, with for each of them, the $n$ extreme points defining the edges, the objectif function of the associated linear program LP, and the optimal value $\rho$ and solution $\omega$ of this latter (if the cone cannot be fathomed, i.e., if $\rho > 1$, we give in addition the expression of $\omega$ in terms of the extreme points defining the cone).*

*The course of the algorithm can be visualized by a directed graph as shown in Figure 7.1: vertices correspond to cones, and there is an edge from a cone to another if the latter is a subcone of the former.*

*We immediately note the cycle $3 - 9 - 18 - 25 - 33 - 41 - 3$: the corresponding cones and their descendants will be generated infinitely many* *time, and hence the algorithm T-G75 will never stop. Note also that some cones have several "fathers cones". For example, the cones* $K^{(6)}$ *is first obtained when subdividing the cone* $K^{(2)}$ *and is obtained again when subdividing the cones* $K^{(23)}$ *and* $K^{(33)}$.

*If in the linear program LP, we add the constraint* $x \in K$ *(or equivalently the constraint* $\lambda \geq 0$ *in LP'), the fathoming procedure described in Section 2.2 is still valid (see, e.g., Bali [1], Zwart [26]). If we solve this more constrained problem for cone* $K^{(25)}$, *we obtain an optimal value of* $\frac{21}{23} < 1$ *which shows that the cone can be fathomed, and hence that the cycle can be broken.*

The following example shows that cycles may occur even if we include the constraint $x \in K$ in the linear programs solved in the fathoming procedure.

**Example 2** *The concave minimization problem that we consider in this example is the following.*

$$\min \quad f(x) = -100(x_1+1)^2 - 5(x_2+8)^2 - 25x_3^2 - 3(x_4+8)^2$$

$$s.t. \quad \left\{ \begin{array}{lll} 15x_1 + 3x_2 + 5x_3 & \leq 0 & (9) \\ -5x_1 + 7x_2 + 5x_3 - 4x_4 & \leq 0 & (10) \\ 5x_1 + x_2 + 10x_3 + x_4 & \leq 0 & (11) \\ -5x_1 - 10x_3 + 2x_4 & \leq 0 & (12) \\ 10x_1 - x_2 - 10x_3 & \leq 10 & (13) \\ -10x_1 - 4x_2 + 40x_3 + 2x_4 & \leq 5 & (14) \\ 15x_1 - x_2 + 5x_3 + 4x_4 & \leq 5 & (15) \\ -15x_1 - 5x_2 - 15x_3 + 6x_4 & \leq 10 & (16) \\ -4x_4 & \leq 35. & (17) \end{array} \right.$$

*The algorithm T-G75 is modified as follows. In Step 2, we solve the linear program LP with the additional constraint* $x \in K$. *If the optimal value is less than or equal to 1, the cone is fathomed. Otherwise, we solve the linear program without the additional constraint (i.e., the linear program as defined in Section 2.2) to get the point* $\omega$ *that will be used to subdivide the cone. With this modification, we prove the optimality of point* $O = (0,0,0,0)$ *after having generated 92 cones (142 for algorithm T-G75; the number of extreme points generated is the same for the two approaches: 19). Again there is a cycle, described in Tables 7.2-7.4. In Table 7.4,* $\rho_1^j$ *is the optimal value of the linear program associated to the cone* $K^{(j)}$, *in which we have added the constraint* $x \in K$. $\rho^j$ *and* $\omega$ *are respectively the optimal value and an optimal solution of the linear program LP as defined in Section 2.2.*

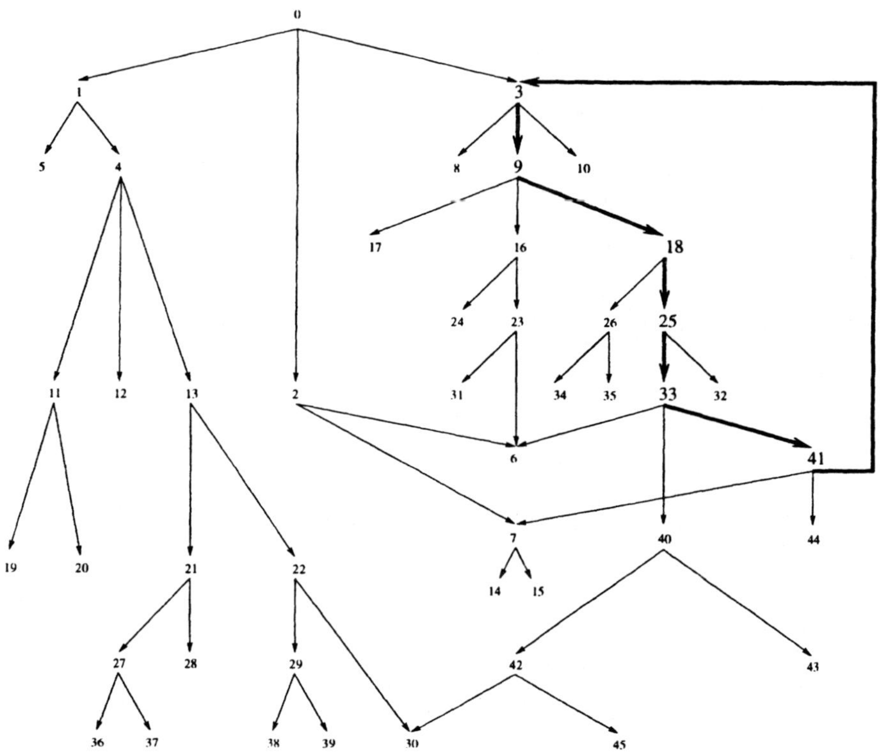


*Figure 7.1.* Exploration graph

*Table 7.1.* Extreme points generated

| $i$ | constraints | $x^i$ | $f(x^i)$ | $\frac{y^i}{x^i}$ |
|---|---|---|---|---|
| 0 | $\{2,3,5,7\}$ | $(0,0,0,0)$ | $-\frac{65}{4}$ | - |
| 1 | $\{3,5,6,7\}$ | $(0,-2,0,-\frac{4}{3})$ | $-\frac{169}{36}$ | 3 |
| 2 | $\{2,5,7,8\}$ | $(0,1,0,-\frac{7}{2})$ | $-9$ | $\frac{82}{53}$ |
| 3 | $\{2,3,4,7\}$ | $(\frac{110}{31},-\frac{22}{31},0,-\frac{88}{31})$ | $-\frac{56481}{3844}$ | $\frac{248}{231}$ |
| 4 | $\{2,3,5,6\}$ | $(0,-2,2.5-0.5)$ | $-\frac{61}{4}$ | $\frac{23}{21}$ |
| 5 | $\{4,5,6,8\}$ | $(0,-2,\frac{8}{7},-\frac{7}{2})$ | $-\frac{64}{49}$ | $\frac{6370}{3441}$ |
| 6 | $\{2,4,5,6\}$ | $(0,-2,\frac{50}{19},-\frac{17}{19})$ | $-\frac{19801}{1444}$ | $\frac{5149}{4233}$ |
| 7 | $\{2,3,4,6\}$ | $(\frac{10}{17},-2,\frac{40}{17},-\frac{16}{17})$ | $-\frac{14369}{1156}$ | $\frac{527}{389}$ |
| 8 | $\{4,6,7,8\}$ | $(\frac{8}{3},-2,0,-\frac{7}{2})$ | $-\frac{64}{9}$ | $\frac{1170}{841}$ |
| 9 | $\{2,4,7,8\}$ | $(\frac{8}{3},-\frac{1}{7},0,-\frac{7}{2})$ | $-\frac{4657}{441}$ | $\frac{44226}{34189}$ |
| 10 | $\{5,6,7,8\}$ | $(0,-2,0,-\frac{7}{2})$ | 0 | 2 |
| 11 | $\{3,4,6,7\}$ | $(\frac{50}{17},-2,0,-\frac{56}{17})$ | $-\frac{10049}{1156}$ | $\frac{374}{283}$ |
| 12 | $\{2,4,5,8\}$ | $(0,\frac{1}{49},\frac{8}{7},-\frac{7}{2})$ | $-\frac{12937}{2401}$ | $\frac{234514}{130197}$ |

*Table 7.2.* Extreme points

| $i$ | constraints | $x^i$ | $f(x^i)$ | $\frac{y^i}{x^i}$ |
|---|---|---|---|---|
| 0 | $\{9,10,11,12\}$ | $(0,0,0,0)$ | $-612$ | - |
| 2 | $\{9,11,12,15\}$ | $(\frac{1}{5},\frac{-15}{16},\frac{-3}{80},\frac{5}{16})$ | $\frac{-153785}{256}$ | $\frac{5120}{2233}$ |
| 3 | $\{9,10,12,13\}$ | $(\frac{16}{49},\frac{-30}{49},\frac{-30}{49},\frac{-110}{49})$ | $\frac{-191256}{343}$ | $\frac{1568}{635}$ |
| 5 | $\{13,14,16,17\}$ | $(\frac{-43}{92},\frac{-220}{23},\frac{-47}{92},\frac{-35}{4})$ | $\frac{-51661}{1058}$ | $\frac{270572}{151405}$ |
| 7 | $\{10,14,16,17\}$ | $(\frac{-229}{132},\frac{-65}{11},\frac{-61}{132},\frac{-35}{4})$ | $\frac{-90263}{1089}$ | $\frac{270006}{154765}$ |
| 8 | $\{13,14,15,16\}$ | $(\frac{39}{92},\frac{-100}{23},\frac{-13}{92}\frac{-5}{4})$ | $\frac{-215108}{529}$ | $\frac{17089}{6225}$ |
| 10 | $\{12,13,15,16\}$ | $(\frac{43}{109},\frac{-320}{109},\frac{-34}{109},\frac{-125}{218})$ | $\frac{-23314763}{47524}$ | $\frac{348800}{118003}$ |
| 12 | $\{9,13,14,17\}$ | $(\frac{157}{188},\frac{-220}{47},\frac{57}{188},\frac{-35}{4})$ | $\frac{-1748781}{4418}$ | $\frac{184804}{121135}$ |
| 14 | $\{10,12,13,17\}$ | $(\frac{-21}{26},\frac{-60}{13},\frac{-35}{26},\frac{-35}{4})$ | $\frac{-291943}{2704}$ | $\frac{514176}{241595}$ |
| 18 | $\{10,12,14,16\}$ | $(\frac{-55}{54},\frac{-95}{36},\frac{-23}{108},\frac{-65}{18})$ | $\frac{-1181929}{5832}$ | $\frac{686016}{208565}$ |

TABLE II. Cones generated

| iter | $j$ | $K^{(j)}$ | objectif | $\rho^j$ | $\omega$ |
|---|---|---|---|---|---|
| 0 | 0 | $(x^1, x^2, x^3, x^4)$ | $\left(\frac{65}{656}, \frac{-3}{82}, \frac{280}{943}, \frac{-8}{41}\right)$ | $\frac{1033}{943}$ | $x^5 = \frac{297}{350}x^1 + \frac{107}{175}x^2 + \frac{16}{35}x^4$ |
| 1 | 1 | $(x^2, x^3, x^4, x^5)$ | $\left(\frac{49130201}{226546320}, \frac{1445666}{6068205}, \frac{7539547}{14159145}, \frac{-4952933}{42477435}\right)$ | $\frac{18439}{17917}$ | $x^6 = \frac{2}{19}x^2 + \frac{20}{19}x^4$ |
| | 2 | $(x^1, x^3, x^4, x^5)$ | $\left(\frac{4378419}{17916080}, \frac{-407333}{2239510}, \frac{251133}{1119755}, \frac{25561}{1119755}\right)$ | $\frac{3171}{3128}$ | $x^7 = \frac{31}{187}x^3 + \frac{16}{17}x^4$ |
| | 3 | $(x^1, x^2, x^3, x^5)$ | $\left(\frac{65}{656}, \frac{-3}{82}, \frac{-56389}{298480}, \frac{-8}{41}\right)$ | $\frac{251}{246}$ | $x^8 = \frac{39}{50}x^1 + \frac{7}{75}x^2 + \frac{124}{165}x^3$ |
| 2 | 4 | $(x^3, x^4, x^5, x^6)$ | $\left(\frac{12545007}{45376240}, \frac{-1904847}{2836015}, \frac{-382971}{2836015}, \frac{525723}{2836015}\right)$ | $\frac{119243931}{77139608}$ | $x^{11} = \frac{155}{187}x^3 + \frac{80}{17}x^4 + \frac{1400}{1683}x^5 - \frac{8132}{1683}x^6$ |
| | 5 | $(x^2, x^3, x^5, x^6)$ | $\left(\frac{15999169}{80888080}, \frac{139918}{722215}, \frac{2101418}{5055505}, \frac{-653759}{5055505}\right)$ | $\frac{20}{21}$ | $x^{12}$ |
| | 6 | $(x^1, x^4, x^5, x^7)$ | $\left(\frac{-3312201}{2239510}, \frac{-407333}{2239510}, \frac{251133}{1119755}, \frac{25561}{1119755}\right)$ | $\frac{3972288}{4255069}$ | $x^8$ |
| | 7 | $(x^1, x^3, x^5, x^7)$ | $\left(\frac{5055443}{24147760}, \frac{-221638}{1509235}, \frac{12040789}{96591040}, \frac{-179407}{6036940}\right)$ | $\frac{137}{136}$ | $x^{11} = \frac{12}{17}x^1 + \frac{155}{187}x^3$ |
| | 8 | $(x^2, x^3, x^5, x^8)$ | $\left(\frac{280487}{1662960}, \frac{1913}{15210}, \frac{2761981}{11640720}, \frac{-46376}{311805}\right)$ | $\frac{20}{21}$ | $x^9$ |
| | 9 | $(x^1, x^3, x^5, x^8)$ | $\left(\frac{13453}{21840}, \frac{-1511}{2730}, \frac{27977}{21840}, \frac{264}{455}\right)$ | $\frac{2701}{672}$ | $x^4 = \frac{261}{28}x^1 + \frac{3317}{308}x^3 + \frac{35}{16}x^5 - \frac{1605}{112}x^6$ |
| | 10 | $(x^1, x^2, x^5, x^8)$ | $\left(\frac{-1789}{127920}, \frac{-3}{82}, \frac{-56389}{298480}, \frac{-8}{41}\right)$ | $\frac{31}{41}$ | $x^{10}$ |
| 3 | 11 | $(x^4, x^5, x^6, x^{11})$ | $\left(\frac{145471}{17826380}, \frac{-1904847}{2836015}, \frac{-382971}{2836015}, \frac{525723}{2836015}\right)$ | $\frac{621746}{567203}$ | $x^1 = \frac{1980}{297}x^4 + \frac{350}{297}x^5 - \frac{2033}{297}x^6$ |
| | 12 | $(x^3, x^5, x^6, x^{11})$ | $\left(\frac{18176453}{86806720}, \frac{522247}{7891520}, \frac{29036593}{86806720}, \frac{-1799767}{21701680}\right)$ | $\frac{231}{248}$ | $x^3$ |
| | 13 | $(x^3, x^4, x^6, x^{11})$ | $\left(\frac{2096145}{4388032}, \frac{-886943}{4388032}, \frac{586725}{2194016}, \frac{701047}{2194016}\right)$ | $\frac{3171}{3128}$ | $x^7 = \frac{31}{187}x^3 + \frac{16}{17}x^4$ |
| | 14 | $(x^3, x^5, x^7, x^{11})$ | $\left(\frac{506956903}{1062501440}, \frac{-214165313}{1062501440}, \frac{159978911}{1062501440}, \frac{84651043}{265625360}\right)$ | $\frac{231}{248}$ | $x^3$ |
| | 15 | $(x^1, x^5, x^7, x^{11})$ | $\left(\frac{787817}{6036940}, \frac{-5087611}{33203170}, \frac{37925501}{265625360}, \frac{-334688}{16601585}\right)$ | $\frac{283}{374}$ | $x^9$ |
| | 16 | $(x^3, x^4, x^5, x^8)$ | $\left(\frac{419483}{1820910}, \frac{280355}{8740368}, \frac{2082621}{5462730}, \frac{-105281}{2185092}\right)$ | $\frac{3171}{3128}$ | $x^7 = \frac{31}{187}x^3 + \frac{16}{17}x^4$ |
| | 17 | $(x^1, x^4, x^5, x^8)$ | $\left(\frac{7122583}{43670445}, \frac{-407333}{2239510}, \frac{251133}{1119755}, \frac{25561}{1119755}\right)$ | $\frac{121235}{125511}$ | $x^7$ |
| | 18 | $(x^1, x^3, x^4, x^8)$ | $\left(\frac{13453}{21840}, \frac{-1511}{2730}, \frac{1208}{31395}, \frac{264}{455}\right)$ | $\frac{3171}{3128}$ | $x^7 = \frac{31}{187}x^3 + \frac{16}{17}x^4$ |
| 4 | 19 | $(x^1, x^5, x^6, x^{11})$ | $\left(\frac{789233}{5425420}, \frac{-41479}{246610}, \frac{45709}{246610}, \frac{283}{123305}\right)$ | $\frac{7894331}{9223214}$ | $x^7$ |
| | 20 | $(x^1, x^4, x^6, x^{11})$ | $\left(\frac{90105}{274252}, \frac{-4379}{12466}, \frac{870}{6233}, \frac{1726}{6233}\right)$ | $\frac{2246319}{2331142}$ | $x^7$ |
| | 21 | $(x^3, x^6, x^7, x^{11})$ | $\left(\frac{54205}{5914304}, \frac{1575589}{5914304}, \frac{2274365}{5914304}, \frac{-566693}{1478576}\right)$ | $\frac{47062857}{26345536}$ | $x^{12} = \frac{3069}{1960}x^3 + \frac{4351}{490}x^6 - \frac{18513}{1960}x^7$ |
| | 22 | $(x^4, x^6, x^7, x^{11})$ | $\left(\frac{-11638405}{34007248}, \frac{-34757337}{34007248}, \frac{-7402965}{17003624}, \frac{1449753}{17003624}\right)$ | $\frac{32824333}{17003624}$ | $x^1 = \frac{20}{3}x^4 - \frac{85}{12}x^7 + \frac{17}{12}x^{11}$ |
| | 23 | $(x^4, x^5, x^7, x^8)$ | $\left(\frac{-153169}{973245}, \frac{-1402855}{1167894}, \frac{-509501}{973245}, \frac{210869}{583947}\right)$ | $\frac{36979}{19251}$ | $x^1 = \frac{20}{29}x^4 - \frac{245}{1044}x^5 - \frac{1819}{261}x^7 + \frac{535}{348}x^8$ |
| | 24 | $(x^3, x^5, x^7, x^8)$ | $\left(\frac{1576299}{8124480}, \frac{141815}{1624896}, \frac{802757}{2708160}, \frac{-43637}{406224}\right)$ | $\frac{231}{248}$ | $x^3$ |
| | 25 | $(x^1, x^4, x^7, x^8)$ | $\left(\frac{-1941046}{973245}, \frac{3091757}{1946490}, \frac{1079723}{973245}, \frac{-2562129}{973245}\right)$ | $\frac{2102666}{194649}$ | $x^2 = -\frac{117}{14}x^1 + \frac{320}{7}x^4 - \frac{340}{7}x^7 + \frac{75}{7}x^8$ |
| | 26 | $(x^1, x^3, x^7, x^8)$ | $\left(\frac{13453}{21840}, \frac{-1511}{2730}, \frac{-17214403}{46038720}, \frac{264}{455}\right)$ | $\frac{137}{136}$ | $x^{11} = \frac{12}{17}x^1 + \frac{155}{187}x^3$ |

TABLE II (continued)

| iter | $j$ | $K^{(j)}$ | objectif | $\rho^j$ | optimal solution |
|---|---|---|---|---|---|
| 5 | 27 | $(x^6, x^7, x^{11}, x^{12})$ | $\left(\frac{-173626378}{1105697615}, \frac{-1516904047}{4422790460}, \frac{-7148183}{2211395230}, \frac{-357533091}{2211395230}\right)$ | $\frac{5536539731}{4422790460}$ | $x^{10} = \frac{443023}{54450}x^6 - \frac{357}{40}x^7 + \frac{357}{200}x^{11} - \frac{392}{1089}x^{12}$ |
| | 28 | $(x^3, x^7, x^{11}, x^{12})$ | $\left(\frac{13150604983}{59797432640}, \frac{3327664807}{59797432640}, \frac{16694020121}{59797432640}, \frac{-2007346279}{29898716320}\right)$ | $\frac{126587307}{135903256}$ | $x^3$ |
| | 29 | $(x^1, x^6, x^7, x^{11})$ | $\left(\frac{-460}{8401}, \frac{11839}{369644}, \frac{348065}{1478576}, \frac{-220339}{739288}\right)$ | $\frac{8646843}{6586384}$ | $x^{12} = -\frac{3267}{2450}x^1 + \frac{4351}{490}x^6 - \frac{18513}{1960}x^7 + \frac{18513}{9800}x^{11}$ |
| | 30 | $(x^1, x^4, x^6, x^7)$ | $\left(\frac{-460}{8401}, \frac{-4379}{12466}, \frac{870}{6233}, \frac{1726}{6233}\right)$ | $\frac{5691}{6233}$ | $x^4$ |
| | 31 | $(x^1, x^5, x^7, x^8)$ | $\left(\frac{9054}{70525}, \frac{-194731}{1269450}, \frac{90934}{634725}, \frac{-4211}{211575}\right)$ | $\frac{3557}{4743}$ | $x^{11}$ |
| | 6 | $(x^1, x^4, x^5, x^7)$ | $\left(\frac{-3312201}{2239510}, \frac{-407333}{2239510}, \frac{251133}{1119755}, \frac{25561}{1119755}\right)$ | $\frac{3972288}{4255069}$ | $x^8$ |
| | 32 | $(x^1, x^2, x^7, x^8)$ | $\left(\frac{-1789}{127920}, \frac{-3}{82}, \frac{3300259}{15862080}, \frac{-8}{41}\right)$ | $\frac{13794379}{13879320}$ | $x^5$ |
| | 33 | $(x^1, x^2, x^4, x^7)$ | $\left(\frac{-10800}{29233}, \frac{-3}{82}, \frac{280}{943}, \frac{-8}{41}\right)$ | $\frac{1033}{943}$ | $x^5 = \frac{297}{350}x^1 + \frac{107}{175}x^2 + \frac{16}{35}x^4$ |
| | 34 | $(x^3, x^7, x^8, x^{11})$ | $\left(\frac{28151}{137280}, \frac{9679}{137280}, \frac{1219697}{4255680}, \frac{-93}{1040}\right)$ | $\frac{231}{248}$ | $x^3$ |
| | 35 | $(x^1, x^7, x^8, x^{11})$ | $\left(\frac{1211}{8580}, \frac{-703}{4290}, \frac{146243}{1063920}, \frac{-3}{715}\right)$ | $\frac{283}{374}$ | $x^{11}$ |
| 6 | 36 | $(x^7, x^{10}, x^{11}, x^{12})$ | $\left(\frac{1023976361}{10872557212}, \frac{-420336643}{5436278606}, \frac{731318161}{3953657168}, \frac{-268209431}{2718139303}\right)$ | $\frac{2255816489}{2981185042}$ | $x^{11}$ |
| | 37 | $(x^6, x^7, x^{10}, x^{12})$ | $\left(\frac{-1995317}{40207186}, \frac{-179702}{1945509}, \frac{137206}{648503}, \frac{-175243}{1945509}\right)$ | $\frac{10129569}{12321557}$ | $x^6$ |
| | 38 | $(x^1, x^7, x^{11}, x^{12})$ | $\left(\frac{25461413}{339758140}, \frac{-182507539}{1868669770}, \frac{2549731889}{14949358160}, \frac{-96703067}{934334885}\right)$ | $\frac{4246111813}{5606009310}$ | $x^8$ |
| | 39 | $(x^1, x^6, x^7, x^{12})$ | $\left(\frac{-1100780}{21003593}, \frac{-135325}{1297006}, \frac{130553}{648503}, \frac{-60632}{648503}\right)$ | $\frac{10129569}{12321557}$ | $x^6$ |
| | 40 | $(x^2, x^4, x^5, x^7)$ | $\left(\frac{-1425405589}{994915922}, \frac{1445666}{6068205}, \frac{7539547}{14159145}, \frac{-4952933}{42477435}\right)$ | $\frac{18439}{17917}$ | $x^6 = \frac{2}{19}x^2 + \frac{20}{19}x^4$ |
| | 6 | $(x^1, x^4, x^5, x^7)$ | $\left(\frac{-3312201}{2239510}, \frac{-407333}{2239510}, \frac{251133}{1119755}, \frac{25561}{1119755}\right)$ | $\frac{3972288}{4255069}$ | $x^8$ |
| | 41 | $(x^1, x^2, x^5, x^7)$ | $\left(\frac{3640859}{2313220}, \frac{-3}{82}, \frac{-56389}{298480}, \frac{-8}{41}\right)$ | $\frac{44207603}{7170982}$ | $x^3 = \frac{3267}{310}x^1 + \frac{1177}{155}x^2 - \frac{385}{31}x^5 + \frac{187}{31}x^7$ |
| 7 | 42 | $(x^4, x^5, x^6, x^7)$ | $\left(\frac{-33771251}{175832930}, \frac{-1904847}{2836015}, \frac{-382971}{2836015}, \frac{525723}{2836015}\right)$ | $\frac{621746}{567203}$ | $x^1 = \frac{20}{3}x^4 + \frac{350}{297}x^5 - \frac{2033}{297}x^6$ |
| | 43 | $(x^2, x^5, x^6, x^7)$ | $\left(\frac{13776733}{313441310}, \frac{139918}{722215}, \frac{2101418}{5055505}, \frac{-653759}{5055505}\right)$ | $\frac{1014411}{1088878}$ | $x^{12}$ |
| | 44 | $(x^2, x^3, x^5, x^7)$ | $\left(\frac{5231677}{27758640}, \frac{1022981}{5948280}, \frac{39812741}{111034560}, \frac{-2821639}{20818980}\right)$ | $\frac{20}{21}$ | $x^9$ |
| | 7 | $(x^1, x^3, x^5, x^7)$ | $\left(\frac{5055443}{24147760}, \frac{-221638}{1509235}, \frac{12040789}{96591040}, \frac{-179407}{6036940}\right)$ | $\frac{137}{136}$ | $x^{11} = \frac{12}{17}x^1 + \frac{155}{187}x^3$ |
| | 3 | $(x^1, x^2, x^3, x^5)$ | $\left(\frac{65}{656}, \frac{-3}{82}, \frac{-56389}{298480}, \frac{-8}{41}\right)$ | $\frac{251}{246}$ | $x^8 = \frac{39}{50}x^1 + \frac{7}{75}x^2 + \frac{124}{165}x^3$ |
| 8 | 45 | $(x^1, x^5, x^6, x^7)$ | $\left(\frac{-460}{8401}, \frac{-41479}{246610}, \frac{45709}{246610}, \frac{283}{123305}\right)$ | $\frac{4233}{5149}$ | $x^6$ |
| | 30 | $(x^1, x^4, x^6, x^7)$ | $\left(\frac{-460}{8401}, \frac{-4379}{12466}, \frac{870}{6233}, \frac{1726}{6233}\right)$ | $\frac{5691}{6233}$ | $x^4$ |

*Table 7.4.* Cones defining a cycle

| $j$ | $K^{(j)}$ | $\rho_1^j$ | $\rho^j$ | $\omega$ |
|---|---|---|---|---|
| 1 | $(x^2, x^3, x^8, x^{14})$ | 1.14491... | 1.14491... | $x^{10} = \frac{140125}{57879}x^2 + \frac{9751343}{157430880}x^3 + \frac{74399}{546635}x^{14}$ |
| 2 | $(x^2, x^8, x^{10}, x^{14})$ | 1.09956... | 21.11789... | $x^8 = \frac{420009200}{6567231}x^2 + \frac{2870423}{2189077}x^8 - \frac{61551646}{2189077}x^{10} + \frac{32437964}{6567231}x^{14}$ |
| 3 | $(x^2, x^7, x^{10}, x^{14})$ | 1.09956... | 16.54557... | $x^{18} = \frac{4569500}{91071}x^2 - \frac{209716}{10119}x^{10} + \frac{324467}{91071}x^{14}$ |
| 4 | $(x^2, x^7, x^{10}, x^{18})$ | 1.00772... | 2.73396... | $x^{12} = \frac{10107520}{1173073}x^2 + \frac{23186911}{3945791}x^7 - \frac{4152028}{1173073}x^{10} - \frac{456092820}{43403701}x^{18}$ |
| 5 | $(x^2, x^{10}, x^{12}, x^{18})$ | 1.59602... | 1.59602... | $x^5 = \frac{65996160}{48481723}x^2 + \frac{43943568}{48481723}x^{10} + \frac{19728955}{48481723}x^{12} + \frac{68403744}{48481723}x^{18}$ |
| 6 | $(x^2, x^5, x^{12}, x^{18})$ | 1.55245... | 1.56301... | $x^8 = \frac{720}{227}x^2 - \frac{7867}{739112}x^5 + \frac{2679}{19976}x^{12} + \frac{2700}{8399}x^{18}$ |
| 7 | $(x^2, x^5, x^8, x^{12})$ | 1.15766... | 21.97873... | $x^{14} = \frac{4144}{65}x^2 + \frac{700961}{114400}x^5 - \frac{35483}{1300}x^8 + \frac{119991}{114400}x^{12}$ |
| 8 | $(x^2, x^8, x^{12}, x^{14})$ | 1.17937... | 3.03015... | $x^3 = \frac{16071200}{2020417}x^2 - \frac{5128080}{2020417}x^8 + \frac{665520}{2020417}x^{12} + \frac{1159392}{2020417}x^{14}$ |

## 5. Conclusions

In this paper, we have shown that, perhaps a little unexpectedly, the first conical algorithm for concave minimization proposed by Tuy in 1964 always finds an optimal solution despite the fact that it can cycle as pointed out by Zwart. It is hoped that this result will contribute to rehabilitate cones covering algorithms.

The most attractive feature of the Tuy-Gallo algorithm is that it only generates extreme points of the polytope. This is interesting for different reasons.

- This algorithm possesses a certain kind of elegance since it is for long time known that it is not necessary to consider other points than the extreme ones when searching for the minimum of a concave function over a polytope.
- It can easily be made finite, for example as proposed in this paper.
- The nonlinear parts of the computation (function evaluations, determinations of $\gamma$-extensions) are rentabilized, at least if we accept to store the extreme points, since it is likely that an extreme point will be generated several times.

In contrast, cones partitioning algorithms generate points on the boundary of the polytope (not necessarily extreme ones) and it is even not known if these algorithms are finite.

On the other hand, the possible presence of cycles in Tuy-Gallo algorithm can be viewed as a serious disadvantage compared to cones partitioning algorithms. A possible direction for future work could be to device new subdivision procedures, in such a way that cycles cannot appear, while preserving the property that edges correspond to extreme points. See [15] for preliminary steps in this direction.

# References


[1] BALI, S. (1973). *Minimization of a Concave Function on a Bounded Convex Polyhedron.* Ph.D. Dissertation, University of California, Los Angeles.

[2] BENSON, H. P. (1996). Concave minimization: Theory, applications and algorithms. In *Handbook of Global Optimization*, R. Horst and P. M. Pardalos, Eds. Kluwer Academic Publishers, Dordrecht, pp. 43–148.

[3] GALLO, G. (1975). On Hoang Tui's concave programming algorithm. Nota scientifica S-76-1, Instituto di Scienze dell'Informazione, University of Pisa, Italie.

[4] GALLO, G., AND ÜLKÜCÜ, A. (1977). Bilinear programming: an exact algorithm. *Mathematical Programming, 12*, 173–194.

[5] HANSEN, P., JAUMARD, B., MEYER, C., AND TUY, H. (1996). Best simplicial and double-simplicial bounds for concave minimization. Les Cahiers du GERAD G-96-17, GERAD, Montréal, Canada.

[6] HAMAMI, M. AND JACOBSEN, S.E. (1988). Exhaustive Nondegenerate Conical Processes for Concave Minimization on Convex Polytopes. *Mathematics of Operations Research, 13*, 479–487.

[7] HORST, R., PHONG T. Q., AND THOAI, N. V. (1990). On solving general reverse convex programming problems by a sequence of linear programs and line searches. *Annals of Operations Research 25*, 1–18.

[8] HORST, R., AND THOAI, N. V. (1989). Modification, implementation and comparison of three algorithms for globally solving linearly constrained concave minimization problems. *Computing, 42*, 271–289.

[9] Horst, R., and Tuy, H. (1991). The Geometric Complementarity Problem and Transcending Stationarity in Global Optimization. *DIMACS Series in Discrete Mathematics and Theoretical Computer Science, 4*, 341–354.

[10] Horst, R., and Tuy, H. (1996). *Global Optimization. Deterministic Approaches*, third, revised and enlarged edition. Springer-Verlag, Berlin.

[11] Horst, R., Thoai, N. V., and Benson, H. P. (1991). Concave minimization via conical partitions and polyhedral outer approximation. *Mathematical Programming, 50*, 259–274.

[12] Horst, R., Thoai, N. V., and de Vries, J. (1992). On geometry and convergence of a class of simplicial covers. *Optimization, 25*, 53–64.

[13] Horst, R., Thoai, N. V., and de Vries, J. (1992). A new simplicial cover technique in constrained global optimization. *Journal of Global Optimization, 2*, 1–19.

[14] Jaumard, B., and Meyer, C. (1996). On the convergence of cone splitting algorithms with $\omega$-subdivisions. Les Cahiers du GERAD G-96-36, GERAD, Montréal, Canada.

[15] Meyer, C. (1996). *Algorithmes coniques pour la minimisation concave.* PhD thesis, Ecole Polytechnique de Montréal, Montréal, Canada.

[16] Thoai, H., and Tuy, H. (1980). Convergent algorithms for minimizing a concave function. *Mathematics of Operations Research, 5*, 556–566.

[17] Tuy, H. (1964). Concave programming under linear constraints, *Soviet Mathematics, 5*, 1437–1440.

[18] Tuy, H. (1990). On Polyhedral Annexation Method for Concave Minimization, in Lev J. Leifman (ed.), *Functional Analysis, Optimization, and Mathematical Economics*, Oxford University Press, New-York, 248–260.

[19] Tuy, H. (1991). Normal Conical Algorithm for Concave Minimization. *Mathematical Programming, 51*, 229–245.

[20] Tuy, H. (1991). Effect of the Subdivision Strategy on Convergence and Efficiency of Some Global Optimization Algorithms. *Journal of Global Optimization, 1*, 23–36.

[21] TUY, H. (1996). D.c. optimization: Theory, methods and algorithms. In *Handbook of Global Optimization*, R. Horst and P. M. Pardalos, Eds. Kluwer Academic Publishers, pp. 149–216.

[22] TUY, H., KHATCHATUROV, V., AND UTKIN, S. (1987). A class of exhaustive cone splitting procedures in conical algorithms for concave minimization. *Optimization, 18* (6), 791–807.

[23] TUY, H., THIEU, T. V., AND THAI, N. Q. (1985). A conical algorithm for globally minimizing a concave function over a closed convex set. *Mathematics of Operations Research, 10*, 498–514.

[24] VAISH, H. (1974). *Nonconvex Programming with Applications to Production and Location Problems.* PhD thesis, School of Industrial and Systems Engineering, Georgia Institute of Technology, Atlanta, Georgia.

[25] ZWART, P. B. (1973). Nonlinear Programming: counterexamples to two global optimization algorithms. *Operations Research, 21*, 1260–1266.

[26] ZWART, P. B. (1974). Global maximization of a convex function with linear inequality constraints. *Operations Research, 22*, 602–609.

# Chapter 8

# THE $\alpha$BB GLOBAL OPTIMIZATION ALGORITHM FOR NONCONVEX PROBLEMS: AN OVERVIEW

Claire S. Adjiman and Christodoulos A. Floudas[1]

*Department of Chemical Engineering*
*Princeton University*
*Princeton, NJ 08544-5263, USA*

[1]Author to whom all correspondence should be addressed

**Abstract** The $\alpha$BB algorithm, a deterministic global optimization algorithm for constrained twice-differentiable NLPs, is presented. It is based on a branch-and-bound approach in which a convex relaxation of the original nonconvex problem is obtained through a dual procedure. First, all nonconvex terms of special structure (i.e., bilinear, trilinear, fractional, fractional trilinear, univariate concave) are replaced by customized tight convex lower bounding functions. Second, valid convex underestimators are generated for general nonconvex terms by using a diagonal shift matrix $\Delta$. $\Delta$ is related to the Hessian matrix of the nonconvex term and four rigorous methods, based on the *interval* Hessian matrix, are proposed for its computation. Branching variable selection strategies that improve the quality of the convex relaxations are presented. Variable bound updates are also found to affect the convergence rate positively. An implementation of the algorithm is used to solve three examples: a small but highly nonconvex problem, a stability problem and a large-scale problem involving the design of batch plant under uncertainty.

**Keywords** : Global optimization; $\alpha$BB algorithm; branch and bound; twice differentiable.



*A. Migdalas et al. (eds.), From Local to Global Optimization,* 155–186.

# 1. Introduction

Optimization is a powerful tool that can potentially be applied to a wide range of problems, such as the determination of phase equilibrium or molecular conformation, the design of separation, reactor or heat exchanger networks, the design or scheduling of batch plants and the design of VLSI chips. The mathematical representation of these systems, however, usually requires the use of highly nonconvex functions in order to achieve a satisfactory degree of accuracy. The nonconvexities that are characteristic of these problems often prevent local optimization algorithms from identifying the global optimum solution, despite its economic or physical signifance. The treatment of nonconvex NLPs has therefore received increasing attention in recent years. Reviews of some recent developments and their applications to process synthesis and design, process control and computational chemistry can be found in Floudas and Grossmann (1995), Floudas and Pardalos (1996), Grossmann (1996) and Floudas (1997).

The deterministic global optimization methods proposed to date fall into several classes such as interval methods (Hansen, 1992; Neumaier, 1990), primal-dual methods (Floudas and Visweswaran, 1993), and branch-and-bound methods (Horst and Tuy, 1990). Many of these appraoches are based on the generation of valid convex underestimators for certain types of nonconvex functions. The GOP algorithm developed by Floudas and Visweswaran (1990, 1993, see also Visweswaran and Floudas, 1990, 1993, 1996a,b) is a primal-dual method which relies on the partitioning of the variables into two sets in order to construct valid underestimators. The branch-and-bound algorithm described in Al-Khayyal and Falk (1983) and Al-Khayyal (1990) uses the convex envelope of bilinear functions to deal with their inherent nonconvexity. Smith and Pantelides (1996) recently presented a branch-and-bound algorithm capable of handling a large number of nonconvex functions thanks to an initial symbolic reformulation which results in the generation of a nonconvex problem containing only terms for which a convex underestimator is known. These include bilinear, univariate concave, convex, linear fractional and simple power terms. The reformulation is achieved through the addition of new variables and constraints to the original problem. Finally, the $\alpha$BB algorithm (Maranas and Floudas, 1994a,b; Androulakis *et al.*, 1995) is applicable to the broad class of twice-differentiable functions (Liu and Floudas, 1993), and has been used to identify all solutions of nonlinearly constrained systems of equations (Maranas and Floudas, 1995). For any twice-differentiable nonconvex function, a convex underestimator is obtained by subtracting a separable positive quadratic term (Maranas and Floudas, 1994a,b). In the general case, the determination of the magnitude of this quadratic term depends on the properties of the Hessian matrix of the nonconvex function and on the variable bounds. For many problems, the completion of this underestimation process in a rigorous way is hindered by the complexity of the Hessian matrix analysis. The availability of procedures that simplify this task is of crucial importance in light of the large number of problems that can potentially be addressed with the $\alpha$BB algorithm.

In this chapter, a review of the recent theoretical and computational results on the $\alpha$BB global optimization algorithm is presented. The basic principles of the $\alpha$BB algorithm are first laid out. This is followed by a discussion of methodologies that have been proposed to resolve the issue of convex underestimator generation for twice-differentiable nonconvex terms. Other important steps in the algorithm, such as branching variable selection and variable bound updates, are then outlined. Finally, the algorithm is tested on three nonconvex constrained examples.

# 2. The $\alpha$BB Global Optimization Algorithm

## 2.1. General Framework

The $\alpha$BB algorithm is a deterministic algorithm which can solve nonconvex optimization problems of the type represented by formulation (1) to global optimality.

$$
\begin{array}{lrl}
\min\limits_{\boldsymbol{x}} & f(\boldsymbol{x}) & \\
s.t. & \boldsymbol{g}(\boldsymbol{x}) & \leq 0 \\
 & \boldsymbol{h}(\boldsymbol{x}) & = 0 \\
 & \boldsymbol{x} & \in X \subseteq \Re^n
\end{array}
\tag{1}
$$

where $f$, $\boldsymbol{g}$ and $\boldsymbol{h}$ belong to $C^2$, the set of twice-differentiable functions, and $\boldsymbol{x}$ is a vector of size $n$.

The theoretical properties of this branch-and-bound algorithm guarantee that the global optimum is identified with finite $\epsilon$-convergence. At each iteration of the algorithm, the solution space is partitioned into smaller regions through a branching step. For each newly created subregion, or node of the branch-and-bound tree, upper and lower bounds on the global solution are generated. The lower bound is obtained by constructing a convex relaxation of the nonconvex problem whose unique solution can be obtained with a standard local optimization technique. The upper bound is calculated either by solving the original problem locally or by performing a problem evaluation at the solution of the lower bounding problem. The most crucial phase of the algorithm is the computation of the lower bound: convergence to the global optimum solution can be guaranteed only if valid convex underestimators are constructed for the nonconvex functions in the problem and if the lower bounding problems become tighter as the size of the solution domain decreases. From the practical standpoint, great care must be taken in deriving underestimators since the rate of convergence of the algorithm depends on the tightness of the relaxation obtained.

## 2.2. Lower Bounding Problem Generation

The derivation of the lower bounding problem in a rigorous and yet efficient way is discussed in detail in the following pages. A determining step in the convexification strategy is the decomposition of each nonlinear function in the problem into a sum of terms belonging to one of several categories: linear, bilinear, trilinear, fractional, fractional trilinear, convex, univariate concave or general nonconvex. Linear and convex terms can be left unchanged, and valid convex underestimators can be derived for each of the other term types.

**Bilinear, Trilinear, Fractional, Fractional Trilinear Terms.** For a bilinear term, McCormick (1976) showed that the convex envelope is obtained by substituting the term with a new variable and adding two linear inequalities to the problem. In addition, two linear inequalities providing upper bounds on the new variable can be included to provide a tighter relaxation. Maranas and Floudas (1995) showed that other nonconvex terms can be similarly underestimated through the replacement of the term by a new variable and the introduction of new constraints. Thus, a trilinear term can be underestimated with a new variable which is bounded by eight linear inequality constraints. For fractional terms, two new constraints are required. Finally, for fractional trilinear terms, eight inequality constraints are appended to the problem. Their analysis is based on general results for products of univariate functions.

**Univariate Concave Terms.** Univariate concave functions are trivially underestimated by their linearization at the lower bound of the variable range. The generation of the best convex underestimator for a univariate concave function does not require the introduction of additional variables or constraints.

**General Nonconvex Terms.** For the most general nonconvexities, an $\alpha$-based underestimator first proposed by Maranas and Floudas (1994b) is used. A general nonconvex term $f_{nc}(\boldsymbol{x}) \in \mathcal{C}^2(\Re^n)$ is underestimated over the entire domain $[\boldsymbol{x}^L, \boldsymbol{x}^U]$ by the function $\mathcal{L}(\boldsymbol{x})$

$$\mathcal{L}(\boldsymbol{x}) = f_{nc}(\boldsymbol{x}) + \sum_{i=1}^{n} \alpha_i (x_i^L - x_i)(x_i^U - x_i) \tag{2}$$

where the $\alpha_i$'s are positive scalars.

Since the summation term in Eq. (2) is negative over the entire region $[\boldsymbol{x}^L, \boldsymbol{x}^U]$, $\mathcal{L}(\boldsymbol{x})$ is a guaranteed underestimator of $f_{nc}(\boldsymbol{x})$. Furthermore, since the quadratic term in the underestimator is convex, the convexity of $\mathcal{L}(\boldsymbol{x})$ can be guaranteed by using sufficiently large values of the $\alpha_i$ parameters. Since $\mathcal{L}(\boldsymbol{x})$ is convex if and only if its Hessian matrix $H_{\mathcal{L}}(\boldsymbol{x})$ is positive semi-definite, a convexity condition is derived by noting that $H_{\mathcal{L}}(\boldsymbol{x})$ is related to the Hessian matrix $H_{f_{nc}}(\boldsymbol{x})$ of $f_{nc}(\boldsymbol{x})$ by

$$H_{\mathcal{L}}(\boldsymbol{x}) = H_{f_{nc}}\boldsymbol{x}) + 2\,\Delta, \tag{3}$$

where $\Delta$ is a diagonal matrix whose diagonal elements are the $\alpha_i$'s. $\Delta$ is referred to as the *diagonal shift matrix*, since the addition of the quadratic term to the function $f_{nc}(\boldsymbol{x})$, as shown in Eq. (2), corresponds to the introduction of a *shift* in the diagonal elements of its Hessian matrix $H_{f_{nc}}(\boldsymbol{x})$. The following theorem can then be used to ensure that $\mathcal{L}(\boldsymbol{x})$ is indeed a convex underestimator:

**Theorem 1** *$\mathcal{L}$ $(\boldsymbol{x})$, as defined in Eq. (2), is convex if and only if $H_{f_{nc}}(\boldsymbol{x}) + 2\,\Delta = H_{f_{nc}}(\boldsymbol{x}) + 2\,\mathrm{diag}\,(\alpha_i)$ is positive semi-definite for all $\boldsymbol{x} \in [\boldsymbol{x}^L, \boldsymbol{x}^U]$.*

Some deterministic methods to automatically derive a diagonal shift matrix in which all diagonal elements are equal to the same $\alpha$ value have been presented in Adjiman *et al.* (1996), Adjiman and Floudas (1996) and Adjiman *et al.* (1997a).

An important property of the type of underestimator proposed in Eq. (2) is that the maximum separation distance between the general nonconvex term and its convex lower bounding function is always bounded and given by

$$d_{max} = \max_{\boldsymbol{x}^L \leq \boldsymbol{x} \leq \boldsymbol{x}^U} (f_{nc}(\boldsymbol{x}) - \mathcal{L}(\boldsymbol{x})) = \frac{1}{4} \sum_{i=1}^{n} \alpha_i (\boldsymbol{x}_i^U - \boldsymbol{x}_i^L)^2. \quad (4)$$

The dependence of $d_{max}$ on the variable ranges shows that, as the size of the solution domain decreases, the underestimator becomes a better approximation of the original term.

**Overall Function Underestimator.** The convex underestimator for a nonlinear function is simply obtained by summing the specialized convex underestimators for all the terms that appear in that function. Equality constraints constitute a special case as they must be linear for a problem to be convex. For equalities in which all the nonlinear terms are replaced by a new variable, the transformed constraint is linear and no additional steps need to be taken. Equality constraints containing other types of nonlinear terms, however, must be replaced by an equivalent set of two inequality constraints which can then be convexified and underestimated.

Since the $\alpha$BB algorithm follows a branch-and-bound approach and all the underestimators presented depend on the variable bounds, the convex lower bounding problem becomes tighter as the size of the solution space is reduced. Thus the lower bounds on the optimum solution form a non-decreasing sequence, as required for convergence.

Theoretically, the decomposition of the functions into terms with a special structure is not necessary for the algorithm to converge to the global optimum, but it can significantly improve its performance. For instance, specialized underestimators are often tighter than general underestimators. In addition, smaller terms tend to depend on a smaller number of variables. This, in turn, reduces the size of the corresponding Hessian matrices and the computational effort required to compute the diagonal shift matrix.

# 3. Computing the Diagonal Shift Matrix

The focus of this section is the development of methods that generate rigorously an appropriate diagonal shift matrix, that is, a set of $\alpha$ parameters satisfying Theorem 1.

The diagonal shift matrix $\Delta$ as defined in Eq. (3) and Theorem 1 is closely linked to the Hessian matrix $H_{f_{nc}}(\boldsymbol{x})$ of the term being underestimated. For general twice-differentiable functions, the elements of the Hessian matrix $H_{f_{nc}}(\boldsymbol{x})$ are likely to be nonlinear functions of the variables, so that the derivation of a matrix $\Delta$ valid over the entire underestimation domain is a very challenging task. The difficulties arising from the presence of the variables in the convexity condition can be alleviated through the transformation of the exact $\boldsymbol{x}$-dependent Hessian matrix to an interval matrix $[H_{f_{nc}}]$ such that $H_{f_{nc}}(\boldsymbol{x}) \subseteq [H_{f_{nc}}]$, $\forall \boldsymbol{x} \in [\boldsymbol{x}^L, \boldsymbol{x}^U]$. The elements of the original Hessian matrix are treated as independent when calculating their natural interval extensions (Ratschek and Rokne, 1988; Neumaier, 1990). The interval Hessian matrix family $[H_{f_{nc}}]$ is then used to relax the $\alpha$ calculation problem.

**Theorem 2** *Consider a general nonconvex function $f_{nc}(\boldsymbol{x})$ with continuous second-order derivatives and its Hessian matrix $H_{f_{nc}}(\boldsymbol{x})$. Let $\mathcal{L}(\boldsymbol{x})$ be defined by Eq. (2). Let $[H_{f_{nc}}]$ be a real symmetric interval matrix such that $H_{f_{nc}}(\boldsymbol{x}) \subseteq [H_{f_{nc}}]$, $\forall \boldsymbol{x} \in [\boldsymbol{x}^L, \boldsymbol{x}^U]$. If the matrix $[H_{\mathcal{L}}]$ defined by $[H_{\mathcal{L}}] = [H_{f_{nc}}] + 2\,\Delta = [H_{f_{nc}}] + 2\,\mathrm{diag}\,(\alpha_i)$ is positive semi-definite, then $\mathcal{L}(\boldsymbol{x})$ is convex over the domain $[\boldsymbol{x}^L, \boldsymbol{x}^U]$.*

The domain of validity of the underestimator, $[\boldsymbol{x}^L, \boldsymbol{x}^U]$, participates in the interval convexity condition implicitly, through the interval matrix. The use of interval arithmetic reduces the computational complexity of the $\alpha$ calculations and allows the preservation and transfer of global information.

Four methods to compute a diagonal shift matrix satisfying Theorem 2 are presented here. The first rigorous method introduced generates a diagonal shift matrix for which all diagonal elements are equal. The second approach is an $\mathcal{O}(n^2)$ method which generates different elements for the diagonal matrix, while the last two methods presented are iterative, with varying degrees of complexity.

## 3.1. *E*-Matrix Method

In the case where the non-zero elements of the diagonal shift matrix are all equal to a single value $\alpha$, Maranas and Floudas (1994b) showed that $\mathcal{L}(\boldsymbol{x})$ as defined by Eq. (2) is convex if and only if

$$\alpha \geq \max\{0, -\frac{1}{2} \min_{i, \boldsymbol{x}^L \leq \boldsymbol{x} \leq \boldsymbol{x}^U} \lambda_i(\boldsymbol{x})\} \tag{5}$$

where the $\lambda_i(\boldsymbol{x})$'s are the eigenvalues of $H_{f_{nc}}(\boldsymbol{x})$, the Hessian matrix of the general nonconvex term $f_{nc}(\boldsymbol{x})$. The condition can be relaxed through the use of interval arithmetic to relate the diagonal elements of $\Delta$ to the minimum eigenvalue of the interval Hessian matrix $[H_{f_{nc}}]$:

$$\alpha \geq \left\{0, -\frac{1}{2}\lambda_{min}\left([H_{f_{nc}}]\right)\right\} \tag{6}$$

where $\lambda_{min}\left([H_{f_{nc}}]\right)$ is the minimum eigenvalue of the interval matrix family $[H_{f_{nc}}]$.

Deif (1991) and Rohn (1996) proposed some methods for the calculation of a lower bound on all the eigenvalues of an interval matrix. The *E*-matrix approach is an extension of their work. In the present case, only the real part of the eigenvalues are of concern since all matrices are symmetric.

The following result and definitions are required to derive the main result for this method.

**Definition 1:** Given an interval matrix $[A]$ with elements $[\underline{a}_{ij}, \overline{a}_{ij}]$, its *modified* radius matrix $\widetilde{\Delta A} = (\widetilde{\Delta a}_{ij})$ is such that

$$\widetilde{\Delta a}_{ij} = \begin{cases} 0 & \text{if } i = j \\ \Delta a_{ij} & \text{otherwise} \end{cases}$$

**Definition 2:** Given an interval matrix $[A]$ with elements $[\underline{a}_{ij}, \overline{a}_{ij}]$, its *modified* midpoint matrix $\tilde{A}_M = (\tilde{a}_{M,ij})$ is such that

$$\tilde{a}_{M,ij} = \begin{cases} \underline{a}_{ij} & \text{if } i = j \\ a_{M,ij} & \text{otherwise} \end{cases}$$

**Theorem 3** *Let $E$ be an arbitrary real symmetric matrix. Given a symmetric interval matrix $[A]$, its modified midpoint matrix $\tilde{A}_M$ and its* *modified radius matrix $\widetilde{\Delta A}$, the minimum eigenvalue $\lambda_{min}(A)$ of any symmetric matrix $A \in [A]$ is such that*

$$\lambda_{min}(A) \geq \lambda_{min}\left(\tilde{A}_M + E\right) - \rho\left(\widetilde{\Delta A} + |E|\right)$$

*where $\rho(M)$ denotes the spectral radius of a matrix $M$, i.e. its maximal absolute eigenvalue, and $|E|$ is the absolute value taken componentwise. (See proof in Adjiman et al., 1997a).*

The optimal choice for the matrix $E$ is a matrix that maximizes the lower bound $\lambda_{min}$ obtained. Unfortunately, this matrix cannot be determined *a priori*. Two choices of the matrix $E$ have been used to illustrate this $\alpha$ calculation method, $E = 0$ and $E = \text{diag}(\Delta A)$. The second choice of $E$ matrix yields a result equivalent to the theorems of Deif (1991) and Rohn (1996), namely

$$\lambda_{min}(A) \geq \lambda_{min}(A_M) - \rho(\Delta A). \tag{7}$$

## 3.2. Scaled Gerschgorin Theorem Method

This second approach allows the calculation of a different $\alpha$ value for each variable in order to construct an underestimator of the form shown in Eq. (2). The non-zero elements of the diagonal shift matrix $\Delta$ can no longer be related to the minimum eigenvalue of the interval Hessian matrix $[H_{f_{nc}}]$. The following theorem is proposed to compute a diagonal shift matrix satisfying Theorem 2.

**Theorem 4** *For any vector $\mathbf{d} > 0$ and a symmetric interval matrix $[A]$, define the vector $\alpha$ as*

$$\alpha_i = \max\left\{0, -\frac{1}{2}\left(\underline{a}_{ii} - \sum_{j \neq i} |a|_{ij} \frac{d_j}{d_i}\right)\right\}$$

*where $|a|_{ij} = \max\{|\underline{a}_{ij}|, |\overline{a}_{ij}|\}$.*

*Then, for all $A \in [A]$, the matrix $A_{\mathcal{L}} = A + 2\Delta$ with $\Delta = \text{diag}(\alpha_i)$ is positive semi-definite. (See proof in Adjiman et al., 1997a).*

The choice of the non-negative vector $\boldsymbol{d}$ in Theorem 4 is arbitrary. In practice, two choices for $\boldsymbol{d}$ have been used: a vector with unit elements and a vector based on the variable ranges, $\boldsymbol{d} = \boldsymbol{x}^U - \boldsymbol{x}^L$. In the second case, the off-diagonal contributions to the value of $\alpha_i$ are divided by $x_i^U - x_i^L$. This reflects the fact that variables with a wide range have a larger effect on the quality of the underestimator than variables with a smaller range.

## 3.3. *H*-Matrix Method

In this method, the properties of $H$-matrices are used to obtain a diagonal shift matrix satisfying Theorem 2.

**Theorem 5** *Consider the symmetric interval matrix $[G]$ and its modified midpoint matrix $\tilde{G}_M$. Let $C = \tilde{G}_M^{-1}$. If the following two sufficient conditions are met*

*(i) $\tilde{G}_M$ is positive definite, and*

*(ii) the pre-conditioned matrix $C[G]$ is an $H$-matrix,*

*then $[G]$ is positive definite. (See proof in Adjiman et al., 1997a).*

This theorem can be used for the identification of an appropriate diagonal shift matrix $\Delta$ for the interval matrix $[A]$ by setting $[G] = [A] + 2\Delta$. The modified midpoint matrix of $[G]$ is then $\tilde{A}_M + \Delta$, where $\tilde{A}_M$ is the modified midpoint matrix of $[A]$. The conditions for the positive definiteness of the interval matrix $[A] + 2\Delta$ are then

(i) $\tilde{A}_M + \Delta$ is positive definite, and

(ii) the pre-conditioned matrix $C[A] + 2C\Delta$ is an $H$-matrix, where $C = \left(\tilde{A}_M + \Delta\right)^{-1}$.

**Remarks On the Sufficient Conditions.**

**Condition (i): Checking the positive definiteness of a real symmetric matrix.** The positive definiteness of some real symmetric matrix $A$ can be tested by constructing a Cholesky decomposition of the matrix. If this decomposition fails, the matrix $A$

is not positive definite. Such an approach does not provide any insight on how far the matrix $A$ is from being positive definite and therefore does not guide the choice of an appropriate diagonal shift matrix $\Delta$. The *modified Cholesky decomposition* of Neumaier (1997) can be used to obtain quantitative information. The results of this modified Cholesky decomposition are a lower triangular matrix $L$ and a vector $\boldsymbol{p}$ such that

$$A = LL^T - \text{diag}(\boldsymbol{p}).$$

The matrix $A + \text{diag}(\boldsymbol{p})$ is therefore positive definite. This result can be used for an initial guess of $\Delta$, namely $\Delta_0 = 1/2\ \text{diag}(\boldsymbol{p})$. Interestingly, while the modified Cholesky decomposition is more conclusive than the traditional Cholesky decomposition, the number of operations it requires is almost the same as that required by the traditional Cholesky decomposition. Moreover, when $A$ is positive definite, the vector $\boldsymbol{p}$ vanishes and the result of the traditional Cholesky decomposition is recovered.

**Condition (ii): Checking the $H$-matrix property.** Whether the matrix $[A] + 2\Delta$ is an $H$-matrix can be tested as follows:

> Construct $[B] := C\,([A] + 2\Delta)$, where $C$ is the pre-conditioning matrix $(\tilde{A}_M + \Delta)^{-1}$. If $0 \in B_{ii}$ for some $i$, $[A] + 2\Delta$ is not an $H$-matrix. Otherwise, compute $[B]\boldsymbol{e}$ where $\boldsymbol{e} = (1, \cdots, 1)^T$ and $[B]\boldsymbol{u}$ where $\boldsymbol{u} = (\langle B_{11}\rangle^{-1}, \cdots, \langle B_{nn}\rangle^{-1})^T$ and $\langle B\rangle$ is the comparison matrix of $[B]$. If all the elements of at least one of the resulting vectors are positive, $[A] + 2\Delta$ is an $H$-matrix.

This test constitutes a *sufficient* condition of the $H$-matrix property. It may therefore fail to recognize that $[A] + 2\Delta$ is indeed an $H$-matrix. The use of the pre-conditioning matrix $C$ is an important factor in reducing the failure rate of this test as it helps to diagonalize the matrix $[A] + 2\Delta$. Choosing the identity matrix for $C$ would not change the theoretical analysis but would have a negative effect on the numerical aspects of the method. An exact check for the $H$-matrix property can be performed using Proposition 3.7.3 of Neumaier (1990). However, this check is much more computationally expensive than the simple matrix and vector multiplications used in the above procedure.

**Bounds on $\Delta$.** Since the fact that a valid diagonal shift matrix $\Delta$ has been identified may not be recognized by the proposed tests, an upper limit should be imposed on the elements of $\Delta$ to ensure that the quality of the constructed underestimator is no worse than that obtained with other techniques. The maximum separation distance between the nonconvex term and its underestimator, $d_{max}$, is an effective measure of the tightness of the underestimator. In practice, the $E$-matrix method with $E = 0$, has been used to determine the largest maximum separation distance $d^E_{max}$ allowed. If no suitable $\Delta$ matrix with a maximum separation distance smaller than $d^E_{max}$ is found by the $H$-matrix method, the diagonal shift matrix found with the $E$-matrix method, $\Delta^E$, is returned.

**Iterative Procedure for $\Delta$ Computation.** Based on the above observations, the following procedure can be set up for the computation of a diagonal shift matrix:

**Step 1**

Given the interval Hessian matrix $[A]$:

Compute a valid $\Delta^E$ using the $E$-matrix method.

All diagonal elements of $\Delta^E$ are equal $\alpha^E \geq 0$.

If $\alpha^E = 0$:

$[A]$ is positive semi-definite.

Return $\Delta = 0$.

Otherwise:

Compute $d^E_{max} = \frac{\alpha^E}{4}||\boldsymbol{x}^U - \boldsymbol{x}^L||^2$.

Set iteration counter $k = 0$.

**Step 2**

Compute the modified midpoint matrix $\tilde{A}_M$ of $[A]$.

Compute the modified Cholesky decomposition of $[A]$, $LL^T - \text{diag}(\boldsymbol{p})$.

Set $\Delta_0 = 1/2\text{diag}(\boldsymbol{p})$.

If $d_{max,0} = \sum\limits_i \frac{\Delta_{0,ii}}{4}(x_i^U - x_i^L)^2 \geq d^E_{max}$.

Return $\Delta = \Delta^E$.

**Step 3**

Compute the pre-conditioning matrix $C_k = \left(\tilde{A}_M + \Delta_k\right)^{-1}$.

Compute the interval matrix $[B_k] = C_k([A] + \Delta_k)$.

If $[B_k]$ is an $H$-matrix:

Return $\Delta = \Delta_k$.

**Step 4**

If the maximum number of iterations has been reached:

Return $\Delta = \Delta^E$.

**Step 5**

Since $[B_k]$ is not an $H$-matrix, a new guess for $\Delta$ is needed.

Augment the elements of $\Delta$:

Compute a step-size $\delta$ such that:

$$\delta = \frac{d^E_{max} - d_{max,k}}{||\boldsymbol{x}^U - \boldsymbol{x}^L||^2}.$$

$$\Delta_{k+1,ii} = \Delta_{k,ii} + \delta.$$

Increase the iteration counter $k$ by 1.

Go to Step 3.

If no matrix $\Delta$ such that $C[A] + 2C\Delta$ is an $H$-matrix has been identified after a fixed number of iterations, the matrix $\Delta^E$ is returned. As a result, the $H$-matrix method is at least as accurate as the $E$-matrix method. It is also more computationally expensive.

## 3.4. Minimization of Maximum Separation Distance

As previously noted, the quality of an underestimator can be measured in terms of the maximum separation distance, $d_{max}$, between the original nonconvex term and the underestimator (Eq. (4)). This method aims to derive a diagonal shift matrix $\Delta$ which is optimal with respect to this measure. This goal can be expressed as an optimization problem of the form

$$\begin{array}{ll} \min & (\boldsymbol{x}^U - \boldsymbol{x}^L)^T \Delta (\boldsymbol{x}^U - \boldsymbol{x}^L) \\ s.t. & H_{f_{nc}}(\boldsymbol{x}) + 2\Delta \geq 0 \\ & \boldsymbol{x} \in [\boldsymbol{x}^L, \boldsymbol{x}^U] \end{array}$$

where $\Delta$ is a diagonal matrix, and $M \geq 0$ means that the matrix $M$ is positive semi-definite.

Due to the nonconvexity of the above problem, the formulation is relaxed to

$$\begin{array}{ll} \min & (\boldsymbol{x}^U - \boldsymbol{x}^L)^T \Delta (\boldsymbol{x}^U - \boldsymbol{x}^L) \\ s.t. & [H_{f_{nc}}] + 2\Delta \geq 0 \end{array}$$

The presence of the interval Hessian matrix in the constraint makes the identification of a solution of this problem difficult. To further simplify it, $[H_{f_{nc}}]$ can be replaced by a real matrix whose minimum eigenvalue is smaller than the minimum eigenvalue of $[H_f]$. Stephens (1997) refers to such matrices as lower bounding Hessians. Such a matrix can be obtained as follows

**Theorem 6** *Given an interval Hessian matrix* $[A] = ([\underline{a}_{ij}, \overline{a}_{ij}])$, *the matrix* $L = (l_{ij})$ *where*

$$l_{ij} = \begin{cases} \underline{a}_{ii} + \sum\limits_{k \neq i} \frac{\underline{a}_{ik} - \overline{a}_{ik}}{2} & i = j \\ \frac{\underline{a}_{ij} + \overline{a}_{ij}}{2} & i \neq j \end{cases}$$

*is a lower bounding Hessian of* $[A]$. *(See proof in Adjiman et al., 1997a).*

The maximum distance minimization problem then becomes

$$\begin{array}{ll} \min & (\boldsymbol{x}^U - \boldsymbol{x}^L)^T \Delta (\boldsymbol{x}^U - \boldsymbol{x}^L) \\ s.t. & L + 2\Delta \geq 0 \end{array} \tag{8}$$

Problem (8), a semi-definite programming problem, is convex and can therefore be solved to global optimality using interior-point methods which have a polynomial worst-case complexity (Vandenberghe and Boyd, 1996). Because of the use of the lower bounding Hessian matrix, the $\Delta$ matrix is rigorously valid but not strictly optimal.

## 4. Branching Variable Selection

At each iteration, a node in the branch-and-bound tree is selected for branching. Since the αBB algorithm is a global optimization technique, all nodes with lower bounds smaller than the current best known value for the optimum solution must eventually be explored. In order to minimize computational expense, the node with the smallest lower bound is therefore chosen for further exploration. A branching variable must then be determined.

Although it does not present any theoretical difficulties, the branching step of any branch-and-bound algorithm often has a significant effect on the rate of convergence. This is especially true of the αBB algorithm as the quality of the underestimator strongly depends on the variable bounds. The convergence rate is particularly sensitive to the bounds on the variables that participate in the problem in a highly nonconvex way. Four selection strategies are therefore proposed to take advantage of this observation.

(i) Use $k$-section on all or some of the variables.

(ii) Use a measure of the quality of each term's underestimator, based on the maximum separation distance between term and underestimator.

(iii) Use a measure of the quality of each term's underestimator, based on the separation distance at the optimum point.

(iv) Use a measure of the overall influence of each variable on the quality of the lower bounding problem.

## 4.1. Strategy 1

By default, bisection on all the variables is used. The user can however specify a restricted list of variables on which to branch. The interval to be bisected is determined through the '*least reduced axis*' rule whose application involves the calculation of a 'current range to original range ratio', $r_i$, for each variable:

$$r_i = \frac{x_i^U - x_i^L}{x_{i,0}^U - x_{i,0}^L} \tag{9}$$

where $x_{i,0}^U$ and $x_{i,0}^L$ are respectively the upper and lower bounds on variable $x_i$ at the first node of the branch-and-bound tree, and $x_i^U$ and $x_i^L$ are respectively the upper and lower bounds on variable $x_i$ at the current node of the tree. The variable with the largest $r_i$ is selected for branching.

## 4.2. Strategy 2

While the second branching option requires additional computational effort, it results in a significant improvement of the convergence time for difficult problems. A different measure is defined for each type of underestimator to facilitate the assessment of the quality of the lower bounding function. The maximum separation distance between the underestimator and the original term is one possible indicator of the degree of accuracy achieved in the construction of the convex approximation. This measure is denoted by $\mu_t^{max}$. The term which appears to have the worst underestimator, that is, the largest $\mu_t^{max}$, is then used as a basis for the selection of the branching variable. Out of the set of variables that participate in that term, the one with the '*least reduced axis*', as defined by Eq. (9), is chosen for $k$-section. With this strategy, the influence of variable bounds on the quality of the underestimators is taken into account, and hence this adaptive branching scheme ensures the effective tightening of the lower bounding problem from iteration to iteration.

## 4.3. Strategy 3

This strategy is a variant of Strategy 2. Instead of computing the maximum separation distance between a term and its underestimator, their separation distance at the *optimum solution* of the lower bounding problem is used. This measure is denoted by $\mu_t^*$.

### 4.4. Strategy 4

The fourth branching procedure takes the approach of Strategy 3 one step further by considering the overall influence of each variable on the convex problem. After the relevant measures $\mu_t^*$ have been calculated for every term, a measure $\mu_v$ of each variable's contribution may be obtained as follows:

$$\mu_v^i = \sum_{t \in T_i} \mu_t^* \tag{10}$$

where $\mu_v^i$ is the measure for the $i$th variable, $T_i$ is the set of terms in which the $i$th variable participates.

## 5. Variable Bound Updates

The quality of the convex lower bounding problem can also be improved by ensuring that the variable bounds are as tight as possible. Variable bound updates can either be performed at the onset of an αBB run or at each iteration.

In both cases, the same procedure is followed in order to construct the bound update problem. Given a solution domain, the convex underestimator for every constraint in the original problem is formulated. The *bound update problem* for variable $x_i$ is then expressed as

$$x_i^{L,NEW} / x_i^{U,NEW} = \left\{ \begin{array}{ll} \min\limits_{\boldsymbol{x}} / \max\limits_{\boldsymbol{x}} & x_i \\ s.t. & \mathcal{G}(\boldsymbol{x}) \leq 0 \\ & \mathcal{H}(\boldsymbol{x}) = 0 \\ & \mathcal{G}_{add}(\boldsymbol{x}) \leq 0 \\ & \boldsymbol{x}^L \leq \boldsymbol{x} \leq \boldsymbol{x}^U \end{array} \right. \tag{11}$$

where $\mathcal{G}(\boldsymbol{x})$ are the convex underestimators for the inequality constraints, $\mathcal{H}(\boldsymbol{x})$ are the linear underestimators for the equality constraints, $\mathcal{G}_{add}(\boldsymbol{x})$ are the additional constraints arising from term substitutions and the bounds on the variables, $\boldsymbol{x}^L$ and $\boldsymbol{x}^U$ are the best calculated bounds.

Thus, once a new lower bound $x_i^{L,NEW}$ on $x_i$ has been computed via a minimization, this value is used in the formulation of the maximization problem for the generation of an upper bound $x_i^{U,NEW}$.

Because of the computational expense incurred by an update of the bounds on all variables, it is often desirable to define a smaller subset of the variables on which this operation is to be performed. The criteria devised for the selection of the branching variables can be used in this instance, since they provide a measure of the sensitivity of the problem to each variable. Bound updates are therefore carried out only for a fraction of the variables with a non-zero $\mu_v$, as calculated in Eq. (10).

# 6. Algorithm and Implementation

## 6.1. Algorithmic Procedure

The strategies described in the previous sections for the generation of a convex relaxation, the selection of a branching variable and the tightening of variable bounds are incorporated in the $\alpha$BB algorithm. The main algorithmic steps are outlined below. The procedures presented in this work have been highlighted.

**Step 1** Initialization

Set absolute tolerance $\epsilon$.

Set lower bound $LBD^* = -infty$ and upper bound $UBD^* = \infty$.

Set node counter $m = 0$.

Initialize set of nodes to be bounded, $J = \{0\}$.

$N$, the list of nodes to be explored, is empty.

The number of new nodes to be created at each iteration is $k$.

**Decompose functions into sum of terms with special structure.**

**Transform nonlinear equalities if needed.**

**Step 2** Bounding

For each node $N_j, j \in J$ {

If **variable bound update** option is on:

**Convexify constraints:**

Generate specialized underestimators.

Compute diagonal shift matrix for general nonconvexterms.

Solve bound update problems (convex NLPs).

**Convexify problem:**

Generate specialized underestimators.

Compute diagonal shift matrix for general nonconvex terms.

Solve convex NLP. Solution is $LBD_j$.

If $LBD_j \leq UBD^*$,

Add the current node to $\boldsymbol{N}$.

Else

Fathom the current node.

}

**Step 3** Convergence Test

Set $LBD^*$ to the lowest lower bound from the list $\boldsymbol{N}$.

If $UBD^* - LBD^* \leq \epsilon$,

Terminate with solution $UBD^*$.

Else,

proceed to Step 4.

**Step 4** Branching Variable Selection

Select the node with the lowest lower bound $LBD^*$ from the list $\boldsymbol{N}$. This node is $N_i$ $(i < m)$

**Compute measures for branching based on selected strategy.**

Select branching variable.

Create $k$ new nodes, $N_{m+1}$ to $N_{m+k}$.

Set $J = \{m+1, \ldots, m+k\}$ and $m = m + k$.

Go back to Step 2.

## 6.2. Implementation of the Algorithm

The $\alpha$BB algorithm has been implemented in C, as is discussed in detail in Adjiman *et al.* (1997b). The implementation includes all the features discussed in this work. A user-friendly interface has been set up to facilitate problem input. All Jacobian and Hessian information is derived automatically from the analytical functions specified in the input file. A C++ interval library, PROFIL/BIAS, developed by Knüppel (1993), has been linked to the package for the computation of interval Hessian matrices.

# 7. Computational Studies

The purpose of this section is to demonstrate the performance of the $\alpha$BB algorithm in identifying the global minimum of nonconvex optimization problems. More extensive studies are presented in Adjiman *et al.* (1997b). Three examples were selected and solved on an HP9000/730.

## 7.1. Example 1: Constrained Nonlinear Optimization Example

This nonconvex problem is taken from Murtagh and Saunders (1983). The $\alpha$BB algorithm is very well suited for this example since it combines bilinear, univariate concave and nonconvex terms.

$$\min \quad (x_1 - 1)^2 + (x_1 - x_2)^2 + (x_2 - x_3)^3 + (x_3 - x_4)^4 + (x_4 - x_5)^4$$

s.t.

$$\begin{aligned} x_1 + x_2^2 + x_3^3 &= 3\sqrt{2} + 2 \\ x_2 - x_3^2 + x_4 &= 2\sqrt{2} - 2 \\ x_1 x_5 &= 2 \\ x_i &\in [-5, 5], \ i = 1, \cdots, 5 \end{aligned}$$

The global minimum as well as some local minima are presented in Table 8.1.

All the diagonal shift matrix computation methods proposed in this work have been used to solve this problem to global optimality. Bound

*Table 8.1.* Global and local minima for Example 1.

| | *obj* | $x_1$ | $x_2$ | $x_3$ | $x_4$ | $x_5$ |
|---|---|---|---|---|---|---|
| Global | 0.0293 | 1.1166 | 1.2204 | 1.5378 | 1.9728 | 1.7911 |
| Local 1 | 27.8719 | -1.2731 | 2.4104 | 1.1949 | -0.1542 | -1.5710 |
| Local 2 | 44.0221 | -0.7034 | 2.6357 | -0.0963 | -1.7980 | -2.8434 |
| Local 3 | 52.9026 | 0.7280 | -2.2452 | 0.7795 | 3.6813 | 2.7472 |
| Local 4 | 64.8740 | 4.5695 | -1.2522 | 0.4718 | 2.3032 | 4.3770 |

updates were performed on a subset of the variables at every iteration for all runs. All branching strategies were tried for this problem and the most successful run is reported in Table 8.2.

*Table 8.2.* Results for Example 1.[a]

| *Method* | | *Iter.* | *CPU sec.* | $t_U$ *(%)* | *Branch* |
|---|---|---|---|---|---|
| $E$-Matrix | $E = 0$ | 337 | 70.7 | 2.8 | 3 |
| $E$-Matrix | $E = \text{diag}(\Delta H)$ | 339 | 72.9 | 2.3 | 2 |
| Scaled G. | $d_i = 1$ | 352 | 76.1 | 2.0 | 4 |
| Scaled G. | $d_i = (x_i^U - x_i^L)$ | 330 | 69.7 | 2.0 | 3 |
| $H$-Matrix | | 347 | 79.6 | 3.1 | 3 |
| Min. Max. distance | | 318 | 100.4 | 30.3 | 4 |

[a] $t_U$ denotes the percentage of total CPU time spent generating convex underestimators.

Branching strategies 3 and 4 are especially well-suited for this problem. If the variable bounds are updated only once at the beginning of the αBB run, the scaled Gerschgorin theorem method requires 2026 iterations and 224.2 CPU seconds. The relatively high degree of nonlinearity of this example problem increases the dependence of the quality of the lower bounding problem on the variable bounds. Variable bound updates should therefore be carried out preferentially on the most nonlinear variables.

The best results are given by the scaled Gerschgorin theorem approach with $d_i = (x_i^U - x_i^L)$. The minimization of the maximum separation distance approach converges with the smallest number of iterations but performs poorly in terms of CPU requirements. For all the methods used, the quality of the lower bound increases sharply from iteration to iteration at the beginning of a run. After a few levels of the branch-and-bound tree have been explored, the rate of improvement tapers off. Figure 8.1 shows the progress of the lower bound for the first 100 iterations using the scaled Gerschgorin theorem method and one can clearly see the dramatic amelioration in the quality of the underestimator.

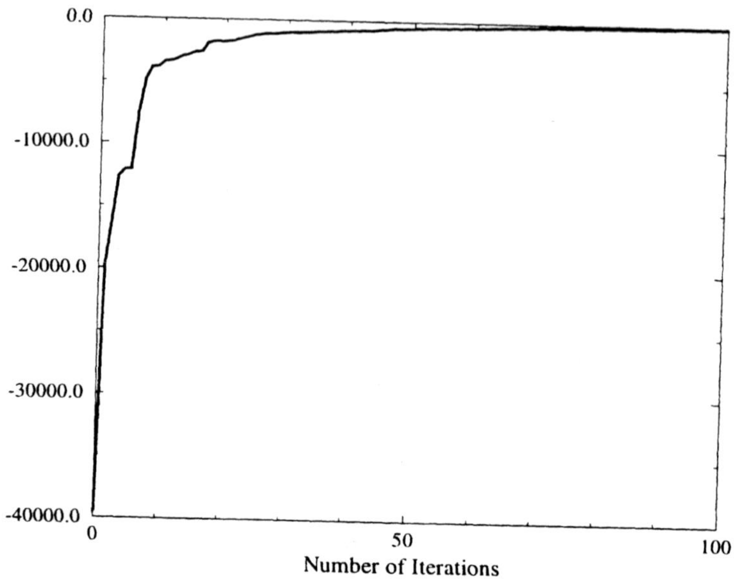


*Figure 8.1.* Progress of the lower bound for Example 1

## 7.2. Example 2: Fiat Engine Stability

This problem was developed to study the stability of the Fiat Dedra spark ignition engine. It belongs to the important class of generalized geometric programming problems, a subclass of the twice-differentiable problems that the αBB algorithm can address. The main property of the functions involved in the formulation is that they are the algebraic sum of posynomials, terms of the form $c \prod_{i=1}^{n} x_i^{d_i}$ where $c$ is a positive real number, the $x_i$'s are positive real variable and the $d_i$'s are scalars. No restrictions of integrality or positivity are imposed on the exponents. Maranas and Floudas (1997) proposed a new approach to tackle such problems, based on a difference of convex functions transformation embedded in a branch-and-bound framework. The applicability of the αBB algorithm is demonstrated on the most challenging example they presented.

Since the goal of this problem is to test the stability of the engine, the stability margin $k$ is minimized. The engine is deemed unstable if there exists a solution with $k < 1$. The problem can therefore be treated as a feasibility study by restricting $k$ to the interval $[0, 1]$.

The problem involves 9 variables and is highly nonlinear, as seen in the formulation:

$$\min \; k$$

$$\begin{array}{lrcl} \text{subject to} & -a_6(\boldsymbol{q})\omega^6 + a_4(\boldsymbol{q})\omega^4 - a_2(\boldsymbol{q})\omega^2 + a_0(\boldsymbol{q}) & = & 0 \\ & a_7(\boldsymbol{q})\omega^6 - a_5(\boldsymbol{q})\omega^4 + a_3(\boldsymbol{q})\omega^2 - a_1(\boldsymbol{q}) & = & 0 \end{array}$$

$$\begin{array}{rcccl} 3.4329 - 1.2721k & \leq & q_1 & \leq & 3.4329 \\ 0.1627 - 0.06k & \leq & q_2 & \leq & 0.1627 \\ 0.1139 - 0.0782k & \leq & q_3 & \leq & 0.1139 \\ 0.2539 & \leq & q_4 & \leq & 0.2539 + 0.3068k \\ 0.0208 - 0.0108k & \leq & q_5 & \leq & 0.0208 \\ 2.0247 & \leq & q_6 & \leq & 2.0247 + 2.4715k \\ 1.0000 & \leq & q_7 & \leq & 1.0000 + 9.0000k \end{array}$$

$$\text{where } a_0(\boldsymbol{q}) = 6.82079\; 10^{-5} q_1 q_3 q_4^2 + 6.82079\; 10^{-5} q_1 q_2 q_4 q_5$$

$$a_1(\boldsymbol{q}) = 7.61760\; 10^{-4} q_2^2 q_5^2 + 7.61760\; 10^{-4} q_3^2 q_4^2 +$$

$$\begin{aligned}
& 4.02141\ 10^{-4} q_1 q_2 q_5^2 + 0.00336706 q_1 q_3 q_4^2 + 6.82079\ 10^{-5} q_1 q_4 q_5 \\
& +5.16120\ 10^{-4} q_2^2 q_5 q_6 + 0.00336706 q_1 q_2 q_4 q_5 \\
& +6.82079\ 10^{-5} q_1 q_2 q_4 q_7 + 6.28987\ 10^{-5} q_1 q_2 q_5 q_6 \\
& +4.02141\ 10^{-4} q_1 q_3 q_4 q_5 + 6.28987\ 10^{-5} q_1 q_3 q_4 q_6 \\
& +0.00152352 q_2 q_3 q_4 q_5 + 5.16120\ 10^{-4} q_2 q_3 q_4 q_6
\end{aligned}$$

$$\begin{aligned}
a_2(\boldsymbol{q}) \quad = \quad & 4.02141\ 10^{-4} q_1 q_5^2 + 0.00152352 q_2 q_5^2 + 0.0552 q_2^2 q_5^2 \\
& +0.0552 q_3^2 q_4^2 + 0.0189477 q_1 q_2 q_5^2 + 0.034862 q_1 q_3 q_4^2 \\
& +0.00336706 q_1 q_4 q_5 + 6.82079\ 10^{-5} q_1 q_4 q_7 \\
& +6.28987\ 10^{-5} q_1 q_5 q_6 + 0.00152352 q_3 q_4 q_5 \\
& +5.16120\ 10^{-4} q_3 q_4 q_6 - 0.00234048 q_3^2 q_4 q_6 \\
& +0.034862 q_1 q_2 q_4 q_5 + 0.0237398 q_2^2 q_5 q_6 \\
& +0.00152352 q_2^2 q_5 q_7 + 5.16120\ 10^{-4} q_2^2 q_6 q_7 \\
& +0.00336706 q_1 q_2 q_4 q_7 + 0.00287416 q_1 q_2 q_5 q_6 \\
& +8.04282\ 10^{-4} q_1 q_2 q_5 q_7 + 6.28987\ 10^{-5} q_1 q_2 q_6 q_7 \\
& +0.0189477 q_1 q_3 q_4 q_5 + 0.00287416 q_1 q_3 q_4 q_6 \\
& +4.02141\ 10^{-4} q_1 q_3 q_4 q_7 + 0.1104 q_2 q_3 q_4 q_5 \\
& +0.0237398 q_2 q_3 q_4 q_6 + 0.00152352 q_2 q_3 q_4 q_7 \\
& -0.00234048 q_2 q_3 q_5 q_6 + 0.00103224 q_2 q_5 q_6
\end{aligned}$$

$$\begin{aligned}
a_3(\boldsymbol{q}) \quad = \quad & 0.0189477 q_1 q_5^2 + 0.1104 q_2 q_5^2 + 5.16120\ 10^{-4} q_5 q_6 + q_2^2 q_5^2 \\
& +7.61760\ 10^{-4} q_2^2 q_7^2 + q_3^2 q_4^2 + 0.1586 q_1 q_2 q_5^2 \\
& +4.02141\ 10^{-4} q_1 q_2 q_7^2 + 0.0872 q_1 q_3 q_4^2 + 0.034862 q_1 q_4 q_5 \\
& +0.00336706 q_1 q_4 q_7 + 0.00287416 q_1 q_5 q_6 \\
& +6.28987\ 10^{-5} q_1 q_6 q_7 + 0.00103224 q_2 q_6 q_7 + 0.1104 q_3 q_4 q_5 \\
& +0.0237398 q_3 q_4 q_6 + 0.00152352 q_3 q_4 q_7 - 0.00234048 q_3 q_5 q_6 \\
& +0.1826 q_2^2 q_5 q_6 + 0.1104 q_2^2 q_5 q_7 + 0.0237398 q_2^2 q_6 q_7 \\
& -0.0848 q_3^2 q_4 q_6 + 0.0872 q_1 q_2 q_4 q_5 + 0.034862 q_1 q_2 q_4 q_7 \\
& +0.0215658 q_1 q_2 q_5 q_6 + 0.0378954 q_1 q_2 q_5 q_7 \\
& +0.00287416 q_1 q_2 q_6 q_7 + 0.1586 q_1 q_3 q_4 q_5 \\
& +0.0215658 q_1 q_3 q_4 q_6 + 0.0189477 q_1 q_3 q_4 q_7 + 2 q_2 q_3 q_4 q_5 \\
& +0.1826 q_2 q_3 q_4 q_6 + 0.1104 q_2 q_3 q_4 q_7 - 0.0848 q_2 q_3 q_5 q_6 \\
& -0.00234048 q_2 q_3 q_6 q_7 + 7.61760\ 10^{-4} q_5^2 + 0.0474795 q_2 q_5 q_6 \\
& +8.04282\ 10^{-4} q_1 q_5 q_7 + 0.00304704 q_2 q_5 q_7
\end{aligned}$$

$$
\begin{aligned}
a_4(\boldsymbol{q}) \;=\;& 0.1586 q_1 q_5^2 + 4.02141\ 10^{-4} q_1 q_7^2 + 2 q_2 q_5^2 + 0.00152352 q_2 q_7^2 \\
& + 0.0237398 q_5 q_6 + 0.00152352 q_5 q_7 + 5.16120\ 10^{-4} q_6 q_7 \\
& + 0.0552 q_2^2 q_7^2 + 0.0189477 q_1 q_2 q_7^2 + 0.0872 q_1 q_4 q_5 \\
& + 0.034862 q_1 q_4 q_7 + 0.0215658 q_1 q_5 q_6 + 0.00287416 q_1 q_6 q_7 \\
& + 0.0474795 q_2 q_6 q_7 + 2 q_3 q_4 q_5 + 0.1826 q_3 q_4 q_6 + 0.1104 q_3 q_4 q_7 \\
& - 0.0848 q_3 q_5 q_6 - 0.00234048 q_3 q_6 q_7 + 2 q_2^2 q_5 q_7 + 0.1826 q_2^2 q_6 q_7 \\
& + 0.0872 q_1 q_2 q_4 q_7 + 0.3172 q_1 q_2 q_5 q_7 + 0.0215658 q_1 q_2 q_6 q_7 \\
& + 0.1586 q_1 q_3 q_4 q_7 + 2 q_2 q_3 q_4 q_7 - 0.0848 q_2 q_3 q_6 q_7 + 0.0552 q_5^2 \\
& + 0.3652 q_2 q_5 q_6 + 0.0378954 q_1 q_5 q_7 + 0.2208 q_2 q_5 q_7
\end{aligned}
$$

$$
\begin{aligned}
a_5(\boldsymbol{q}) \;=\;& 0.0189477 q_1 q_7^2 + 0.1104 q_2 q_7^2 + 0.1826 q_5 q_6 + 0.1104 q_5 q_7 \\
& + 0.0237398 q_6 q_7 + q_2^2 q_7^2 + 0.1586 q_1 q_2 q_7^2 + 0.0872 q_1 q_4 q_7 \\
& + 0.0215658 q_1 q_6 q_7 + 0.3652 q_2 q_6 q_7 + 2 q_3 q_4 q_7 - 0.0848 q_3 q_6 q_7 \\
& + q_5^2 + 7.61760\ 10^{-4} q_7^2 + 0.3172 q_1 q_5 q_7 \\
& + 4 q_2 q_5 q_7
\end{aligned}
$$

$$
a_6(\boldsymbol{q}) \;=\; 0.1586 q_1 q_7^2 + 2 q_2 q_7^2 + 2 q_5 q_7 + 0.1826 q_6 q_7 + 0.0552 q_7^2
$$

$$
a_7(\boldsymbol{q}) \;=\; q_7^2
$$

The solution of the stability problem shows that this system is stable and computational results, obtained with $\omega \in [0, 10]$, are reported in Table 8.3. The best runs were obtained without updating variable bounds and with branching strategy 4.

*Table 8.3.* Results for Example 2.[a,b]

| *Method* | | *Iter.* | *CPU sec.* | $t_U$ *(%)* |
|---|---|---|---|---|
| $E$-Matrix | $E = 0$ | 140 | 44.4 | 64.4 |
| $E$-Matrix | $E = \mathrm{diag}(\Delta H)$ | 186 | 52.0 | 62.9 |
| Scaled Ger. | $d_i = 1$ | 163 | 48.9 | 63.7 |
| Scaled Ger. | $d_i = (x_i^U - x_i^L)$ | 3944 | 1129.2 | 64.0 |
| $H$-Matrix | | 143 | 45.1 | 71.3 |
| Min. Max. distance | | 246 | 1528.4 | 96.5 |

[a] $t_U$ denotes the percentage of total CPU time spent generating convex underestimators.
[b] $\omega \in [0, 10]$.

The $E$-matrix method with $E = 0$ outperforms other underestimation approaches in this case. The $H$-matrix method also yields good results, but the increase in computational expense between the $E$-matrix and the $H$-matrix methods is not sufficiently compensated by the improvement in lower bounding function quality. The minimization of maximum separation distance approach performs well in terms of number of iterations but does poorly in terms of CPU time. The additional time requirement arises from the need to solve a semi-definite programming problem for each nonconvex term at each node. Thus $t_U$, the percentage of time spent generating convex underestimators, is much larger for this method than for any other approach.

## 7.3. Example 3: Batch Process Design and Scheduling under Uncertainty

This class of problems addresses the design of multiproduct batch plants given uncertain demands and processing parameters. The aim of the problem is the identification of the equipment sizes, batch sizes and production rates which maximize profit, given a fixed number of stages. The uncertainties are taken into account in two ways: the demands are assigned probability distributions, while the size factors and the processing times are assigned a set of discrete values. Each of the P potential combinations of these values gives rise to a different scenario and optimization must be carried out over all P possibilities. The derivation of the NLP formulation used to represent batch design problems is presented in detail by Harding and Floudas (1997). Its final form is given by:

$$
\begin{aligned}
\min_{\boldsymbol{v},\boldsymbol{b},\boldsymbol{N}} \quad & \delta \sum_{j=1}^{M} \alpha_j N_j \exp(\beta_j v_j) \\
& - \sum_{p=1}^{P} \frac{1}{\omega^p} \sum_{q=1}^{Q} \omega^q J^q \sum_{i=1}^{N} p_i Q_i^{qp} \\
\text{s.t.} \quad & v_j \;-\; b_i \;\geq\; \ln(S_{ij}^p) \qquad \forall\, i, \forall\, j, \forall\, p \qquad (12) \\
& \sum_{i=1}^{N} Q_i^{qp} \cdot \exp(t_{Li}^p - b_i) \;\leq\; H \qquad \forall\, q, \forall\, p \\
& \theta_i^L \;\leq\; Q_i^{qp} \;\leq\; \theta_i^q \qquad \forall\, i, \forall\, q, \forall\, p \\
& \ln(V_j^L) \;\leq\; v_j \;\leq\; \ln(V_j^U)
\end{aligned}
$$

where the parameters are defined as follows: $\delta$ is the coefficient used for capital cost annualization, $M$ denotes the number of stages, $N$ is the number of products, $N_j$ is the number of identical pieces of equipment in stage $j$, $\alpha_j$ and $\beta_j$ are the fixed charge cost coefficients for the equipment in stage $j$, $p_i$ is the market price of product $i$, $\omega^p$ is the weighting factor for scenario $p$, $S_{ij}^p$ is the equipment volume used to produce one mass unit of product $i$ in stage $j$ and for scenario $p$, $t_{Li}^p$ is the natural logarithm of the longest processing time for product $i$ over all stages in scenario $p$, $H$ is the time horizon for the campaign, $\theta_i^L$ is the lower bound for the uncertain demand for product $i$, $V_j^L$ and $V_j^U$ are the lower and upper bounds on the equipment size for stage $j$. The second term in the objective function is an approximation of the expected revenue based on a Gaussian quadrature formula. $Q$ is the total number of quadrature points and $J^q$ is the probability of quadrature point $q$, $\omega^q$ is its weighting factor, $\theta_i^q$ is the upper bound on the uncertain demand of product $i$ associated with quadrature point $q$ in scenario $p$. The variables are $Q_i^{pq}$, the amount of product $i$ produced in scenario $p$ and associated with quadrature point $q$, $v_j$, the natural logarithm of the equipment size for stage $j$, and $b_i$, the natural logarithm of the size of the batch for product $i$.

The size of the problem increases dramatically with the number of scenarios, stages, products and quadrature points. The presence in the constraints of nonconvex terms involving many of the variables renders these problems difficult.

The example solved here represents a four-product plant with six stages and a single unit per stage. There is a single scenario since only the demands are considered uncertain. Five quadrature points are used for each demand with a normal distribution. As a result, the problem involves 2510 variables and 649 constraints. There are 1250 general nonconvex terms to be underestimated. The algorithm converges to the global optimum solution after 3 iterations and 3055 CPU seconds with the scaled Gerschgorin theorem and $d_i = (x_i^U - x_i^L)$. Other underestimation techniques, however, fail to identify the optimum solution in reasonable CPU time.

## 8. Conclusions

As demonstrated in this work, the $\alpha$BB algorithm can rigorously identify the global solution of twice-differentiable nonconvex programming problems based on a general convex underestimating scheme. The decomposition of the nonlinear functions into a sum of terms constitutes a central concept for the generation of tight underestimators. Convex underestimators for bilinear, fractional, trilinear, fractional trilinear, univariate concave and general nonconvex terms are used to generate convex lower bounding functions for the objective function and constraints. Several reliable methods have been developed for the calculation of the $\alpha$ parameters necessary for the construction of tight valid convex underestimators. Other algorithmic issues such as variable bound updates and branching strategies have been addressed in a way that improves the convergence rate of the algorithm.

The $\alpha$BB algorithm has been fully implemented and successfully tested on a number of nonconvex problems including a small constrained problem, a stability problem and the design of a batch plant under uncertainty.

## Acknowledgments

The authors gratefully acknowledge financial support from the National Science Foundation, the Air Force Office of Scientific Research, the National Institutes of Health, Exxon Foundation and Mobil Technology Company.

# References


[1] C. S. Adjiman, I. P. Androulakis, C. D. Maranas and C. A. Floudas. A Global Optimization Method, $\alpha$BB, for Process Design. *Computers & Chem. Engng, Suppl.*, 20:S419–S424, 1996.

[2] C. S. Adjiman, I. P. Androulakis and C. A. Floudas. A Global Optimization Method, $\alpha$BB, for General Twice–Differentiable NLPs – II. Implementation and Computational Results. *Computers & Chem. Engng.*, Accepted for publication, 1997b.

[3] C. S. Adjiman, S. Dallwig, A. Neumaier and C. A. Floudas. A Global Optimization Method, $\alpha$BB, for General Twice–Differentiable NLPs – I. Theoretical Advances. *Computers & Chem. Engng.*, Accepted for publication, 1997a.

[4] C. S. Adjiman and C. A. Floudas. Rigorous Convex Underestimators for General Twice–Differentiable Problems. *J. of Glob. Opt.*, 9:23–40, 1996.

[5] F. A. Al-Khayyal. Jointly Constrained Bilinear Programs and Related Problems : An Overview. *Computers Math. Applic.*, 19(11):53–62, 1990.

[6] F. A. Al-Khayyal and J. E. Falk. Jointly Constrained Biconvex Programming. *Maths Ops Res.*, 8:273–286, 1983.

[7] I. P. Androulakis, C. D. Maranas and C. A. Floudas. $\alpha$BB : A Global Optimization Method for General Constrained Nonconvex Problems. *J. of Glob. Opt.*, 7:337–363, 1995.

[8] A. S. Deif. The Interval Eigenvalue Problem. *Z. angew. Math. Mech*, 71(1):61–64, 1991.

[9] C. A. Floudas. Deterministic Global Optimization in Design, Control, and Computational Chemistry. volume 93 of *IMA Volumes in*

*Mathematics and its Applications : Large Scale Optimization with Applications, Part II*, L. T. Biegler, T. F. Coleman, A. R. Conn and F. N. Santosa (Eds.), pp. 129–184, Springer-Verlag, New York, 1997.

[10] C. A. Floudas and I. E. Grossmann. Algorithmic Approaches to Process Synthesis : Logic and Global Optimization. *FOCAPD'94*, volume 91 of *AICHE Symposium Series*, pp. 198–221, 1995.

[11] C. A. Floudas and P. M. Pardalos (Eds). *State of the Art in Global Optimization*, Kluwer Academic Publishers, 1996.

[12] C. A. Floudas and V. Visweswaran. A Global Optimization Algorithm (GOP) for Certain Classes of Nonconvex NLPs : I. Theory. *Computers & Chem. Engng.*, 14:1397–1417, 1990.

[13] C. A. Floudas and V. Visweswaran. A Primal-Relaxed Dual Global Optimization Approach. *J. Optim. Theory and Appl.*, 78(2):187–225, 1993.

[14] I. E. Grossmann (Ed.). *Global Optimization in Engineering Design*, Kluwer Academic Publishers, 1996.

[15] E. Hansen. *Global Optimization Using Interval Analysis. Pure and Applied Mathematics*, Marcel Dekker, Inc., 1992.

[16] S. T. Harding and C. A. Floudas. Global Optimization in Multiproduct and Multipurpose Batch Design under Uncertainty, *Ind. Eng. Chem. Res.*, 36(5):1644–1664, 1997.

[17] R. Horst and H. Tuy. *Global Optimization: Deterministic Approaches.* Springer-Verlag, Berlin, Germany, 1990.

[18] O. Knüppel. *PROFIL – Programmer's Runtime Optimized Fast Interval Library.* Technische Informatik III, Technische Universität Hamburg–Harburg, 1993.

[19] W. B. Liu and C. A. Floudas. A Remark on the GOP Algorithm for Global Optimization. *J. of Glob. Opt.*, 3(3):519–521, 1993.

[20] C. D. Maranas and C. A. Floudas. A deteministic Global Optimization Approach for Molecular Structure Determination. *J. Chem. Phys.*, 100(2):1247–1261, 1994a.

[21] C. D. Maranas and C. A. Floudas. Global Minimum Potential Energy Conformations of Small Molecules. *J. of Glob. Opt.*, 4:135–170, 1994b.

[22] C. D. Maranas and C. A. Floudas. Finding All Solutions of Nonlinearly Constrained Systems of Equations. *J. of Glob. Opt.*, 7(2):143–182, 1995.

[23] C. D. Maranas and C. A. Floudas. Global Optimization in Generalized Geometric Programming. *Computers & Chem. Engng.*, 21(4):351–370, 1997.

[24] G. P. McCormick. Computability of Global Solutions to Factorable Nonconvex Programs : Part I – Convex Underestimating Problems. *Math. Programming*, 10:147–175, 1976.

[25] B. A. Murtagh and M. A. Saunders. *MINOS 5.4 User's Guide.* Dept. of Operations Research, Stanford University, CA, Systems Optimization Laboratory, p94, 1983.

[26] A. Neumaier. *Interval Methods for Systems of Equations. Encyclopedia of Mathematics and its Applications*, Cambridge University Press, 1990.

[27] A. Neumaier. An Optimality Criterion for Global Quadratic Optimization. *J. of Glob. Opt.*, 2:201–208, 1992.

[28] A. Neumaier. On Satisfying Second–Order Optimality Conditions Using Modified Cholesky Factorizations. *SIAM J. Optimization*, Submitted for publication, 1997.

[29] H. Ratschek and J. Rokne. *Computer Methods for the Range of Functions. series Ellis Horwood Series in Mathematics and its Applications*, Halsted Press, 1988.

[30] J. Rohn. *Bounds on Eigenvalues of Interval Matrices.* Technical Report no. 688, Institute of Computer Science, Academy of Sciences, Prague, 1996.

[31] C. Stephens. Interval and Bounding Hessians. *Developments in Global Optimization*, I. M. Bomze et al. (Eds.), Kluwer Academic Publishers, pp. 109–199, 1997.

[32] E. M. B. Smith and C. C. Pantelides. Global Optimisation of General Process Models. *Global Optimization in Engineering Design*, I. E. Grossmann (Ed.), Kluwer Academic Publishers, pp. 355–386, 1996.

[33] L. Vandenberghe and S. Boyd. Semidefinite Programming. *SIAM Review*, 38:49–95, 1996.

[34] V. Visweswaran and C. A. Floudas. A Global Optimization Algorithm GOP for Certain Classes of Nonconvex NLPs : II. Application of Theory and Test Problems. *Computers & Chem. Engng.*, 14:1419–1434, 1990.

[35] V. Visweswaran and C. A. Floudas. New Properties and Computational Improvement of the GOP Algorithm for Problems with Quadratic Objective Function and Constraints. *J. of Glob. Opt.*, 3(3):439–462, 1993.

[36] V. Visweswaran and C. A. Floudas. New Formulations and Branching Strategies for the GOP Algorithm. *Global Optimization in Engineering Design*, I. E. Grossmann (Ed.), Kluwer Book Series in Nonconvex Optimization and its Applications, Chapter 3, pp. 75–100, 1996a.

[37] V. Visweswaran and C. A. Floudas. Computational Results for an Efficient Implementation of the GOP Algorithm and its Variants. *Global Optimization in Engineering Design*, I. E. Grossmann (Ed.), Kluwer Book Series in Nonconvex Optimization and its Applications, Chapter 4, pp. 111–153, 1996b.

# Chapter 9

# INFIMUM OF POLYNOMIALS AND SINGULARITY AT INFINITY

HA Huy Vui

*Institute of Mathematics, P.O. Box 631, Bo Ho, Hanoi, Vietnam*
hhvui@thevinh.ncst.vn

**Abstract** We prove that if $c$ is the infimum value of a polynomial of two variables $f$ and if $f$ does not attain $c$, then $c$ is a critical value of singularties at infinity of the global Milnor fibration of $f$. This provides a method of complex geometry for finding the infimum values of real polynomials

**Keywords:** Singularity at infinity, Milnor's number, Palais-Smale condition.

## 1. Introduction

Let $c$ be the infimum value of a differentiable function $F$ from a space $E$ into the real line $\mathbb{R}$. If $F$ attains its minimum $c$ at a point $x_0$, then grad $F(x_0) = 0$ and the value $c$ is critical for $F$. Suppose that $F$ does not attain its infimum value $c$. What can we say about $c$ in this case? It is not hard to imagine that the value $c$ must be somehow related to the behavior of $F$ in a neighborhood of infinity. Roughly speaking, *$c$ must be a critical value of singularities at infinity of $F$.*

However, the exact meaning of the latter is not clear. In fact, one might think about *singularities at infinity* in many different ways. In this paper we try to answer the question raised above for the simplest case, when the function under consideration is a polynomial of several real variables. We hope to clarify the nature of infimum values of real polynomials by using



*A. Migdalas et al. (eds.), From Local to Global Optimization,* 187–204.

global Milnor's fibrations, defined by the same polynomials, considered as functions in complex variables. These fibrations are a subject of very intensive study in the singularity theory during the last years [2] [5] [6] [9] [10] [11-13] [17] [19] [20] [21] [22]. For the global Milnor fibrations the notion of singularities at infinity has a precise meaning. But even in this very restrictive setting, we cannot justify our hope completely. What we can do is only to study the case of 2 variables.

In brief, our main result says that if $f$ is a polynomial of 2 real variables and if $c \neq -\infty$ is its infimum, then $c$ is either critical value of $f$, or critical value of singularities at infinity of $f$, considered as a map from $\mathbb{C}^2$ to $\mathbb{C}$.

Since the critical values of singularities at infinity of polynomials in 2 complex variables can be characterized by topological and algebraic invariants, our result provides a method of complex geometry for finding the infimum values of real polynomials. At the moment we do not know, practically, how useful this method can be. But it seems to us that in some situations this method is very easy to handle.

The paper is organized as follows

§2. Statement of results

§3. A method of complex geometry for finding infimum values

§4. An example

§5. Proofs.

This work was done while the author was visiting Emile Picard's Laboratory at the University of Paul Sabatier, Toulouse. He thanks this institute for hospitality and financial support. He also thanks N.V. Dung for his help in the preparation of this paper.

## 2. Statement of results

We first recall some facts about the topology of complex polynomials.

Let $f : \mathbb{C}^n \to \mathbb{C}$ be a polynomial of $n$ complex variables. According to R. Thom [24], there exists a finite set $A \subset \mathbb{C}$ such that the restriction

$$f_| : \mathbb{C}^n \setminus f^{-1}(A) \to \mathbb{C} \setminus A$$

is an $C^\infty$ - locally trivial fibration. Let us denote by $A_f$ the smallest set with this property. It is clear that $A_f$ contains the set of all the critical values of $f$, but in general, does not coincide with it. For example, if $f(x,y) = y^2x + y$, then $f$ has no critical values, but $A_f = \{0\}$, since $f^{-1}(t)$ is homeomorphic to $\mathbb{C}^*$, while $f^{-1}(0) \approx \mathbb{C} \amalg \mathbb{C}^*$ [5].

**2.1. Definition** [20] *(i) a value $t_0 \in \mathbb{C}$ is said to be regular at infinity for $f$, if there exist a compact set $K \subset \mathbb{C}^n$ and a disc $D_\delta$ centered at $t_0$, of radius $\delta$, such that the restriction*

$$f_| : f^{-1}(D_\delta) \setminus K \to D_\delta$$

*defines a $C^\infty$ - trivial fibration.*

*(ii) If $t_0$ is not regular at infinity for $f$, we say that $t_0$ is a critical value of singularities at infinity of $f$.*

Denote by crit $(f)$ and $\text{cri}_\infty(f)$ the set of critical values and the set of critical values of singularities at infinity, respectively, of $f$. In [12] we have shown that

$$A_f = \text{crit}(f) \cup \text{crit}_\infty(f).$$

Our main result now is

**Theorem A.** *Let $f(x,y)$ be a polynomial of 2 real variables. Suppose that $c \neq -\infty$ is the infimum value of $f$. Then $c \in A_f$.*

**2.2. Definition** *[4] We say that the polynomial $f : \mathbb{C}^n \to \mathbb{C}$ satisfies the Palais-Smale condition at the value $t_0$, if any sequence $\{x_n\} \subset \mathbb{C}^n$, such that $f(x_n) \to t_0$ and $\|\text{grad} f(x_n)\| \to 0$, is relatively compact.*

Theorem A is a consequence of Ekeland's variational principle and the following result

**Theorem B.** *Let $f(x,y)$ be a polynomial of 2 complex variables. Then the following 2 conditions are equivalent:*

*(i)* $t_0 \in \text{crit}_\infty(f)$

*(ii)* *$f$ does not satisfy the Palais-Smale condition at $t_0$*

## 3. A method of complex geometry for finding infimum values

By Theorem A, if $c = \inf f(x,y)$, then $c \in A_f = \mathrm{crit}(f) \cup \mathrm{crit}_\infty(f)$. Suppose that $f$ does not attain $c$, then $c \in \mathrm{crit}_\infty(f)$. More precisely, $c \in \mathrm{crit}_\infty(f) \cap \mathbb{R}$. So, in order to find $c$, we have to find $\mathrm{crit}_\infty(f)$.

Our method is based on the result of [12], characterizing critical values of singularities at infinity of polynomials in 2 complex variables.

Let $C$ be an analytic curve, defined by the equation $g(x,y) = 0$. Suppose that $g(0,0) = 0$ and that $(0,0)$ is a singular point of $C$. Then Milnor's number is defined by

$$\mu(C,0) := \dim_{\mathbb{C}} \mathbb{C}\{x,y\}/(\frac{\partial g}{\partial x}\frac{\partial g}{\partial y})$$

where $\mathbb{C}\{x,y\}$ is the ring of germs at $(0,0)$ of analytic functions in 2 complex variables, and $(\frac{\partial g}{\partial x}, \frac{\partial g}{\partial y})$ is the ideal generated by $\frac{\partial g}{\partial x}$ and $\frac{\partial g}{\partial y}$ ([18]).

Let $f(x,y)$ be a polynomial and let

$$f(x,y) = \sum_{\alpha+\beta \le d} c_{\alpha\beta} x^\alpha y^\beta,$$

The homogeneization of $f(x,y)$ is the homogeneous complex polynomial $F(X,Y,Z)$ of degree $d$ :

$$F(X,Y,Z) = \sum_{\alpha+\beta \le d} c_{\alpha\beta} X^a Y^\beta Z^{d-\alpha-\beta}$$

The compactification of the fibers $f = t$ in the complex projective plane $\mathbb{C}\,\mathbf{P}^2$ are the projective curves $C_t$ with projective equations $F - tZ^d = 0$. These curves pass through the same points on the line at infinity $Z = 0$, namely through

$$\{A_1, ..., A_r\} = \{F = Z^d = 0\},$$

which correspond to the asymptotic directions of the level curves $f = t$.

For each $A_i$, $1 \le i \le r$, there is an integer

$$\mu_i = \inf_{t \in \mathbb{C}} \mu(C_t, A_i)$$

where $\mu(C_t, A_i)$ is the Milnor number of the curve $c_t$ at the point $A_i$. We have

**3.1. Theorem** [12]

$$\text{crit}_\infty(f) = \{t \in \mathbb{C} \mid \textit{there exists } i, \textit{ with } 1 \leq i \leq r, \textit{ such that } \mu(c_t, A_i) \neq \mu_i\}$$

**3.2. Remark**

Thus, in order to compute the set $\text{crit}_\infty(f)$, we have to look for the values $t_0 \in \mathbb{C}$, for which the Milnor numbers $\mu(C_t, A_i)$ are not constant for some $i$, $1 \leq i \leq r$. In other words, we have to find the values $t_0$, at which the family of curves with singular points $A_i$, is not equisingular in the sense of Zariski. [15]-[16], [23], [26],... .

This remark is practically useful. It permits us to use the result of equisingularity theory.

In conclusion, our method for finding the infimum values of polynomials in 2 real variables consists of the following steps

1) Find all the critical values of $f$ (they are finitely many, by Bertini-Sard Theorem), and take the minimum of these values. Denote this minimum by $c_1$.

2) Determine the set $\text{crit}_\infty(f)$ by finding those values $t \in \mathbb{C}$, for which $\mu(C_t, A_i) \neq \mu_i$ for some $i$ and take the set

$$\text{crit}_\infty(f)^{\mathbb{R}} = \text{crit}_\infty(f) \cap \mathbb{R}.$$

The infimum value of $f$ must be contained in the set $\text{crit}_\infty(f)^{\mathbb{R}} \cup \{c_1\}$.

**3.3. Remark.** In ([10], [17]) one has an algorithmic way to carry out Step 2) (via resolution of singularities).

## 4. An example

Let

$$f(x, y) = 2x^4y^4 + x^2y^2 + 2xy + x^2.$$

We want to find $c = \inf_{\mathbb{R}^2} f(x, y)$ by the method proposed in §3.

1) Finding the set crit $(f)$:

We see that the only critical point of $f$ is $(0,0)$. Thus crit $(f) = \{f(0,0)\} = \{0\}$

2) Finding $\text{crit}_\infty(f)$:

- Homogeneization of $f$:

$$F(X,Y,Z) = 2X^4Y^4 + X^2Y^2Z^6 + 2XYZ^6 + X^2Z^6$$

-

$$C_t = \{(X:Y:Z) \in \mathbb{P}^2(\mathbb{C}),\quad F(X,Y,Z) = tZ^8\}$$

- $C_t \cap \{Z=0\} = \{A_1, A_2\},\quad A_1 = (0:1:0),\quad A_2 = (1:0:0)$

- Locally in a neighborhood of $A_1, C_t$ determine a family of analytic curves

$$C_t^1 = \{(X,Z) \in U_1 | 2X^4 + X^2Z^4 + 2XZ^6 + X^2Z^6 - tZ^8 = 0\}$$

where $U_1$ is a sufficiently small neighborhood of $(0,0) \in \mathbb{C}^2$

- Analoguously, $C_t$ determine in a neighborhood of $A_2$ a family of analytic curves

$$C_t^2 = \{(Y,Z) \in U_2 \mid 2Y^2 + Y^2Z^4 + 2YZ^6 + Z^6 - tZ^8 = 0\}$$

- We find now the set of point $t \in \mathbb{C}$ such that

$$\mu(C_t^1, A_1) \neq \mu_1 = \inf_{t \in \mathbb{C}} \mu(C_t^1, A_1).$$

We first notice that $C_t^1$ is the family of analytic curves with $(0,0)$ as a semi-quasihomogeneous singularity in the sense of ([1]).

The quasihomogeneous part is

$$g_t(X,Z) = 2X^4 + X^4Z^4 + 2XZ^6 - tZ^8.$$

For generic $t$, $g_t(X,Z)$ is nondegenerated ([1]) and has an isolated singularity in $(0,0)$. Thus, according to an well known fact, we have, for generic $t$

$$\mu_1 = \mu(C_t^1, A_1) = \mu(\widetilde{C}_t^1, A_1) = 21,$$

where $\widetilde{C}_t^1$ is a germ of analytic curve, defined by the equation $g_t(X,Z) = 0$.

The values $t \in \mathbb{C}$, for which $\mu(C_t^1, A_1) \neq \mu_1$ are exactly those $t$, for which $g_t(X, Z)$ has non-isolated singularity at $(0,0)$. An elementary computation shows that these values are $h(-\frac{1}{2}) = -\frac{5}{8}$, $h(\frac{1 \pm \sqrt{7}i}{4})$, where $h$ is the polynomial $h(u) = 2u^4 + u^2 + 2u$ and $i = \sqrt{-1}$.

- In the same way we can see that

$$\{t \in \mathbb{C} \mid \mu(C_t^2, A_2) \neq \mu_2\} = \emptyset$$

- So, we have

$$\text{crit}_\infty(f) = \{-\frac{5}{8},\ h(\frac{1-\sqrt{7}i}{4}),\ h(\frac{1+\sqrt{7}i}{4})\}$$

and $\text{crit}_\infty(f)^{\mathbb{R}} = \{-\frac{5}{8}\}$

- Since crit $(f) = \{0\}$ and $0 > -\frac{5}{8}$, it remains to show that $-\frac{5}{8}$ is really the infimum value. In fact, taking $x_n = -\frac{1}{2n}$, $y_n = n$, we see that $f(x_n, y_n) \to -\frac{5}{8}$.

Thus $\inf_{\mathbb{R}^2} f = -\frac{5}{8}$ and the polynomial $f$ does not attain its infimum value.

**4.1. Remark.** It is easy to solve this example "by hand", and the result (certainly) is the same ($\inf f = -\frac{5}{8}$).

## 5. Proofs

**5.1. Proof of Theorem B:** (i) $\Rightarrow$ (ii). Suppose the contrary, that $c \in \text{crit}_\infty(f)$ but the Palais-Smale condition at $c$ holds. Then there exists $R \gg 1$ and $\delta > 0$, $\epsilon > 0$ such that : $\forall x \in f^{-1}(D_\delta) \setminus B_R$, we have

$$\|\text{grad} f(x)\| \geq \epsilon$$

Assertion: There exist $\delta_0 \leq \delta$, $R_0 \geq R$, such that for any $x \in f^{-1}(D_{\delta_0})/B_{R_0}$, the vectors grad $f(x)$ and $\bar{x}$ are linearly independent over $\mathbb{C}$. In fact, if it is not so, the set

$$V = \{x \in f^{-1}(D_\delta),\ \exists \lambda \in \mathbb{C},\ \ \text{grad} f(x) = \lambda \bar{x}\}$$

is not bounded. It is easy to see that $V$ is a semi-algebraic set. By a version at infinity of the Curve Selection Lemma [19], there exists a meromorphic map $\varphi : (0, \epsilon_0] \to V$ such that

$$|\varphi(\tau)| \to \infty \quad \text{as} \quad \tau \to 0.$$

Since $\varphi(\tau) \in V$, grad $f(\varphi(\tau)) = \lambda(\tau)\overline{\varphi}(\tau)$ for $\lambda(\tau) \in \mathbb{C}$.

We have

$$\frac{df(\varphi(\tau))}{d\tau} = < \text{grad} f(\varphi(\tau)), \frac{d\varphi(\tau)}{d\tau} > = \lambda(\tau) < \overline{\varphi}(\tau), \frac{d\varphi(\tau)}{d\tau} > .$$

Then

$$\frac{1}{\lambda(\tau)} \frac{df(\varphi(\tau))}{d\tau} = < \overline{\varphi}(\tau), \frac{d\varphi(\tau)}{d\tau} >$$

and

$$\frac{1}{\overline{\lambda}(\tau)} \frac{d\bar{f}(\varphi(\tau))}{d\tau} = < \varphi(\tau), \frac{d\overline{\varphi}(\tau)}{d\tau} > .$$

Then

$$\frac{1}{\lambda(\tau)} \frac{df(\varphi(\tau))}{d\tau} + \frac{1}{\overline{\lambda}(\tau)} \frac{df(\varphi(\tau))}{d\tau} = \frac{d}{d\tau} \|\varphi(\tau)\|^2$$

It follows that

$$|\frac{df(\varphi(\tau))}{d\tau}| \geq \frac{|\lambda(\tau)|}{2} \frac{d}{d\tau} \|\varphi(\tau)\|^2 = \frac{\|\text{grad} f(\varphi(\tau))\|}{2\|\varphi(\tau)\|} \frac{d}{d\tau} \|\varphi(\tau)\|^2.$$

Suppose that

$$\|\varphi(\tau)\| = \tau^q \|u_1(\tau)\|, \ \|\text{grad} f(\varphi(\tau))\| = \tau^k \|\varphi_2(\tau)\|, \ \|f(\varphi(\tau))\| = \tau^s \|u_3(\tau)\|,$$

where $u_1(\tau), u_2(\tau), u_3(\tau)$ are real analytical mappings and $\|u_i(0)\| \neq 0, \ i = 1, 2, 3$.

It follows from the last inequality that

$$s \leq k + q.$$

On the other hand, since $\|\text{grad} f(\varphi(\tau))\| \geq \epsilon, \ \ k \leq 0$. Since $|f(\varphi(x)| < \delta, \ s \geq 0$, since $\|\varphi(\tau)\| \to \infty, \ q < 0$. Then the inequality $s \leq k + q$ cannot be true and the assertion is proved.

Using this assertion, it is easy to construct the vector field trivializing the fibration

$$f_| : f^{-1}(D_{\delta_0}) \setminus B_{R_0} \to D_{\delta_0}$$

In fact, since the vector grad $f(x)$ and $\bar{x}$ are linearly independent in $f^{-1}(D_{\delta_0}) \setminus B_{R_0}$, there exist 2 vector fields $w(x)$ and $\tilde{w}(x)$ such that

$$\begin{cases} \langle \mathrm{grad} f(x), w(x) \rangle &= 1, \\ \langle \bar{x}, w(x) \rangle &= 0, \end{cases}$$

and

$$\begin{cases} \langle \mathrm{grad} f(x), \tilde{w}(x) \rangle &= +\sqrt{-1}, \\ \langle \bar{x}, \tilde{w}(x) \rangle &= 0. \end{cases}$$

Let $x(\tau)$ be the integral curve of $w(x)$, starting form $x(0)$. We see that

$$\frac{d}{d\tau}(x(\tau), x(\tau)) = 0.$$

This implies that $\|x(\tau)\| = \|x(0)\| \quad \forall \tau$, and we can extend the curve $x(\tau)$ for any $\tau \in \mathbb{R}$.

Moreover, since $\dfrac{df(x(\tau))}{d\tau} = 1$, we have

$$\begin{cases} \mathrm{Re}(f(x(\tau))) &= \mathrm{Re}(f(x(0))) + \tau \\ \mathrm{Im}(f(x(\tau))) &= \mathrm{Im}(f(x(0))). \end{cases}$$

Similarily, if $\tilde{x}(\tau)$ is the integral curve of $\tilde{w}$ then

$$\begin{cases} \mathrm{Re}(f(\tilde{x}(\tau))) &= \mathrm{Re}(f(\tilde{x}(0))) \\ \mathrm{Im}(f(\tilde{x}(\tau))) &= \mathrm{Im}(f(\tilde{x}(0))) + \tau. \end{cases}$$

Let $\psi(x^0, \tau)$ and $\tilde{\psi}(x^0, \tau)$ be the flows of $w$ and $\tilde{w}$ respectively, we define the map

$$h : (f^{-1}(c) \times D_{\delta_0}) \setminus B_{R_0} \to f^{-1}(D_{\delta_0}) \setminus B_{R_0}$$

$$(x, t) \mapsto \psi(\tilde{\psi}(x, \mathrm{Re}(t) - \mathrm{Re}(c), \ \mathrm{Im}(t) - \mathrm{Im}(c))$$

Then $h$ is a diffeomorphism trivializing the fibration.

**5.2. Proof of Theorem B:** (ii) $\Rightarrow$ (i). This part of the proof is not trivial. We begin with some preliminaries. First of all we need the following version at infinity of the Puiseux expansion of affine curves.

Let $C = f^{-1}(0) \subset \mathbb{C}^2$ be an affine plane curve. Let $f = f_d(x,y) + g(x,y)$, where the degree of $g$ is less than $d$, $f_d$ is the homogeneous part of degree $d$, and $d =$ dgree of $f$

$$f_d(x,y) = \prod_{i=1}^{s}(a_i x + b_i y)^{k_i}, \quad \sum_{i=1}^{s} k_i = d$$

**5.2.1. Proposition.** (Puiseux's expansion at infinity)

*There exist d roots $a_i(x)$, $i = 1, ..., d$ of the equation $f(x, y) = 0$. Each of these roots is of the form*

$$\begin{aligned} a_i(x) &= c_i x + \sum_{j=-1}^{-k_0(i)} c_{0,j}(i) x^j + c(i)_1 x^{\frac{m_1(i)}{n_1(i)}} + \sum_{j=-1}^{-k_1(i)} c(i)_{1,j} x^{\frac{m_1(i)+j}{n_1(i)}} \\ + \cdots &+ c(i)_{g(i)} x^{\frac{m_{g(i)}(i)}{n_1(i)\cdots n_{g(i)}}} + \sum_{j=-1}^{-\infty} c(i)_{g(i),j} x^{\frac{m_{g(i)}(i)+j}{n_1(i)\cdots n_{g(i)}}(i)} \end{aligned}$$

*where $c_i \in \{-\frac{a_i}{b_i},\ i = 1, ..., s\}$. Moreover, the series $a_i(\xi^{n_1(i)\cdots n_{g(i)}(i)})$ converge in $|\xi| \gg r_0 > 1$.*

**Proof:** Let $\overline{C}$ be a compactification of $C = f^{-1}(0)$ in $\mathbb{C}\,\mathbb{P}^2$. $\overline{C}$ intersects with "the line at infinity" $z = 0$ of $\mathbb{C}\,\mathbb{P}^2$ at the points

$$A_i = (1 : -\frac{a_i}{b_i} : 0), \quad i = 1, ..., s.$$

We consider the equation $z^d f(-\frac{1}{z}, \frac{y}{z}) = 0$ locally in a neighborhood of $A_i$, $i = 1, ..., s$. The change of variables $u_i = y + \frac{a_i}{b_i}$ sends the point $(-\frac{a_i}{b_i}, 0)$ to the point $(0, 0)$.

Set

$$f_i(u_i, z) = z^d f(\frac{1}{z}, u_i - \frac{a_i}{b_i}).$$

Then $f_i(u_i, z)$ is of the form

$$f_i(u_i, z) = g_i(u_i) + z h_i(u_i, z)$$

and the multiplicity of $g_i(u_i)$ at $u_i = 0$ is $s_i$. Using Newton's algorithm described in [25], one can find $s_i$ roots $u_i = u_{i,j}$, $j = 1, ..., s_i$ of the equation $f_i(u_i, z) = 0$ (locally in a neighborhood of $(0, 0)$). Let $u(z)$ be one of these roots. Then $u(z)$ is of the form

$$\begin{aligned} u(z) &= \sum_{j=1}^{k_0} c_{0,j} z^j + c_1 z^{\frac{m'_1}{n'_1}} + \sum_{j=1}^{k_1} c_{1,j} z^{\frac{(m'_1+j)}{n'_1}} + \cdots + \\ &+ c_g z^{\frac{m'_g}{n'_1\cdots n'_g}} + \sum_{j=1}^{\infty} c_{g,j} z^{\frac{(m'_g+j)}{n'_1\cdots n'_g}}. \end{aligned}$$

Set $a'(z) = -\frac{a_i}{b_i} + u(z)$, then $a'(z)$ is one of the roots of $z^d f(\frac{1}{z}, \frac{y}{z}) = 0$, locally at $A_i$. That is

$$z^d f(\frac{1}{z}, \frac{a'(z)}{z}) \equiv 0 \quad \text{for} \quad |z| \le \epsilon,\ \epsilon > 0.$$

Set $a(x) = xa'(\frac{1}{x})$. Then

$$f(x, a(x)) \equiv 0 \quad \text{for} \quad |x| > \frac{1}{\epsilon}.$$

It is easy to see that $a(x)$ is of the form stated in the Proposition.

**5.2.2. Remark.** Let $a'(z)$ and $a(z)$ be as in the proof of Proposition 3.2.1. Then among the exponents $m_j(i)$, $n_j(i)$ and $m'_j(i)$, $n'_j(i)$ the following relations holds

$$\begin{aligned} m_j(i) &= n'_1(i) \cdots n'_j(i) - m'_j(i) \\ n_j(i) &= n'_j(i), \quad j = 1, ..., g(i). \end{aligned}$$

Let $v$ be the valuation in the ring of formal series, with rational exponents, and each serie has only a finite number of terms of positive exponents. The following result is an analogue of a local result of Kuo-Lu [14].

**5.2.3. Lemma:** *Let $a_1(x), ..., a_d(x)$ be the Puiseux expansions at $\infty$ of $V = f^{-1}(0)$ and $b_1(x), ..., b_{d-1}(x)$ those of $f_y^{-1}(0)$. Then, for every $v, j$, $i \neq j$ there exists $b_s(x)$ such that*

$$v(a_i(x) - b_s(x)) = v(a_j(x) - b_s(x)) = v(a_i(x) - a_j(x)).$$

**Proof.** This follows from the above Remark 5.2.2 and Lemma 3.3 of [14].

Let $X'$ be a ring of formal series with rational exponents, involving only finitely many terms with negative exponents. Let $v'$ be the valuation in $X'$.

**5.2.4. Lemma.** *Let $a, , b', c' \in X'$. At least two numbers among $v'(a' - b')$, $v'(a' - c')$, $v'(b' - c')$ are equal to each other and they are no greater than the third.*

**Proof.** Evident.

**5.2.5. Lemma.** ([7], Proposition 2.2)

*Let $a'_1, ..., a'_n$, $b'_1, ..., b'_m$, $c' \in X'$. Then, at least one of the following statements holds.*

*(i)* *There exists* $s \in \{1, ..., m\}$ *such that*

$$v'(a'_r - b'_s) \geq v'(a'_r - c') \quad \forall r \in \{1, ..., n\}$$

*(ii)* *There exists* $r \in \{1, ..., n\}$ *such that*

$$v'(b'_s - a'_r) \geq v'(b'_s - c') \quad \forall s \in \{1, ..., m\}$$

**5.2.6. Lemma.** *Let* $a_1(x), ..., a_d(x)$ *be the Puiseux expansions at infinity of* $f^{-1}(t_0)$ *and* $b_1(x), ..., b_{d-1}(x)$ *those of* $f_y^{-1}(0)$. *Let* $\varphi(\tau), \psi(\tau)$ *be meromorphic functions in a neighborhood of* $\tau = 0$. *Set*

$$\begin{aligned}\alpha &= v'(f(\varphi(\tau), \psi(\tau)) - t_0)\\ \beta &= v'(f_y(\varphi(\tau), \psi(\tau)))\\ \gamma &= v'(f_x(\varphi(\tau), \psi(\tau))).\end{aligned}$$

*Then, there exists* $b_{s_0}(x)$ *such that*

$$v'(f(\varphi(\tau), b_{s_0}(\varphi(\tau)) - t_0) \geq \min\{\alpha, \beta + \frac{\beta}{d-1}, \gamma + \frac{\gamma}{d-1}\}.$$

**Proof.** Consider

$$\begin{aligned}&a_1(\varphi(\tau)), ..., a_d(\varphi(\tau)) \in X'\\ &b_1(\varphi(\tau)), ..., b_{d-1}(\varphi(\tau)) \in X'\\ &\psi(\tau) \in X'.\end{aligned}$$

By Lemma 5.2.5, at least one of the following possibilities occurs

(i) There exists $b_{s_0}(\varphi(\tau))$ such that

$$v'(a_r(\varphi(\tau)) - b_{s_0}(\varphi(\tau))) \geq v'(a_r(\varphi(\tau)) - \psi(\tau)) \quad \forall r = 1, ..., d$$

(ii) There exists $a_{r_0}(\varphi(\tau))$ such that

$$v'(b_s(\varphi(\tau)) - a_{r_0}(\varphi(\tau))) \geq v'(b_s(\varphi(\tau)) - \psi(\tau)) \quad \forall s = 1, ..., d$$

In the case (i), we have

$$\begin{aligned}v'(f(\varphi(\tau), \psi(\tau)) - t_0) &= \textstyle\sum_{r=1}^d v'(a_r(\varphi(\tau)) - b_{s_0}(\varphi(\tau))\\ &\geq \textstyle\sum_{r=1}^d v'(a_r(\varphi(\tau)) - \psi(\tau))\\ &= v'(f(\varphi(\tau), \psi(\tau)) - t_0) = \alpha\end{aligned}$$

and the Lemma holds.

Suppose we have (ii). Then

$$\begin{aligned}v'(f_y(\varphi(\tau), a_{r_0}(\varphi(\tau))) &= \textstyle\sum_{s=1}^{d-1} v'(a_{r_0}(\varphi(\tau)) - b_s(\varphi(\tau)))\\ &\geq \textstyle\sum_{s=1}^{d-1} v'(b_s(\varphi(\tau)) - \psi(\tau)) = \beta.\end{aligned}$$

This means that

$$\sum_{r \neq r_0} v'(a_r(\varphi(\tau)) - a_{r_0}(\varphi(\tau))) \geq \beta.$$

We shall show that there exists $b_{s_0}(x)$ such that

$$\sum_{r \neq r_0} v'(a_r(\varphi(\tau)) - b_{s_0}(\varphi(\tau))) \geq \sum_{r \neq r_0} v'(a_r(\varphi(\tau)) - a_{r_0}(\varphi(\tau)))$$

(and therefore $\geq \beta$).

In fact, it follows from Lemma 5.2.3, that for every $a_r(x)$, $r \neq r_0$, there exist $b_{s(r)}(x)$ such that

$$\begin{aligned} v'(a_r(\varphi(\tau)) - a_{r_0}(\varphi(\tau))) &= v'(a_r(\varphi(\tau)) - b_{s(r)}(\varphi(\tau))) \\ &= v'(a_{r_0}(\varphi(\tau)) - b_{s(r)}(\varphi(\tau))). \end{aligned}$$

We choose $b_{s_0}(x)$ such that

$$v'(a_{r_0}(\varphi(\tau)) - b_{s_0}(\varphi(\tau))) = \max_s(v'(a_{r_0}(\varphi(\tau)) - b_s(\varphi(\tau)))$$

Then

$$\begin{aligned} v'(a_r(\varphi(\tau)) - a_{r_0}(\varphi(\tau))) &= v'(a_r(\varphi(\tau)) - b_{s(r)}(\varphi(\tau))) \\ &\leq v'(a_{r_0}(\varphi(\tau)) - b_{s_0}(\varphi(\tau))). \end{aligned}$$

It follows from Lemma 5.2.4 that

$$v'(a_r(\varphi)) - a_{r_0}(\varphi(\tau))) \leq v'(a_r(\varphi(\tau)) - b_{s_0}(\varphi(\tau))).$$

We obtain

$$\sum_{r \neq r_0} v'(a_r(\varphi(\tau)) - b_{s_0}(\varphi(\tau))) \geq \beta.$$

By (ii), and by the choice of $b_{s_0}(x)$ we have

$$\begin{aligned} v'(a_{r_0}(\varphi(\tau)) - b_{s_0}(\varphi(\tau))) &\geq \tfrac{1}{d-1}\textstyle\sum_{s=1}^{d-1} v'(b_s(\varphi(\tau)) - a_{r_0}(\varphi(\tau))) \\ &\geq \tfrac{1}{d-1}\textstyle\sum_{s=1}^{d-1} v'(b_s(\varphi(\tau)) - \psi(\tau)) = \tfrac{\beta}{d-1}. \end{aligned}$$

These inequalities imply that

$$\sum_{r=1}^{d} v'(a_r(\varphi(\tau)) - b_{s_0}(\varphi(\tau))) \geq \beta + \frac{\beta}{d-1}.$$

This means that

$$v'(f(\varphi(\tau),\ b_{s_0}(\varphi(\tau)) - t_0) \geq \beta + \frac{\beta}{d-1}.$$

The Lemma is proved.

Now we are ready to complete the proof of Theorem B. Suppose that $f$ does not satisfy the Palais-Smale condition at $c$. Take any $\delta > 0$, where $\delta$ is close enough to 0 and consider the set

$$V_\delta = \{x \in f^{-1}(\overline{D}_\delta \mid \|\text{grad} f(x)\| = \inf\{\|\text{grad} f(x')\| \mid \|x'| = \|x|,$$
$$x' \in f^{-1}(\overline{D}_\delta)\}\}.$$

It is not hard to see that $V_\delta$ is a semi-algebraic set ([3]) and obviously unbounded. Let $(\varphi(\tau), \psi(\tau))$ be a real meomorphic curve in $V_\delta$ such that $\|(\varphi(\tau), \psi(\tau))\| \to \infty$ as $\tau \to 0$. (The existence of such a curve follows from the version at infinity of Curve Selection Lemma). Since $f$ does not satisfy the Palais-Smale condition at $c$, $v(\text{grad} f(\varphi(\tau), \psi(\tau))) = \beta > 0$ and $v(f(\varphi(\tau), \psi(\tau) - c)) > 0$.

By Lemma 5.2.6, there exists a root $b_{s_0}(x)$ of the equation $f_y(x, y) = 0$, such that

$$v'(f(\varphi(\tau), b_{s_0}(\varphi(\tau)) - c) \geq \min\{\alpha, \beta + \frac{\beta}{d-1}\} > .0$$

Thus

$$f(\varphi(\tau), b_{s_0}(\varphi(\tau)) - c \;=\; c_0\tau^\varrho + \cdots \to 0 \qquad (1)$$
$$f_y(\varphi(\tau), b_{s_0}(\varphi(\tau))) \;\equiv\; 0 \qquad (2)$$

for $\tau \sim 0$ and $\|(\varphi(\tau))\| \to \infty$ as $\tau \to 0$.

Set $t = c + c_0\tau^\varrho + \cdots$ we shall show that for any compact $K \subset \mathbb{C}^2$, there exists $\tau$ sufficiently close to 0 (therefore $t$ sufficiently close to $c$) such that

$$\chi(f^{-1}(c) \setminus K) \;>\; \chi(f^{-1}(t) \setminus K),$$

where $\chi(.)$ denotes the Euler characteristic. It is obvious that the Theorem will be proved if this inequality is established.

We see that the points $(\varphi(\tau), b_{s_0}(\varphi(\tau)))$ satisfying (1) (2) are exactly points of ramification of the Riemann surface $f^{-1}(t)$. As we have mentioned before, $\|(\varphi(\tau), b_{s_0}(\varphi(\tau)))\| \to \infty$ as $\tau \to 0$. Hence, for any compact $K \subset \mathbb{C}^2$ and for any $\delta > 0$, there exists $t \in D_\delta = \{t \mid |t - c| < \delta\}$ such that the Riemann surface $f^{-1}(t)$ has some points of ramifications lying outside $K$. Choose the compact $K$ to be so large that all points of ramifications of $f^{-1}(c)$ are contained in $K$.

We consider the map

$$\pi : \mathbb{C}^2 \to \mathbb{C}$$
$$(x, y) \mapsto x$$

By the choice of $K$, the restriction

$$\pi_{c_|} : f^{-1}(c) \setminus K \to C$$

is a covering without ramification. Nevertheless, for $t \neq c$, close to $c$, $\pi_{t_|}$ has some ramification points. Using the Mayer-Vietoris sequence, we see that

$$\chi(f^{-1}(c) \setminus K) - \chi(f^{-1}(t) \setminus K) = m$$

where $m$ is the sum of multiplicities of ramification points of $\pi_{t_|}$.

The proof of Theorem B is complete.

## 5.3. Proof of Theorem A

Let $c$ be the infimum value of the polynomial $f(x, y)$. If $f$ attains $c$ at some point $(x_0, y_0)$, then grad $f(x_0, y_0) = 0$ and $c = f(x_0, y_0) \in \text{crit}(f) \subset A_f$.

Suppose that $f$ does not attain $c$. It follows then from Ekeland's variational principle [8] that there exists a sequence of points $(x_n, y_n)$ such that

$$f(x_n, y_n) \to c$$

and $\|\text{grad}(x_n, y_n)\| \to 0$, as $n \to \infty$.

If the sequence $(x_n, y_n)$ is bounded, then there exists a subsequence $(x_{n'}, y_{n'})$ such that $(x_{n'}, y_{n'}) \to (x_0, y_0) \in \mathbb{R}^2$. This implies that $c = f(x_0, y_0)$, a contradiction. Thus one can suppose that $(x_n, y_n) \to \infty$, as $n \to \infty$. This means that $f$ does not satisfy the Palais-Smale condition at $c$. By Theorem B, $c \in A_f$. The proof of Theorem A is complete.

# References


[1] Arnold, V.I., Gussein-Zade, S.M., Varchenko, A.N., (1985 and 1988), *Singularites of differentiable maps*, Vols 1 and 2, Monographs in Mathematics 82 and 83, Birkhäuser, Boston.

[2] Artal Bartolo, E., (1995), *Une démonstration géometrique du théorème d'Abhyankar - Moh*, J. rein angew. Math. 464, 97-108.

[3] Bochnak, J., Lojasiewicz, S., *A converse of the Kuiper-Kuo Theorem.* Proc. of Liverpool Singularities Symposiums, 1993 LNM, 246-254.

[4] Bresis, H., *Points critiques dans les problèmes variationels sans compacité.* Seminaire Bourbaki, 1987-88, No 698.

[5] Broughton, S., (1988) *Milnor numbers and Topology of polynomial hypersurfaces.* Invent. Math. 92, 217-241.

[6] Cassou - Noguès, P., Dimca, A., *Sur la topologie des polynômes complexes.* Preprint Univ. Bordeaux 7/1996.

[7] Chadzynski, J., Ploski, A., (1988), *An inequality for the Intersection multiplicity of analytic curves.* Bul. Pol. Acad. Sci., Math. No. 3-4, 12-17.

[8] Clarke, F.H., (1989), *Optimization and nonsmooth analysis.* Les publications CRM, Univ. de Montréal.

[9] Durfee, A.H., (1997), *Five definitions of critical point at infinity.* Proceedings of the Oberwolfach Conference in honor of Brieskorn's 60th birthday, to appear.

[10] Fourrier, L., (1996), *Topologie d'un polynôme de deux variables complexes au voisinage de l'infini.* Annales de L'Institut Fourier, T. 46, 645-687.

[11] Hà, H.V., (1989), *Sur la fibration globale des polynômes de deux variables complexes.* C.R. Acad. Sci. Paris, Série I, 309, 231-234.

[12] Hà, H.V., Lê, D.T., (1984), *Sur la topologie des polynômes complexes.* Acta Math. Vietnamica 9, 21-32.

[13] Hà, H.V., Zaharia, A., (1996), *Families of polynomials with total Milnor number constant.* Math. Annalen, 304, 481-488.

[14] Kuo, T.C., Lu, Y.C., (1977), *On analytic function germs of two complex variables.* Topology, Vol. 16, 299-310.

[15] Lê, D.T., Ramanujam, C.P., (1976), *The invariance of Milnor's number implies the invariance of the topological type.* Am. J. of Math. 98, N.1, 67-78.

[16] Lê, D.T., Teissier, B., (1983), *Cycles évanescents et conditions de Whitney II.* Proceedings of Symp. Pure Math. 40, Part 2, 65-103.

[17] Lê, D.T., Weber, C., (1994), *A geometrical approach to the Jacobian conjecture for* $n = 2$. Kodai Math. J., 17, 347-381.

[18] Milnor, J., (1968), *Singular points of complex hypersurfaces.* Ann. of Math. Studies, 61, Princeton Univ. Press.

[19] Némethi, A., Zaharia, A., (1992), *Milnor fibration at infinity.* Indag. Math., N.S. 3(3), 323-335.

[20] Neumann, W., (1989), *Complex algebraic plane curves via their links at infinity.* Invent. Math. 98, 445-489.

[21] Parusinski, A., (1995), *On the bifurcation set of complex polynomial with isolated singularities at infinity.* Compositio Math. 97, 369-384.

[22] Siersma, J.D., Tibar, M., (1995), *Singularities at infinity and their vanishing cycles.* Duke Math. J., 80, 771-783.

[23] Teissier, B., (1973), *Cycles évanescents, sections planes et conditions de Whitney.* Asterisque 718, 285-362.

[24] Thom, R., (1969), *Ensembles et morphismes stratifiés.* Bull. Amer. Math. Soc. 75, 240-284.

[25] Walker, R., (1950), *Algebraic curves.* Princeton Univ. Press.

[26] Zariski, O., (1965), *Studies in equisingularity II: Equisingularity in codimension 1.* Am. J. of Math., 87, 1324-1351.

Chapter 10

# SOLVING GLOBAL OPTIMIZATION PROBLEMS WITH BARON

Vinay Ghildyal[1] and Nikolaos V. Sahinidis[2]

[1] *Sabre Decision Technologies,*
*1 E. Kirkwood Dr.*
*Southlake, TX 76092, USA*

[2] *University of Illinois at Urbana-Champaign,*
*Department of Chemical Engineering,*
*600 South Mathews Avenue, Urbana, Illinois 61801, USA*

[2]Corresponding author. nikos@uiuc.edu

**Abstract** The purpose of this paper is to report our progress to date towards the development of a general purpose global optimization system. The system has evolved from a 1800 line code initially written in the GAMS modeling language to a more than 30,000 line code written in FORTRAN and C. It provides eleven modules for the solution of structured global optimization problems, a parser for inputing and solving factorable nonlinear programs, and a core component for maintaining the branch and bound data structures of arbitrary nonconvex problems.

**Keywords:** branch and reduce methods, the +BARON software, global optimization, factorable nonlinear programming,



*A. Migdalas et al. (eds.), From Local to Global Optimization,* 205–230.

# 1. Introduction

Optimization problems with multiple local optima are encountered in all areas of engineering and the sciences. Determining global optima finds numerous applications in fields such as structural and shape optimization, mechanical equipment and parts design, analysis and design of control systems, integrated circuit design, prediction of molecular structures and molecular design, and chemical process synthesis and operations.

Motivated by a large number of potential applications, we have developed BARON, a computational system for facilitating the solution of nonconvex optimization problems to global optimality. The Branch And Reduce Optimization Navigator derives its name from its combining interval analysis and duality in its "reduce" arsenal with enhanced "branch and bound" concepts as it winds its way through the hills and valleys of complex optimization problems in search of global solutions.

The first version of BARON was merely 1800 lines of code written in the GAMS modeling language [1] in 1991-93 when duality-based range reduction techniques were developed [15]. The code was initially applied to a collection of standard engineering design problems [15], design of just-in-time manufacturing systems [6b], circuit layout and compaction [3], and chemical process planning [9].

The second version was approximately 10,000 lines of code written in FORTRAN 77 in 1994-95. This code incorporated some additional range reduction techniques [16] and was applied to polynomial and multiplicative programs [16], and robust controller design [20]. This code also included specialized algorithms for separable concave minimization problems [17], and fixed-charge and other concave minimization problems arising in multiperiod planning problems [10]. At that time, heuristic techniques were added for feasibility-based range reduction as well as branching schemes to ensure finiteness for certain problem classes [17]. These specialized codes and FORTRAN version of BARON were first made available through anonymous ftp on 2 March, 1995.

The current version of the code was developed in 1996-97. It is a much improved version as it offers, among other new features, more efficient memory management techniques, a specialized code for factorable nonlinear programming problems, an easy-to-use parser, and a detailed manual. This version of BARON is currently approximately 23,000 lines of FORTRAN 90 code and 10,000 lines of code written in C.

This report describes the algorithms used by the software and details the problem classes for which we have developed specialized solution algorithms and solver routines. Numerous examples are provided to illustrate the current capabilities of the system.

## 2. Algorithms Used

The global optimization problem we address is:

$$
\begin{array}{lrl}
(P): & \min & f(x) \\
 & \text{s.t.} & g(x) \leq 0 \\
 & & x \in X
\end{array}
$$

where $f : X \to \Re$, $g : X \to \Re^m$, and $X \subset \Re^n$. The functions $f(x)$ and $g(x)$ can be arbitrarily nonlinear and the set $X$ may include integrality restrictions. The basic algorithmic assumption made here is that there exists a "relaxation", $R$, of $P$. The optimal solution of $R$ provides a lower bound for the problem $P$. It is also assumed that a standard algorithm is available to solve the relaxation $R$ to optimality.

The algorithm used here is of the branch and bound type (see Horst and Tuy [7] for detailed discussions). Its key features are two: (a) the inclusion of range contraction techniques that reduce the range of the feasible space [16], and (b) the incorporation of branching schemes that guarantee finite termination with the exact global optimum for certain problem classes [17].

At a given iteration, the search domain is partitioned and lower and upper bounds are developed over each partition. The partitions are refined until the least lower and upper bounds taken over all partitions come within a predetermined tolerance. During the course of this algorithm, regions that are infeasible or inferior are fathomed. The flowchart in Figure 10.1 describes the branch and reduce algorithm.

The steps of the algorithm are as follows.

### 2.1. Initialization

This step involves setting the upper bound of the problem objective to $+\infty$. The relaxation, $R$, of the original problem is placed in the list of active nodes and assigned a lower bound of $-\infty$.

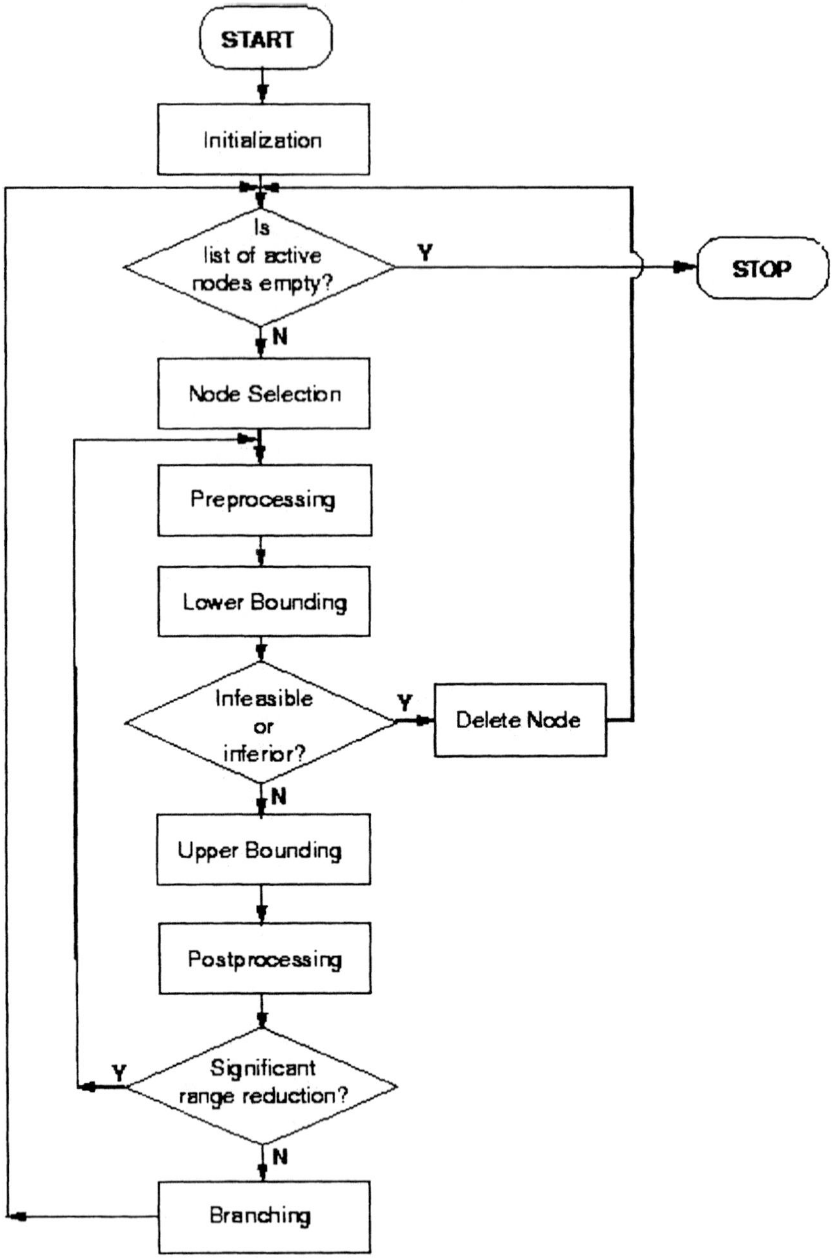


*Figure 10.1.* Branch and reduce algorithm.

From here, we proceed to the iterative step. In a typical iteration, the following steps are encountered.

## 2.2. Node Selection

This involves selecting a node or subproblem from the list of active nodes. If this list is empty, the algorithm terminates. If the list is not empty, the node with the best (lowest) lower bound is selected. Note that other node selection rules might be used here. The node selection rule is not needed to guarantee finite termination of the algorithm in case the search space is discrete. However, the least lower bound selection rule is, in general, required to guarantee convergence for arbitrary search spaces. Let the selected node be denoted by $R_i$.

## 2.3. Preprocessing

The variable bounds for the node $R_i$, selected in the previous step, are reduced using feasibility-based tightening tests. These tests contract variable ranges based on the constraints of the original problem. If a finite upper bound on the problem exists at this time, an "objective function cut" can be introduced to further contract the variable ranges. If linear constraints are present, variable bounds can be tightned through the solution of linear programs. It is often beneficial to solve only these LPs through heuristic techniques, a process we refer to as "poor man's LPs" [17]. Nonlinear terms can also be subject to interval analysis techniques to tighten variable bounds. These tests can be repeated a prespecified number of times for effective reduction in the ranges of the variables. The tighter variable bounds aid in the construction of tighter relaxations and hence facilitate the development of a tighter lower bound for the subproblem.

## 2.4. Lower Bounding

The relaxed problem for the node $R_i$ is constructed using the newly tightened variable bounds. Typically, the relaxed problem is constructed using factorable programming techniques (McCormick [11], [12], [13]), so that the relaxations are exact at the variable bounds. The tightness of the relaxation depends on the tightness of the variable bounds. Construction of underestimators for certain nonconvex terms is described below.

(i) Bilinearities of the form $x_i x_j$ can be underestimated by two constraints, depending on the sign of the inequality in which they appear and the sign of the bilinear term. If the inequality is of the type $x_i x_j + g(x) \leq 0$, the following constraints may be used:

$$w_{ij} + g(x) \leq 0$$

$$w_{ij} \geq x_j^U x_i + x_i^U x_j - x_i^U x_j^U \quad (1)$$

$$w_{ij} \geq x_j^L x_i + x_i^L x_j - x_i^L x_j^L \quad (2)$$

where, in order to maintain differentiability, a new variable, $w_{ij}$, was substituted for $x_i x_j$.

If the inequality in which bilinear terms appear is of the type $x_i x_j + g(x) \geq 0$, the following constraints may be employed:

$$w_{ij} + g(x) \geq 0$$

$$w_{ij} \leq x_j^L x_i + x_i^U x_j - x_i^L x_j^U \quad (3)$$

$$w_{ij} \leq x_j^U x_i + x_i^L x_j - x_i^U x_j^L. \quad (4)$$

(ii) Similarly, linear fractional terms of the form $\frac{x_i}{x_j}$, with $x_j \geq x_j^L > 0$, can also be underestimated by either the first or last two of the following four constraints, depending on the sign of the inequality in which they appear:

$$\frac{x_i}{x_j} \geq \frac{x_i^U}{x_j} + \frac{x_i}{x_j^L} - \frac{x_i^U}{x_j^L}$$

$$\frac{x_i}{x_j} \geq \frac{x_i^L}{x_j} + \frac{x_i}{x_j^U} - \frac{x_i^L}{x_j^U}$$

$$\frac{x_i}{x_j} \leq \frac{x_i^U}{x_j} + \frac{x_i}{x_j^U} - \frac{x_i^U}{x_j^U}$$

$$\frac{x_i}{x_j} \leq \frac{x_i^L}{x_j} + \frac{x_i}{x_j^L} - \frac{x_i^L}{x_j^L}.$$

(iii) Univariate terms which are concave over their entire domain can be underestimated by their secant. Such concave terms include $x^n$ with $0 \leq n \leq 1, x \geq 0$ and $x^n$ with $n = 2k+1, k \in \{1, 2, \ldots\}, x \leq 0$. Consider any such function, $g(x), x \in [l, u]$. This concave function can be underestimated by the term $\alpha x + \beta$, where

$$\alpha = \frac{g(u) - g(l)}{u - l},$$

$$\beta = g(u) - \alpha u.$$

(iv) Terms of the form $x^n$ where $n$ is odd and $0 \in [l, u]$, can be treated in two different ways.

(a) The interval $[l, u]$ can be partitioned at $x = 0$ resulting into two subproblems such that $x^n$ is convex in one subproblem and concave in the other. The concave part can then be treated as above. This process increases the number of open nodes in the branch and bound tree.

(b) If branching is not desired, a convex envelope, $\gamma(x)$, of the term can be obtained as shown in Figure 10.2:

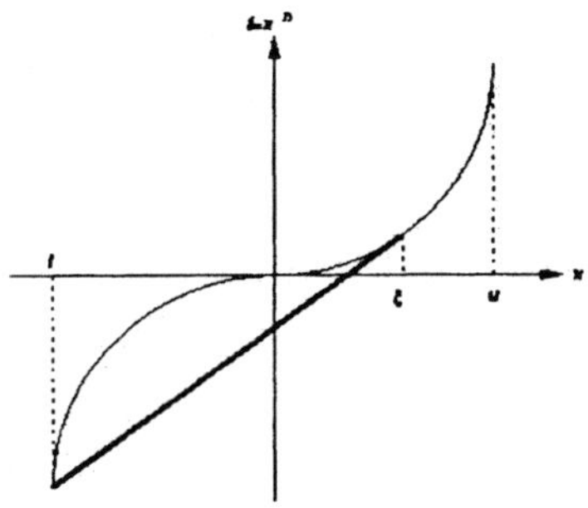

*Figure 10.2.* Monomial underestimator.

$$\gamma(x) = \begin{cases} x_n, & \text{if } x \geq \xi, \\ \alpha x + \beta, & \text{if } x < \xi. \end{cases}$$

The point $\xi$ is where the monomial and its underestimator have the same slope and is, therefore, determined by solving the equation, $(n-1)\xi^n - nl\xi^{n-1} + l^n = 0$. The coefficients $\alpha$ and $\beta$ are then given by:

$$\alpha = n\xi^{n-1},$$
$$\beta = l^n - \alpha l.$$

With the relaxation at hand, the lower bounding problem is then solved. If the problem is infeasible, this node is deleted form the active list and a new node is selected using the node selection rule. If the lower bound of this node is greater than the best known upper bound on the nonconvex problem, the current node is deleted from the list of active nodes and a new node is selected. Otherwise, the upper bounding step is performed.

## 2.5. Upper Bounding

If the solution of the lower bounding problem is feasible to the original problem, the upper bound can be updated. Additionally, a local search is conducted about this point and a randomized heuristic is employed for upper bounding in the subregion $R_i$. If an improved feasible solution is found, the upper bound is updated. All nodes with a lower bound higher than the best known upper bound are fathomed at this step and, if the list of active nodes becomes empty, the algorithm terminates.

After upper bounding, the current node is subjected to postprocessing.

## 2.6. Postprocessing

The variable ranges are further reduced using optimality-based and feasibility-based range reduction tests. In addition to the tests outlined in the preprocessing section, optimality-based tightening is done using the marginal values obtained from the lower bounding step. Variable ranges can also be reduced by probing, i.e., temporarily fixing a variable at its bound and using the marginal values obtained to tighten the variable bounds. The theoretical justification behind these operations is provided in [16].

If the range of any variable is reduced by a factor more than a prespecified value, the lower bounding step is performed again. Note that a better upper bound makes the optimality-based range reduction more effective [16]. If the reduction in variable ranges is significant, the lower bounding, upper bounding and postprocessing steps are repeated. Otherwise, the current node is branched.

## 2.7. Branching

Here, the domain of $R_i$ is partitioned into a finite number of subregions. Rectangular, conical or simplicial partitioning schemes can be used for this purpose (Tuy [18], Tuy *et al.* [19]). In our implementation, the rectangular scheme is employed for the sake of simplicity. This involves selecting a branching variable and a branching point in its domain. Typically, the branching variable is the one which "contributes" the most to the relaxation gap, i.e., the one with the largest "violation" between a nonconvex term and its relaxation. Measures of the violation include the distance between the underestimator and the nonconvex term at the solution of $R_i$, the ranges of the variables, and the distance of the nearest variable bound from its solution value. In most cases, the violations are calculated by combining all these criteria.

Once the branching variable is identified, the branching point is chosen as follows:

(i) Once every predetermined number of times, the variable is branched at its midpoint (bisection). This ensures a significant reduction in the domains of the created subproblems and hence exhaustiveness [18], [19], which is a key to finiteness as shown in [17].

(ii) If the incumbent solution is present in $R_i$, the value of the branching variable at the incubment is used as the branching point provided that this process leads to two new subproblems that are distinct from their father node. This partitioning rule makes the incumbent gapless and is also key to the finiteness of the algorithm [17].

(iii) If the above two conditions do not apply, branching is done at the point that solves the lower bounding problem ("omega" branching).

At the end of the branching step, two subproblems are created and a new node is selected for preprocessing.

## 2.8. Illustrative Example

To illustrate the branch and reduce algorithm, we solve the following concave quadratic programming problem (Floudas and Pardalos [5]).

$$\begin{aligned} \min \quad & 42x_1 - 50x_1^2 + 44x_2 - 50x_2^2 + 45x_3 - 50x_3^2 \\ & +47x_4 - 50x_4^2 + 47.5x_5 - 50x_5^2 \\ \text{s.t.} \quad & 20x_1 + 12x_2 + 11x_3 + 7x_4 + 4x_5 \le 40 \\ & \mathbf{0} \le (x_1, x_2, x_3, x_4, x_5) \le \mathbf{1}. \end{aligned}$$

The global minimum for this problem occurs at $\mathbf{x} = (1, 1, 0, 1, 0)$, with an objective function value of $-17$.

A standard branch and bound algorithm using midpoint branching was first used to solve this problem. This algorithm requires 17 iterations for the upper and lower bounds to converge with a tolerance of $10^{-6}$. The branch and reduce algorithm, applied to the problem, requires 7 iterations. The branch and reduce tree is outlined in Figure 10.3. At each node, $L_i$ and $U_i$ represent the lower and upper bounds for that node. $L$ and $U$ are the global lower and upper bounds. A superscript is used if the node was repeated due to significant bounds tightening during postprocessing.

In this process, preprocessing included feasibility based range reduction (poor man's LPs) and range reduction based on an objective function cut. In the postprocessing step, bounds were tightened using the marginal values obtained during lower bounding. Probing was employed for variables which were strictly within their bounds. The relaxed problem was constructed by underestimating each concave quadratic term in the objective by a linear term, within the variable bounds.

At the root node, preprocessing is performed by minimizing and maximizing each variable subject to the problem constraints. Any feasible solution obtained is used to update the upper bound. Even though this step did not yield improved variable bounds, an upper bound of $-14$ was obtained.

**Iteration 1**: The lower bounding problem yields a solution of $\mathbf{x} = (0.3, 1, 1, 1, 1)$, with $L_1 = -18.9$. No bounds tightening was possible by postprocessing. The node is branched at $x_1 = 0.3$. At this time, the global bounds are $L = -18.9$ and $U = -14.0$.

**Iteration 2**: Here, the node with $(0,0,0,0,0) \leq \mathbf{x} \leq (0.3,1,1,1,1)$ is considered. After preprocessing, the variable ranges become $(0, 0.953, 0.952, 0.950, 0) \leq \mathbf{x} \leq (0.064, 1, 1, 1, 1)$. The relaxed problem yields $L_2 = -16.5$, with $\mathbf{x} = (0, 1, 1, 1, 1)$. This value of $\mathbf{x}$ is also feasible to the original problem with $U_2 = -16.5$. As the upper and lower bounds in this node have collapsed, this node is fathomed. After this iteration, the global bounds are $L = -18.9$ and $U = -16.5$.

**Iteration 3**: Now the node with $(0.3, 0, 0, 0, 0) \leq \mathbf{x} \leq (1, 1, 1, 1, 1)$ is solved. In preprocessing, the bounds are tightened to $(0.84, 0, 0, 0, 0) \leq \mathbf{x} \leq (1, 1, 1, 1, 1)$. The relaxed problem is solved with $\mathbf{x} = (1, 1, 0.364, 0, 1)$,

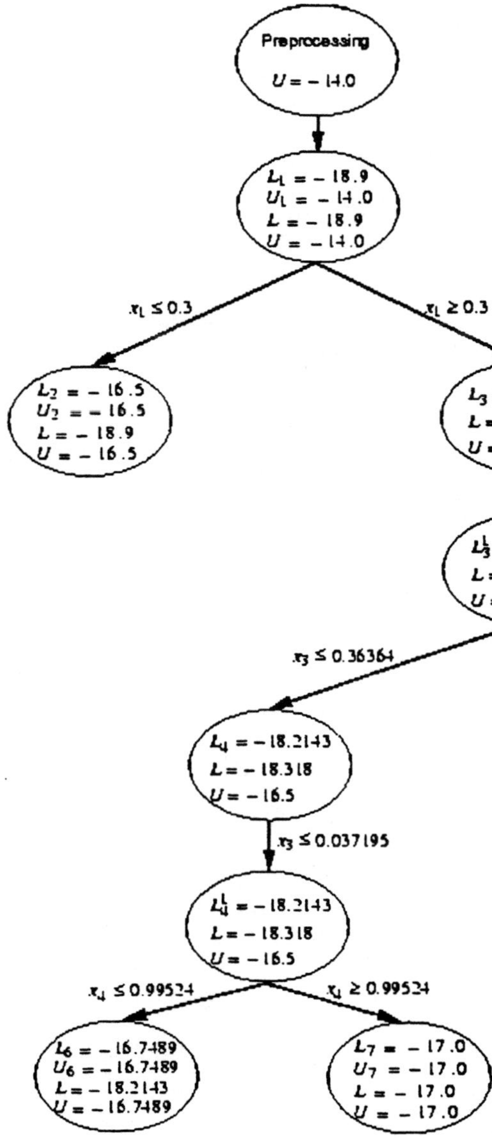


*Figure 10.3.* **Branch and reduce tree for the illustrative example**

and $L_3 = -18.318$. Now the dual multiplier corresponding to the variable bound constraint $x_1 - 1 \leq 0$ is $\lambda_1 = 40.909$. This gives the bound $x_1 \geq x_1^u - (U - L_3)/\lambda_1$, yielding $0.956 \leq x_1 \leq 1$. As the bounds on $x_1$ have been tightened significantly, this node is solved again with the updated bounds. This again yields the solution $\mathbf{x} = (1, 1, 0.364, 0, 1)$ and $L_3^1 = -18.318$. Tightening using the marginals during postprocessing yields $0.961 \leq x_1 \leq 1$. At this stage, the node is branched at $x_3 = 0.364$. After this iteration, the bounds are $L = -18.318$ and $U = -16.5$.

**Iteration 4**: The node with $(0.961, 0, 0, 0, 0) \leq \mathbf{x} \leq (1, 1, 0.364, 1, 1)$ is solved here. After preprocessing, the variable bounds are $(0.961, 0.944, 0, 0, 0) \leq \mathbf{x} \leq (1, 1, 0.0725, 1, 1)$. The relaxed problem yields $\mathbf{x} = (1, 1, 0, 0.571, 1)$ and $L_4 = -18.214$. Postprocessing using marginals and probing contracts the bounds to $(0.964, 0.964, 0, 0, 0) \leq \mathbf{x} \leq (1, 1, 0.0372, 1, 1)$. The node is solved again with the new bounds. The solution obtained is $\mathbf{x} = (1, 1, 0, 0.571,\ 1)$ and $L_4^1 = -18.214$. Postprocessing on the node yields $(0.991, 0.991, 0, 0.990,\ 0) \leq \mathbf{x} \leq (1, 1, 0.0112, 1, 0.0106)$. The current node is branched at $x_4 = 0.99524$. After this iteration, the bounds are $L = -18.318$ and $U = -16.5$.

**Iteration 5**: The node with $(0.961, 0, 0.364, 0, 0) \leq \mathbf{x} \leq (1, 1, 1, 1, 1)$ is solved here. During preprocessing, this node becomes infeasible, indicating that all points in it are "inferior" to the current best upper bound. Hence, this node is fathomed. At this stage, the bounds are $L = -18.214$ and $U = -16.5$.

**Iteration 6**: The subregion of the tree with $(0.991, 0.991, 0, 0.990, 0) \leq \mathbf{x} \leq (1, 1, 0.0112, 0.995, 0.0106)$ is solved here. After preprocessing, the bounds become, $(0.996, 0.996, 0, 0.990, 0) \leq \mathbf{x} \leq (1, 1, 0.00556, 0.995, 0.00527)$. The solution of the relaxed problem is $\mathbf{x} = (1, 1, 0, 0.995, 0)$ with $L_6 = -16.7489$. This value of $\mathbf{x}$ is also feasible to the original problem with $U_6 = -16.749$. Since the lower and upper bounds are same for this node, it is fathomed. After this iteration, $L = -18.214$ and $U = -16.74$.

**Iteration 7**: The subregion of the tree with $(0.991, 0.991, 0, 0.995, 0) \leq \mathbf{x} \leq (1, 1, 0.0112, 1, 0.0106)$ is solved now. After preprocessing, the variable bounds become $(0.996, 0.996, 0, 0.995, 0) \leq \mathbf{x} \leq (1, 1, 0.00562, 1, 0.00532)$. The minimum for the relaxed problem occurs at $\mathbf{x} = (1, 1, 0, 1, 0)$ with $L_7 = -17.0$. This point is also feasible to the original quadratic problem with $U_7 = -17.0$. As the bounds for this node have collapsed, this node is fathomed. After this iteration, $L = -17.0$ and $U = -17.0$.

As there are no more active nodes, the algorithm terminates at this point with $\mathbf{x} = (1, 1, 0, 1, 0)$ and $f = -17.0$.

As can be seen, range reduction and other acceleration techniques greatly enhance the effectiveness of the branch and bound algorithm.

## 3. System Features

As exploiting distinct mathematical structure in a problem can significantly expedite the solution process, different specialized modules have been developed around a core component that maintains the branch and bound datastructures. In addition, a module has been developed for solving arbitrary factorable nonlinear programs. These modules are ready to use and take input in the form of simple data files or in the form of a high level modeling language. The modules are also available as callable subroutines.

### 3.1. Specialized Modules

All modules can handle integer variables even when not explicitly stated in the description that follows.

### 3.2. Separable Concave Quadratic Programming

$$
\begin{aligned}
(SCQP): \qquad & \min \quad f(x) = \sum_{i=1}^{n} (c_i x_i + q_i x_i^2) \\
& \text{s.t.} \quad a \leq Ax \leq b \\
& \qquad\quad l \leq x \leq u
\end{aligned}
$$

where $x \in \Re^n$, $c \in \Re^n$, $q \in \Re^n_-$, $a \in \Re^m$, $b \in \Re^m$, $l \in \Re^n$, $u \in \Re^n$ and $A \in \Re^{m \times n}$.

In this module, for the purpose of lower bounding, each concave quadratic term is underestimated by its secant and all integrality restrictions are relaxed. In addition to feasibility-based tightening, bounds are also tightened by including an objective function cut described in [17].

## 3.3. Separable Concave Programming

$$
\begin{aligned}
(SCP): \qquad & \min \quad f(x) = \sum_{i=1}^{n} f_i(x_i) \\
& \text{s.t.} \quad a \le Ax \le b \\
& \qquad\quad l \le x \le u
\end{aligned}
$$

where $x \in \Re^n$, $a \in \Re^m$, $b \in \Re^m$, $l \in \Re^n$, $u \in \Re^n$ ,$A \in \Re^{m\times n}$ and it is assumed that $f_i(x_i)$ are concave and bounded over $l_i \le x_i \le u_i$ for $i = 1 \ldots n$.

This module is handled in a way similar to the $(SCQP)$ module, except that the concave terms need not be quadratic. The concave term corresponding to each variable is provided by means of a user-written subroutine.

## 3.4. Problems with Power Economies of Scale (Cobb-Douglas functions)

$$
\begin{aligned}
(PES): \qquad & \min \quad f(x) = \sum_{i=1}^{n} c_i x_i^{q_i} \\
& \text{s.t.} \quad a \le Ax \le b \\
& \qquad\quad l \le x \le u
\end{aligned}
$$

where $x \in \Re^n$, $c \in \Re^n_+$, $q \in \Re^n$, $0 < q \le 1$, $a \in \Re^m$, $b \in \Re^m$, $l \in \Re^n_+$, $u \in \Re^n_+$ and $A \in \Re^{m\times n}$.

Each term in the objective function is either linear $(q = 1)$ or concave $(q < 1)$. The concave terms are underestimated by their secants (linear underestimator) and all integrality restrictions are relaxed for lower bounding purposes.

## 3.5. Fixed-Charge Programming

$$
(FCP): \qquad \min \quad f(x) = \sum_{i=1}^{n} f_i(x_i)
$$

$$\begin{aligned} \text{s.t.} \quad & a \le Ax \le b \\ & l \le x \le u \end{aligned}$$

where $x \in \Re^n$, $f_i(x_i)$ equals $c_i + q_i x_i$ if $x_i > 0$ and 0 otherwise, $c \in \Re^n_+$, $q \in \Re^n$, $a \in \Re^m$, $b \in \Re^m$, $l \in \Re^n_+$, $u \in \Re^n_+$ and $A \in \Re^{m \times n}$.

For each variable $x_i$ with $0 \in [l_i, u_i]$, the objective function is treated as a concave term. Although these problems can be linearized through the introduction of 0-1 variables and variable upper bounding constraints associated with each concave variable, the above formulation has the advantage of avoiding the introduction of these variables and constraints. At the same time, the lower bound used by linearizing the concve terms is as tight as the corresponding of the LP relaxation of the corresponding 0-1 formulation.

## 3.6. Fractional Programming

$$\begin{aligned} (FP): \quad \min \quad & f(x) = \frac{cx + \alpha}{qx + \beta} \\ \text{s.t.} \quad & a \le Ax \le b \\ & l \le x \le u \end{aligned}$$

where $x \in \Re^n$, $c \in \Re^n_+$, $q \in \Re^n$, $\alpha \in \Re$, $\beta \in \Re$, $a \in \Re^m$, $b \in \Re^m$, $l \in \Re^n_+$, $u \in \Re^n_+$, $A \in \Re^{m \times n}$. Note that this problem can be solved using linear programming techniques if no integer variables are present.

The lower bound at each iteration is obtained by relaxing the integrality restriction and solving the resulting linear fractional program by the method of Charnes and Cooper [2]. This involves substituting $z = \frac{1}{qx+\beta}$ and $y = zx$. The linear fractional program is then equivalent to the following linear program:

$$\begin{aligned} (RFP): \quad \min \quad & cx + \alpha z \\ \text{s.t.} \quad & az \le Ay \le bz \\ & qy + \beta z = 1 \\ & \mathbf{y} \ge \mathbf{0} \\ & z \ge 1. \end{aligned}$$

## 3.7. Univariate Polynomial Programming

$$
\begin{array}{lll}
(POLY): & \min & f(x) = \sum_{i=0}^{k} c_i x^i \\
 & \text{s.t.} & l \le x \le u
\end{array}
$$

where $x \in \Re$, $c \in \Re^k$, $l \in \Re$, $u \in \Re$.

## 3.8. Linear Multiplicative Programming

$$
\begin{array}{lll}
(LMP): & \min & f(x) = \prod_{i=1}^{p} f_i(x) = \prod_{i=1}^{p} \left(c_i^t x + c_{i0}\right) \\
 & \text{s.t.} & a \le Ax \le b \\
 & & c_i^t x + c_{i0} \ge 0 \quad (i = 1, \ldots, p)
\end{array}
$$

where $x \in \Re^n$, $c_i \in \Re^n$ and $c_{i0} \in \Re$ $(i = 1, \ldots, p)$, $a \in \Re^m$, $b \in \Re^m$, and $A \in \Re^{m \times n}$.

## 3.9. General Linear Multiplicative Programming

$$
\begin{array}{lll}
(GLMP): & \min & f(x) = \sum_{i=1}^{nt} \prod_{j=1}^{p_i} \left(c0_{ij} + c_{ij}^t x\right) \\
 & \text{s.t.} & a \le Ax \le b
\end{array}
$$

where $x \in \Re^n$, $c_{ij} \in \Re^n$ and $c0_{ij} \in \Re$ $(i = 1, \ldots, nt; j = 1, \ldots, p_i)$, $a \in \Re^m$, $b \in \Re^m$, and $A \in \Re^{m \times n}$.

## 3.10. Mixed Integer Convex Quadratic Programming

$$
(MIQP): \quad \min \quad f(x) = \sum_{i=1}^{n} (c_i x_i + q_i x_i^2)
$$

$$\begin{aligned} \text{s.t.} \quad & a \leq Ax \leq b \\ & l \leq x \leq u \end{aligned}$$

where $x \in \Re^n$, $c \in \Re^n$, $q \in \Re^n_+$, $a \in \Re^m$, $b \in \Re^m$, $l \in \Re^n$, $u \in \Re^n$ and $A \in \Re^{m \times n}$.

The lower bounding in this module can be done in two ways. Firstly, with the integrality restrictions relaxed, the resulting convex quadratic program can be solved. Alternatively, each quadratic convex term $x_i^2$ can be underestimated by the two inequalities (1-2) by treating $x_i^2$ as a bilinear term of the form $x_i x_i$.

## 3.11. Mixed Integer Linear Programming

$$\begin{aligned} (MILP): \qquad \min \quad & f(x) = \sum_{i=1}^{n} c_i x_i \\ \text{s.t.} \quad & a \leq Ax \leq b \\ & l \leq x \leq u \end{aligned}$$

where $x \in \Re^n$, $c \in \Re^n$, $a \in \Re^m$, $b \in \Re^m$, $l \in \Re^n$, $u \in \Re^n$, $A \in \Re^{m \times n}$, and some (or all) of the $x$-variables are required to be integers.

The relaxed linear program is solved for lower bounding at each iteration. For upper bounding, a randomized rounding scheme is employed to fix the integer variables and the resulting LP is solved to restore constraint feasibility.

## 3.12. Indefinite Quadratic Programming

$$\begin{aligned} (IQP): \qquad \min \quad & f(x) = \sum_{i=1}^{n} \sum_{j=1}^{n} q_{ij} x_i x_j + \sum_{i=1}^{n} c_i x_i \\ \text{s.t.} \quad & a \leq Ax \leq b \\ & l \leq x \leq u \end{aligned}$$

where $x \in \Re^n$, $q \in \Re^{n \times n}$, $c \in \Re^n$, $a, b \in \Re^m$, $l, u \in \Re^n$ and $A \in \Re^{m \times n}$.

Note that this module subsumes the $(SCQP)$ and $(MIQP)$ modules by allowing for an objective function that is indefinite.

## 3.13. Factorable NLP Module

This is the most general of the supplied modules that do not require any coding from the user. The input to this module is through the parser, which is described in the next section.

The module solves fairly general nonlinear programs:

$$
\begin{aligned}
(NLP): \quad & \min && f(x) \\
& \text{s.t.} && a \leq g(x) \leq b \\
& && l \leq x \leq u
\end{aligned}
$$

where, $x \in \Re^n$, $f : \Re^n \to \Re$, $g : \Re^n \to \Re^m$, $a \in \Re^m$, $b \in \Re^m$, $l \in \Re^n$, $u \in \Re^n$ and $f, g$ are factorable functions.

Factorable functions are recursive compositions of sums and products of functions of single variables. Most functions of several variables used in nonlinear optimization are factorable and can be easily brought into separable form [13].

Currently, the types of functions allowed in this module include $x^\alpha, \beta^x$, for $\alpha, \beta \in \Re$, $exp(x)$ and $ln(x)$. In this way, the capabilities of this module subsume those of all other specialized modules. However, this module is not designed to take advantage of the special structure arising in previously described modules.

The relaxed problem is constructed using separable programming techniques. During the course of parsing, nonlinearities are "removed" from expressions by introducing new variables for the nonlinear terms. This process can best be visualized by viewing each expression as a binary tree [8]. The nonterminal nodes of this tree contain the mathematical operators: binary nodes contain the binary operators like "+", "−", "×", "÷" and "^" (power), where the unary nodes contain unary operators such as "−" (unary minus) and "*ln*". The terminal nodes of the tree are comprised of variables and constants. As an example, the expression $3x^2 + 4xy$ is represented by means of a binary tree in Figure 10.4.

For the reformulation, the tree is traversed from below. As and when nonlinearities are encountered, a new variable can be introduced in the tree. The traversed part of the tree can then be delinked and used to form a new constraint. A detailed reformulation algorithm presented by Smith and Pantelides [4] has been integrated with the parsing strategy for this purpose.

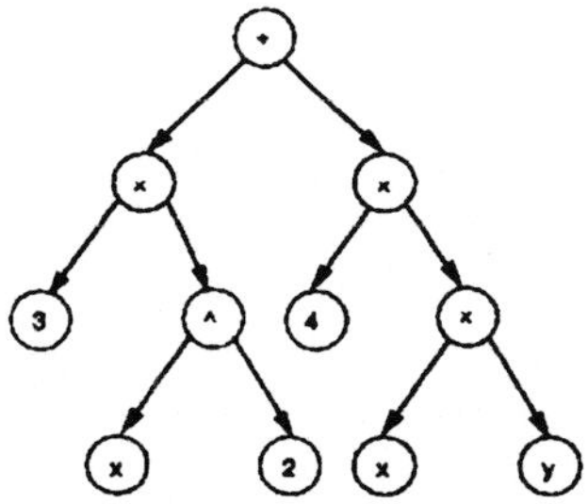

*Figure 10.4.* Binary tree form of an expression

This leads to an equivalent problem with an increase in the number of variables and the number of constraints. The distinct characteristic of this formulation is that, all the nonlinearities of the problem have been identified in terms of "simple" constraints. Underestimators for these constraints can then be constructed using the standard techniques outlined below.

(i) **Bilinear Constraints**: These constraints are of the form $x_i x_j = x_k$. The convex envelope for this bilinear term is given by the inequalities (1)-(4).

(ii) **Linear Fractional Terms**: These terms are of the form $\frac{x_i}{x_j} = x_k$. This can be rearranged as $x_j x_k = x_i$, $(0 \notin [x_j^L, x_j^U])$ and treated in the same way as bilinear terms.

(iii) **Concave Univariate Terms**: These terms can be underestimated by their secant (i.e., the linear underestimator between the variable bounds). For example, the inequality, $f(x_i) - x_j \leq 0$, with $f(x_i)$ concave in $[x_i^L, x_i^U]$, can be underestimated by $\alpha x_i + \beta - x_j \leq 0$, where,

$$\alpha = \frac{f(x_i^U) - f(x_i^L)}{x_i^U - x_i^L},$$
$$\beta = f(x_i^U) - \alpha x_i^U.$$

(iv) Other **polynomial terms** can be treated using techniques described in section 2.4.

Feasibility based tightening on the above nonlinear terms can be done by interval analysis techniques. In addition, linearity-based tightening can be performed on the linear constraints of the reformulated problem. Since the objective function is linear at the end of the reformulation process, an objective function cut can be introduced and included in the linearity based tightening.

For the purpose of upper bounding, the original nonconvex problem is solved as it is of a smaller size compared to the reformulated problem. The original constraints are evaluated by evaluating the corresponding binary tree at the given point. The partial derivative at a given point can also be evaluated by applying the chain rule of differentiation to the binary tree. Both the upper bounding and the lower bounding problems are solved using MINOS [14].

## 4. Usage and Input/Output

A problem can be input to BARON by three different formats: through data files, callable subroutines or through the parser. The input can be broadly classified into algorithmic options and problem related data. If data files or subroutines are used, the options and the problem data are input separately. If the parser input is used, both the options and the problem data can be input together.

We will first illustrate use of the specialized module for $(SCQP)$ and then describe the parser that can be used to input factorable NLPs and all the problem classes described above. All other specialized modules are accessible similar to $(SCQP)$.

### 4.1. Module barscqp: Separable Concave Quadratic Programming

$$\begin{aligned}
\min \quad & -10.5x_1 - 0.5x_1^2 - 7.5x_2 - 0.5x_2^2 - 3.5x_3 - 0.5x_3^2 \\
& -2.5x_4 - 0.5x_4^2 - 1.5x_5 - 0.5x_5^2 - 10x_6 \\
\text{s.t.} \quad & 6x_1 + 3x_2 + 3x_3 + 2x_4 + x_5 \leq 6.5 \\
& 10x_1 + 10x_3 + x_6 \leq 20 \\
& 0 \leq x_i \leq 1 \text{for } i = 1, \ldots, 6.
\end{aligned}$$

The above problem can be entered by inputing all elements (including zeros) of the constraint matrix by speficying **type** $= 0$, in the following input format.

```
2                                                 SCQP problem name
2        6        0                               m, n, numint
0                                                 type
        0    0    0    0    0    0                xlo
-1e31   6    3    3    2    1    0     6.5        lr, a, ur
-1e31  10    0   10    0    0    1     20         lr, a, ur
        1    1    1    1    1    1e31             xup
      -10.5 -7.5 -3.5 -2.5 -1.5 -10               c
      -.5   -.5  -.5  -.5  -.5   0                q
```

Alternatively, the constraint matrix may be input in the sparse column major form using type = 1 as follows:

```
2                                                 SCQP problem name
2        6        0                               m, n, numint
1                                                 type
8                                                 ne
6    10     3    3    10    2    1    1           a(i), i = 1,ne
1     2     1    1     2    1    1    2           ia(i), i=1,ne
1     3     4    6     7    8    9                ja(i), i=1,n+1
-1e31    -1e31                                    lr(i), i=1,m
  6.5      20                                     ur(i), i=1,m
        0    0    0    0    0    0                xlo
        1    1    1    1    1    1e31             xup
      -10.5 -7.5 -3.5 -2.5 -1.5 -10               c
      -.5   -.5  -.5  -.5  -.5   0                q
```

For either value of type, a starting point may be optionally supplied by providing an array of length n in the last row of the input file.

Problem data can also be passed directly in memory by calling subroutine **barscqp** as shown below. The arrays should first be declared as follows:

The real*8 arrays:

$$c = (\,-10.5 \quad -7.5 \quad -3.5 \quad -2.5 \quad -1.5 \quad -10\,)$$
$$q = (\,-0.5 \quad -0.5 \quad -0.5 \quad -0.5 \quad -0.5 \quad -0.5\,)$$
$$a = (\,6 \quad 10 \quad 3 \quad 3 \quad 10 \quad 2 \quad 1 \quad 1\,)$$
$$lr = (\,-1e31 \quad -1e31\,)$$
$$ur = (\,6.5 \quad 20\,)$$
$$xlo = (\,0 \quad 0 \quad 0 \quad 0 \quad 0 \quad 0\,)$$

$$xup = (\,1 \quad 1 \quad 1 \quad 1 \quad 1 \quad 1e31\,)$$

Note that **a** is the array of nonzero matrix elements (length $ne$), **ia** is the array of row indices (length $ne$) and **ja** is the array of column starts (length $n+1$).

The `integer*4` arrays:

$$ia = (\,1 \quad 2 \quad 1 \quad 1 \quad 2 \quad 1 \quad 1 \quad 2\,)$$
$$ja = (\,1 \quad 3 \quad 4 \quad 6 \quad 7 \quad 8 \quad 9\,)$$

The subroutine call is:

```
      call barscqp(
     $      xlo, xup, n, barspace, work, filename,
     $      barstatus, zlbest, zub, xbest, totaltime,
     $      proname, c, q, lr, ur, a, ia, ja, m, numint, ne)
```

The screen output for the above input is shown below.

```
===========================================================================
                    Welcome to BARON v. 3.0
            Global Optimization by BRANCH-AND-REDUCE
===========================================================================
             Separable Concave Quadratic Programming
===========================================================================
 Starting solution is feasible with a value of        .000000D+00
 Preprocessing found feasible solution with value    -.110000D+02
 Preprocessing found feasible solution with value    -.142951D+02
===========================================================================
  Itn. no.      Open Nodes      Total Time      Lower Bound      Upper Bound
*        0               0       000:00:00     -.100000D+21     -.142951D+02
*        1               1       000:00:00     -.100000D+21     -.213000D+03
         1               0       000:00:00     -.213000D+03     -.213000D+03


                 *** Successful Termination ***


  Total time elapsed  :      000:00:00,     in seconds :          .07
       on preprocessing:     000:00:00,     in seconds :          .06
```

```
        on navigating    :     000:00:00,    in seconds :          .01
        on relaxed       :     000:00:00,    in seconds :          .00
        on local         :     000:00:00,    in seconds :          .00
        on tightening    :     000:00:00,    in seconds :          .00
        on marginals     :     000:00:00,    in seconds :          .00
        on probing       :     000:00:00,    in seconds :          .00


    Total no. of BaR iterations:        1
    Best solution found at node:        1
    Max. no. of nodes in memory:        1

 All done with problem 2
===========================================================================
```

## 4.2. Parser

In addition to data files and direct subroutine calls described in the previous subsection, input to BARON can also be in the form of a "high level" program/file. The contents of this file are parsed and the problem is reformulated according to the module being specified in the file. In this way, all specialized modules described in the previous section are accessible through the parser, including the module for solving factorable nonlinear programs.

A sample input file for the NLP module is shown below. It contains a section for speficying the algorithmic and system options, and sections for declaring variables and constraints followed by a rather straightforward description of the problem constraints and objective:

```
// This is input file ex1.bar for the NLP module
OPTIONS{
contol: 1.0e-5; // constraint tolerance
//maxiter: 4;
}
MODULE: NLP;
POSITIVE_VARIABLES x1, x2;
UPPER_BOUNDS{
x1: 6;
x2: 4;
}
EQUATION e1;
e1: x1*x2 <= 4;
OBJ: minimize -(x1 + x2);
```

## 5. Concluding Remarks

We have used BARON over the last 6 years to solve linearly constrained nonconvex problems with up to a few thousands of variables and constraints. Problems from the literature are typically solved within at most minutes on standard engineering workstations [15, 16, 17]. We are currently in the process of a systematic computational experimentation to identify search strategies that improve performance, especially for nonlinearly constrained problems. On the theoretical side, we are exploring the development of novel lower bounding schemes and branching strategies that expedite convergence.

## Acknowledgments

Acknowledgment is made to the donors of the Petroleum Research Fund, administered by the ACS for partial support of this research. The work of the authors was also partially supported by National Science Foundation under CAREER award DMII 95-02722.

# References


[1] A. Brooke, D. Kendrik and A. Meeraus. *GAMS – A User's Guide.* The Scientific Press, Redwood City, CA, 1988.

[2] A. Charnes and W. W. Cooper. Programming with Linear Fractionals. *Naval Research Logistics Quarterly, 9:181-185, 1962*

[3] M. C. Dorneich and N. V. Sahinidis. Global Optimization algorithms for Chip Layout and Compaction. *Engineering Optimization, 25:131-154, 1995.*

[4] E. M. B. Smith and C. C. Pantelides. Global Optimization of General Process Models. In Grossmann [6], pages 355-386.

[5] C. A. Floudas and P. M. Pardalos. *A Collection of Test Problems for Constrained Global Optimization Algorithms*, Springer-Verlag, Berlin, 1990.

[6a] I. E. Grossmann, editor. *Global Optimization in Engineering Design*, Kluwer Academic Publishers, Boston, MA, 1996.

[6b] R. A. Gutierrez and N. V. Sahinidis. A Branch-and-bound Approach for machine Selection in Just-in-Time Manufacturing Systems. *International Journal of Production Research, 31:797-818, 1996.*

[7] R. Horst and H. Tuy. *Global Optimization: Deterministic Approaches.* Springer-Verlag, Berlin, second edition, 1996.

[8] D. E. Knuth. *The Art of Computer Programming - 1. Fundamental Algorithms.* Addison-Wesley, Reading, Mass. 1973.

[9] M. L. Lin and N. V. Sahinidis. Process Planning in a Fuzzy Environment. *European Journal of Operational Research, 100:142-169*, 1997.

[10] M. L. Lin, N. V. Sahinidis and J. P. Shectman. Planning of Chemical Process Networks via Global Concave Minimization. In Gross mann [6]: pages 195-230.

[11] G. P. McCormick. *Converting General Nonlinear Programming Problems to Separable Nonlinear Programming Problems.* Technical Report Serial T-267, The George Washington University, Washington, D.C., 1972.

[12] G. P. McCormick. Computability of Global Solutions to Factorable Nonconvex Programs: Part I - Convex Underestimating Problems. *Mathematical Programming*, 10:147-175, 1976.

[13] G. P. McCormick. *Nonlinear Programming. Theory, Algorithms, and Applications.* Wiley Interscience, New York, 1983.

[14] B. A. Murtagh and M. A. Saunders, *MINOS 5.4 User's Guide.* Technical Report SOL 83-20R, Systems Optimization Laboratory, Department of Operations Research, Stanford University, CA, 1995.

[15] H. S. Ryoo and N. V. Sahinidis. Global Optimization of Nonconvex NLPs and MINLPs with Applications in Process Design. *Cpmputers & Chemical Engineering*, 19:551-566, 1995.

[16] H.S. Ryoo and N. V. Sahinidis. A branch-and-Reduce Approach to Global optimization. *Journal of Global Optimization*, 8:107-139, 1996.

[17] J.P.Sheetman and N. V. Sahinidis. A Finite Algorithm for Global Minimization of Separable Concave Programs. *Journal of Global Optimization*, 1997.

[18] H. Tuy. Effect of the Subdivision Strategy on Convergenec and Efficiency of Some Global Optimization Algorithms. *Journal of Global Optimization*, 1:23-36,1991.

[19] H. Tuy, V. Khatchaturov, and S. Utkin. A class of exhaustive cone splitting procedures in conical algorithms for concave minimization. *Optimization*, 18:791-807, 1987.

[20] J. G. Van Antwerp, R. D. Braatz, and N. V. Sahinidis. Globally Optimal Robust Control for Systems via Nonlinear Time-varying Perturbations. *Computers & Chemichal Engineering*, 21:S125-S130, 1997.

Chapter 11

# D.C. PROGRAMMING APPROACH TO THE MULTIDIMENSIONAL SCALING PROBLEM*

Le Thi Hoai An and Pham Dinh Tao

lethi@insa-rouen.fr pham@insa-rouen.fr

*Mathematical Modelling and Applied Optimization Group,*
*Laboratory of Mathematics - CNRS UPRES-A 60 85,*
*National Institute for Applied Sciences-Rouen,*
*BP 8, F 76 131 Mont Saint Aignan Cedex,*
*France.*

**Abstract** The paper is devoted to the solution of the metric Euclidean Multidimensional Scaling Problem (MDS) and the Euclidean distance geometry problem by methods of d.c. programming developed earlier by the authors. The algorithms, called DCA (DC Algorithms), are based on the duality theory and local optimality conditions for d.c. programming. Different regularization techniques are used in order to improve the qualities (robustness, stability, convergence rate and globality of computed solutions) of DCA. Lagrangian duality without gap allows us to find interesting equivalent forms of (MDS) and a very simple expression of the dual objective function which can be used for checking globality of solutions computed by DCA. DCA (with and without regularization techniques) is described for solving both MDS problems. Requiring only

*This work was partially supported by the computer resources financed by Contrat de Plan Interregional du Bassin Parisien-Pole Interregional de Modelisation en Sciences pour Ingenieurs



*A. Migdalas et al. (eds.), From Local to Global Optimization*, 231–276.

matrix-vector products and only one Cholesky factorization, it is quite simple, and allows exploiting sparsity in large-scale problems. Moreover the well-known majorization method by de Leeuw can be viewed as a special case of DCA. DCA (which globally solves the trust region problem) is also presented in its parametric version to deal with the Euclidean metric MDS problem. Finally extensive numerical experiments are reported which show the robustness and efficiency of DCA for solving the Euclidean metric MDS problem, especially in the large-scale setting. They also show the globality of computed solutions in the case where the dissimilarities are the Euclidean distances between the objects.

**Keywords:** d.c. programming, d.c. duality, local and global optimality, d.c. algorithm (DCA), Lagrangian duality with zero gap, regularization techniques, metric Multidimensional Scaling problem (MDS), subdifferential calculations, distance geometry. problem.

## 1. Introduction

The Multidimensional Scaling Problem, to be denoted by MDS in the sequel, plays a fundamental role in statistics due to various applications in different fields such as social sciences ([7]), biochemistry ([9], [10], [11]), psychology ([29]), mathematical psychology ([6], [57]), data visualization and pattern recognition ([14], [54], [65])... Its mathematical formulation, due to Kruskal ([24], [26]), is as follows:

Let $\varphi$ be a norm on $\mathbb{R}^p$ and $d$ be its corresponding distance. For an $n \times p$ matrix, $X_i$ and $X^j$ denote the $i^{th}$ row and the $j^{th}$ column of $X$ respectively. The transpose of $X$ is denoted by $X^T$, and $(X_i)^T$ will be written as $X_i^T$ for simplicity.
Let two matrices $\Delta = (\delta_{ij})$ and $W = (w_{ij})$ symmetric of order $n$ be given such that

$$\delta_{ij} = \delta_{ji} \geq 0, \quad w_{ij} = w_{ji} \geq 0, \quad \forall i \neq j; \quad \delta_{ii} = w_{ii} = 0, \quad \forall i, j = 1, \ldots, n.$$

The normal case corresponds to $w_{ij} = 1$, $\forall i \neq j$.
We shall refer to $\Delta$ as the dissimilarity matrix and $W$ as the weight matrix. Metric MDS problem consists of finding in $\mathbb{R}^p$ $n$ objects $X_1^{*T}, \ldots, X_n^{*T}$, represented by an $n \times p$ matrix $X^*$, such that the quantity

$$\sum_{i<j} w_{ij} \left( d(X_i^{*T}, X_j^{*T}) - \delta_{ij} \right)^2$$

is smallest.

Let $\mathcal{M}_{n,p}(\mathbb{R})$ denote the space of real matrices of order $n \times p$. We can identify by rows (resp. columns) each matrix $X \in \mathcal{M}_{n,p}(\mathbb{R})$ with a row-vector (resp. column-vector) in $(\mathbb{R}^p)^n$ (resp. $(\mathbb{R}^n)^p$) by writing respectively

$$X \longleftrightarrow \mathcal{X} = (X_1, \dots, X_n),\ X_i^T \in \mathbb{R}^p,\ \mathcal{X}^T \in (\mathbb{R}^p)^n, \quad (1)$$

and

$$X \longleftrightarrow \overline{\mathcal{X}} = \begin{pmatrix} X^1 \\ \ddots \\ X^p \end{pmatrix},\ X^j \in \mathbb{R}^n, \overline{\mathcal{X}} \in (\mathbb{R}^n)^p. \quad (2)$$

The inner product in $\mathcal{M}_{n,p}(\mathbb{R})$ is defined as the inner product in $(\mathbb{R}^p)^n$ or $(\mathbb{R}^n)^p$. That is

$$\begin{aligned} \langle X, Y\rangle_{\mathcal{M}_{n,p}(\mathbb{R})} &= \langle \mathcal{X}^T, \mathcal{Y}^T\rangle_{(\mathbb{R}^p)^n} = \sum_{i=1}^{n} \langle X_i^T, Y_i^T\rangle_{\mathbb{R}^p} = \sum_{i=1}^{n} X_i Y_i^T \quad (3)\\ &= \langle \bar{\mathcal{X}}, \bar{\mathcal{Y}}\rangle_{(\mathbb{R}^n)^p} = \sum_{k=1}^{p} \langle X^{(k)}, Y^{(k)}\rangle_{\mathbb{R}^n} = \sum_{k=1}^{p} (X^{(k)})^T Y^{(k)} \\ &= Tr(X^T Y). \quad (4) \end{aligned}$$

$Tr(X^TY)$ denotes the trace of the matrix $X^TY$. In the sequel, for simplicity, we shall suppress, if no ambiguity is likely, the indices in the inner product and denote by $\|.\|$ the corresponding Euclidean norm on $\mathcal{M}_{n,p}(\mathbb{R})$. Evidently we must choose either representation in a convenient way.

In fact the most general metric MDS problems are of the form

$$(\text{MDS}) \quad \sigma_* = \min\left\{\sigma(X) = \frac{1}{2}\sum_{i<j} w_{ij}\theta(\delta_{ij} - d_{ij}(X)) : \quad X \in \mathcal{M}_{n,p}(\mathbb{R})\right\}$$

where $d_{ij}(X) = d(X_i^T, X_j^T) = \varphi(X_i^T - X_j^T)$ and $\theta(t) = |t|$ (resp. $t^2$) for $t \in \mathbb{R}$.

In this formulation, for $i \neq j$, $w_{ij} = 0$ if and only if $\delta_{ij}$ is missing. More precisely we can assume, without loss of generality, that the weight-matrix $W$ is irreducible (i.e., the graph associated $G(N, \mathcal{S})$ with $N = \{1, \dots, n\}$ and $\mathcal{S} = \{(i,j) \in N \times N : w_{ij} \neq 0\}$ is connected), since (MDS) can be decomposed into a number of smaller problems otherwise. It includes so the general case where some dissimilarity data $\delta_{ij}$ are missing but the weight matrix $W$ is still irreducible.

Other mathematical models have been discussed in [26], [59]. The classical Euclidean metric MDS problem corresponds to $\theta(t) = t^2$ and $\varphi$

being the Euclidean norm on $\mathbb{R}^p$ ([26]). The function $\sigma$ is called the loss function. Some methods of gradient-type were developed by Kruskal [24], [25], Guttman [16], Lingoes, Roskam [30] for solving this nondifferentiable nonconvex optimization problem. But there is no convergence proof, and these gradient-type methods fail when the loss function $\sigma$ is nondifferentiable ([12], [13]).

If the dissimilarity matrix $\Delta$ is a distance matrix, i.e., the $\delta_{ij}$ represent the distances between pairs of points in $\mathbb{R}^p$, then the optimal value $\sigma_*$ of the corresponding (MDS) is zero. In this case the so-called Euclidean distance geometry problem, is stated in one of the following equivalent forms.

$$(\mathrm{EDP})_1 \qquad 0 = \min\left\{\sum_{i<j} w_{ij}(\delta_{ij} - d_{ij}(X))^2 : \quad X \in \mathcal{M}_{n,p}(\mathbb{R})\right\},$$

$$(\mathrm{EDP})_2 \qquad 0 = \min\left\{\sum_{i<j} w_{ij}(\delta_{ij}^2 - d_{ij}^2(X))^2 : \quad X \in \mathcal{M}_{n,p}(\mathbb{R})\right\},$$

where the summation is taken over all pairs $(i,j)$, $i<j$ for which the distances $\delta_{ij}$ are available, i.e. $(i,j) \in \mathcal{S}$ as defined above. It is worth noting that $(\mathrm{EDP})_2$ is a differentiable nonconvex optimization problem.

Euclidean distance matrices have received a lot of attention in recent years both for their elegance and for their many important applications in molecular conformation problems in chemistry (see for instance [15], [17], [18], [19]). Saxe shows that p-dimensional distance geometry problems are strongly NP-hard for all $p \geq 1$. Moré and Wu [36] show that determining an $\varepsilon-$ solution to the distance geometry problem is also NP-hard. Algorithmic difficulties concerning this problem arise from the missing distance data. Indeed, if all distances are given, then the distance geometry problem $(\mathrm{EDP})_2$ can be solved by computing the largest p eigenvalues and eigenvectors of the positive semidefinite rank-p matrix $-\frac{1}{2}(I - \frac{ee^T}{n})(\Delta \otimes \Delta)(I - \frac{ee^T}{n})$ where $e$ is the vector of all one and $\otimes$ denotes Hadamard product, i.e., $\Delta \otimes \Delta = (\delta_{ij}^2)$ ([11]). Alternatively, we could use the Cholesky decomposition with diagonal pivoting. In other words, the "complete" distance geometry problem $(\mathrm{EDP})_2$ can be solved in polynomial time. This explains why the distance geometry problem has been formulated under the form $(\mathrm{EDP})_2$ rather than $(\mathrm{EDP})_1$, and special optimization algorithms have been developed for solving $(\mathrm{EDP})_2$. These algorithms rely on general techniques such as multistarts or simulated annealing to claim convergence to a global minimizer. Hendrickson

[18], [19] used some recently published results in graph theory to develop algorithms that test the uniqueness and rigidity of the distance graph and an adapted trust region algorithm for local optimization. The embed algorithm (Crippen and Havel [11] and Havel [17]) solves the Euclidean distance geometry with lower and upper bound which consists in finding $X$ such that

$$l_{ij} \leq \| X_i^T - X_j^T \| \leq u_{ij}, \; (i,j) \in \mathcal{S} \tag{5}$$

where $l_{ij}$ and $u_{ij}$ are lower and upper bounds on the distance constraints, respectively. This algorithm approaches problem (5) by solving a sequence of distance geometry problems $(\text{EDP})_2$ where all pairwise distances are included. Current implementations of the embed algorithm use a local minimizer of $(\text{EDP})_2$ as a starting point for a simulated annealing procedure. In deterministic approaches, Glunt, Hayden and Raydan([15])have developed a special gradient method for determining a local minimizer of $(\text{EDP})_1$. Moré and Wu ([35]) used global smoothing and continuation technique for solving $(\text{EDP})_2$. In this approach the Gaussian transform is used to map the objective function into a smoother function with fewer local minimizers, and an optimization procedure (more precisely, a trust region based algorithm) is applied to the transformed function, tracing the minimizer back to the original function. Computational experiments with up to $216 \times 3 = 648$ variables showed that the global continuation method is more reliable and efficient than the multistart approach, a standard procedure for finding the global minimizer to $(\text{EDP})_2$. Zou, Bird and Schnabel [66] proposed a stochastic/perturbation algorithm for both $(\text{EDP})_2$ and problem (5). This is a combination of a stochastic phase that identifies an initial set of local minimizers and a more deterministic phase that moves from low to even lower local minimizer. The numerical experiments presented there (with the same data as in Moré and Wu [35] and Hendrickson [19]) showed that this approach is promising. It is worth noting that $(\text{EDP})_2$ is intimately related to the Euclidean distance matrix completion problem ( [1], [28]). This problem has been formulated as a semidefinite programming problem and solved by Alfakih, Khandami and Wolkowicz [1] with an adapted interior-point method. So far the above algorithms seem to be expensive and so, not suitable to large-scale problems.

De Leeuw ([12]) proved that the classical metric MDS problem is equivalent to maximizing the ratio of two seminorms

$$(\text{MDS}_{\text{max}}) \qquad \max \left\{ \frac{\xi(X)}{\eta(X)} : \eta(X) \neq 0 \right\},$$

where

$$\eta(X) = \left[\sum_{i<j} w_{ij} d_{ij}^2(X)\right]^{1/2} \quad \text{and} \quad \xi(X) = \sum_{i<j} w_{ij}\delta_{ij} d_{ij}(X) \tag{6}$$

are two seminorms on $\mathcal{M}_{n,p}(\mathbb{R})$. For solving the Euclidean metric MDS with $\theta(t) = t^2$ de Leeuw proposed an algorithm [12] (see also [13]), called the majorization method which takes advantage of the special structure of the seminorms $\xi$ and $\eta$ and relationships between problems (MDS) and ($\text{MDS}_{\text{max}}$). The majorization method is in fact an adapted version to problem (MDS) (via ($\text{MDS}_{\text{max}}$)) of the aforesaid subgradient method ([40]-[45]) for maximizing a seminorm on the unit ball of another seminorm:

$$(\text{S}_{\text{max}}) \qquad \max\{\xi(X) : \quad \eta(X) \le 1\}.$$

In the Euclidean case de Leeuw observed judiciously that his elegant adapted algorithm is unlike the subgradient method since $\lim_{k\to+\infty} \|X^{(k+1)} - X^{(k)}\|$ is not necessarily zero in the latter. In fact such a property can be obtained by applying regularization techniques to subgradient methods (see Theorem 2). The majorization method was tremendously successful in solving the classical Euclidean metric MDS problem. Note that de Leeuw's approach fails in the case where $\theta(t) = |t|$.

Pham Dinh Tao extended subgradient methods to d.c. programs

$$(\text{P}) \qquad \min\{f(x) = g(x) - h(x) : x \in \mathbb{R}^n\}$$

where $g$ and $h$ are lower semicontinuous proper convex functions on $\mathrm{R}^n$. Later on algorithms for solving d.c. programs, called henceforth DCA, have been extensively developed by Le Thi Hoai An and Pham Dinh Tao [2], [50] and references therein). (MDS) with $\theta(t) = t^2$ is actually a d.c. optimization problem of the form

$$(\text{S}_1) \qquad \min\left\{\frac{1}{2}\eta^2(X) - \xi(X) : \quad X \in \mathcal{M}_{n,p}(\mathbb{R})\right\}, \tag{7}$$

since

$$\sigma(X) = \frac{1}{2}\sum_{i<j} w_{ij} d_{ij}^2(X) + \frac{1}{2}\sum_{i<j} w_{ij}\delta_{ij}^2 - \sum_{i<j} w_{ij}\delta_{ij} d_{ij}(X) = \frac{1}{2}\eta^2(X) - \xi(X) + \frac{1}{2}\sum_{i<j} w_{ij}\delta_{ij}^2.$$

This d.c. programming formulation allows a new insight into the majorization method. Indeed, as will be seen later (Subsection 4.7, Remark 4), a special case of DCA applied to ($\text{S}_1$) in the Euclidean case gives rise, in a natural and elegant way, to the majorization method. Convergence results for the latter then are immediate consequences of the general convergence theorem of DCA. If $\theta(t) = |t|$, it is easily seen ([2]) that the corresponding (MDS) remains a d.c. program (which is a polyhedral d.c. program if the norm $\varphi$ is polyhedral) while the argument used by de Leeuw in [12] is no longer valid in this case. Remark that $(\text{EDP})_2$ is still a d.c. program to which DCA can be applied. Nevertheless DCA is simpler, more elegant and more efficient when it is applied to $(\text{EDP})_1$.

The paper is divided into six sections. In the next section we will briefly review DCA's and their basic properties whose proofs can be found in [2], [47], [50], [51]. In Section 3 we discuss the zero gap property of the Lagrangian duality relative to the problem of maximization of a gauge $\psi$ on the unit ball defined by a gauge $\phi$. These results are used to find various d.c. decompositions for Problem $(\text{S}_{\text{max}})$ and in particular to establish the equivalence of $(\text{S}_{\text{max}})$ and $(\text{S}_1)$ as well as relationships between primal and dual solutions. Moreover the expression of the dual objective function $\gamma(\lambda)$ is very simple and can be used to check globality of solutions computed by DCA. Section 4 constitutes the core of the paper and is devoted to the solution of the Euclidean metric MDS problem and in particular the distance geometry problem $(\text{EDP})_1$. A thorough study of MDS problem in its elegant matrix formulation (including substantial subdifferential calculus relative to the seminorms $\eta$ and $\xi$ on the matrix space $\mathcal{M}_{n,p}(\mathbb{R})$) is crucial because it allows us to compute quite simply sequences of iterates in DCA. It actually requires matrix-vector products (resp. and only one Cholesky factorization) in the normal (resp. general) case. In Section 5, it is pointed out that in the Euclidean metric case, Problem $(\text{S}_{\text{max}})$ can be formulated as a parametric trust region problem. Parametric DCA is then described for solving this problem. Extensive numerical experiments (Section 6) on problem instances with up to $2500 \times 3 = 7500$ variables proved the robustness and the efficiency of DCA for solving the metric Euclidean MDS problem (in particular the Euclidean distance geometry problem $(\text{EDP})_1$), especially in the large scale setting. They also showed that $\text{DCAP}_{\text{max}}$ (DCA applied to $(\text{S}_{\text{max}})$) is more efficient than $\text{DCAP}_1$ and DCA always converges to a global solution of the Euclidean distance geometry problem (i.e. $\sigma_* = 0$).

## 2. D.c. optimization.

The space $E = \mathbb{R}^n$ is equipped with the canonical inner product $\langle x, y \rangle$, thus the dual space $F$ of $E$ can be identified with $E$ itself. Denote by $\Gamma_o(E)$ the set of all lower semicontinuous proper convex functions on $E$. The conjugate function $g^*$ of $g \in \Gamma_o(E)$ is a function belonging to $\Gamma_o(F)$ and defined by

$$g^*(y) = \sup\{\langle x, y \rangle - g(x) : x \in E\}.$$

The Euclidean norm of $E$ is denoted by $\|x\| = \langle x, x \rangle^{1/2}$. For a convex set $C$ in $E$ the indicator function of $C$ is denoted by $\chi_C(x) = 0$ if $x \in C, +\infty$ otherwise; the interior of $C$ is denoted by int $C$ while ri $C$ stands for the relative interior of $C$.We shall use the following usual notations of [53]

$$\text{dom } g = \{x \in E : g(x) < +\infty\}.$$

For $\epsilon > 0$ and $x^0 \in \text{dom } g$, the symbol $\partial_\epsilon g(x^0)$ denotes the $\epsilon$-subdifferential of $g$ at $x^0$, i.e.

$$\partial_\epsilon g(x^0) = \{y \in F : g(x) \geq g(x^0) + \langle x - x^0, y \rangle - \epsilon, \quad \forall x \in E\},$$

while $\partial g(x^0)$ stands for the usual (or exact) subdifferential of $g$ at $x^0$.

Also, $\text{dom } \partial g = \{x \in E : \partial g(x) \neq \emptyset\}$ and $\text{range}\, \partial g = \cup\{\partial g(x) : x \in \text{dom } \partial g\}$. We adopt in the sequel the convention $+\infty - (+\infty) = +\infty$.

A d.c. program is of the form

$$\text{(P)} \quad \alpha = \inf\{f(x) := g(x) - h(x) : x \in E\},$$

where $g$ and $h$ belong to $\Gamma_o(E)$.
Such a function $f$ is called d.c. function on $E$ and $g$, $h$ are called its d.c. components. If $g$ and $h$ are in addition finite on all $E$, then one says that $f = g - h$ is finite d.c. function on $E$. The set of d.c. functions (resp. finite d.c. functions) on $E$ is denoted by $\mathcal{DC}(E)$ (resp. $\mathcal{DC}_f(E)$). By a well known result of Toland [60], the optimal value in (P) is equal to the optimal value in its dual

$$\text{(D)} \quad \alpha = \inf\{h^*(y) - g^*(y) : y \in F\}.$$

Note that the finiteness of $\alpha$ merely implies that

$$\text{dom } g \subset \text{dom } h \quad \text{and} \quad \text{dom } h^* \subset \text{dom } g^*. \tag{8}$$

Such inclusions will be assumed throughout the paper.
A point $x^*$ is said to be *a local minimizer* of $g - h$ if $g(x^*) - h(x^*)$ is

finite (i.e., $x^* \in \text{dom } g \cap \text{dom } h$) and there exists a neighbourhood $U$ of $x^*$ such that

$$g(x^*) - h(x^*) \leq g(x) - h(x), \quad \forall x \in U. \tag{9}$$

Under the convention $+\infty - (+\infty) = +\infty$, the property (9) is equivalent to $g(x^*) - h(x^*) \leq g(x) - h(x), \quad \forall x \in U \cap \text{dom } g$.

$x^*$ is said to be *a critical point* of $g - h$ if $\partial g(x^*) \cap \partial h(x^*) \neq \emptyset$.

A function $\theta \in \Gamma_o(E)$ is said to be polyheral convex if ( [53])

$$\theta(x) = \max\{\langle a^i, x\rangle - \alpha_i : i = 1, \ldots, m\} + \chi_S(x), \quad \forall x \in E$$

where $a^i \in E$, $\alpha_i \in \mathbb{R}$ for $i = 1, \ldots, m$ and $S$ is a nonempty polyhedral convex set in $E$.

Let $\rho \geq 0$ and $C$ be a convex subset of $E$. One says that a function $\theta : C \longrightarrow \mathbb{R} \cup \{+\infty\}$ is $\rho -$ *convex* if

$$\theta[\lambda x + (1-\lambda)x'] \leq \lambda\theta(x) + (1-\lambda)\theta(x') - \frac{\lambda(1-\lambda)}{2}\rho\|x - x'\|^2, \forall \lambda \in ]0,1[,$$

$$\forall x, x' \in C.$$

It amounts to saying that $\theta - (\rho/2)\|\cdot\|^2$ is convex on $C$. The modulus of strong convexity of $\theta$ on $C$, denoted by $\rho(\theta, C)$ or $\rho(\theta)$ if $C = E$, is given by:

$$\rho(\theta, C) = \sup\{\rho \geq 0 : \theta - (\rho/2)\|\cdot\|^2 : \text{ is convex on } C\}. \tag{10}$$

Clearly, $\theta$ is convex on $C$ if and only if $\rho(\theta, C) = 0$. One says that $\theta$ is *strongly convex* on $C$ if $\rho(\theta, C) > 0$.

For $f \in \Gamma_o(E)$ and $\lambda > 0$ the Moreau-Yosida regularization of $f$ with parameter $\lambda$, denoted by $f_\lambda$, is the inf-convolution of $f$ and $\frac{1}{2\lambda}\|\cdot\|^2$ (see Apendix). The function $f_\lambda$ is continuously differentiable, underapproximates $f$ without changing the set of minimizers and $(f_\lambda)_\mu = f_{\lambda+\mu}$ . More precisely, $\nabla f_\lambda = \frac{1}{\lambda}[I - (I + \lambda\partial f)^{-1}]$ is Lipschitzian with ratio $\frac{1}{\lambda}$. The operator $(I + \lambda\partial f)^{-1}$ is called the proximal mapping associated with $\lambda f$ ([32], [33], [55]).

Let $\mathcal{P}$ and $\mathcal{D}$ denote the solution sets of problems (P) and (D), respectively, and let

$$\mathcal{P}_l = \{x^* \in E : \partial h(x^*) \subset \partial g(x^*)\}, \quad \mathcal{D}_l = \{y^* \in F : \partial g^*(y^*) \subset \partial h^*(y^*)\}.$$

We present below some fundamental results of d.c. optimization which constitute the basis of DCA presented in Subsection 2.1.

**Theorem 1** *([2], [51]) (i)* $\cup\{\partial h(x) : x \in \mathcal{P}\} \subset \mathcal{D} \subset dom\, h^*$.
*The first inclusion becomes equality if* $g^*$ *is subdifferentiable in* $\mathcal{D}$ *(in particular if* $\mathcal{D} \subset ri(dom\, g^*)$ *or if* $g^*$ *is subdifferentiable in* $dom\, h^*$*). In this case* $\mathcal{D} \subset (dom\, \partial g^* \cap dom\, \partial h^*)$.
*(ii) If* $x^*$ *is a local minimum of* $g - h$*, then* $x^* \in \mathcal{P}_l$.
*(iii) Let* $x^*$ *be a critical point of* $g - h$ *and* $y^* \in \partial g(x^*) \cap \partial h(x^*)$*. Let* $U$ *be a neighbourhood of* $x^*$ *such that* $(U \cap dom\ g) \subset dom\ \partial h$ *. If for any* $x \in U \cap dom\ g$ *there is* $y \in \partial h(x)$ *such that* $h^*(y) - g^*(y) \geq h^*(y^*) - g^*(y^*)$ *then* $x^*$ *is a local minimum of* $g - h$*. More precisely,*

$$g(x) - h(x) \geq g(x^*) - h(x^*), \quad \forall x \in U \cap dom\ g.$$

*(iv) Let* $x^* \in dom\ \partial h$ *be a local minimum of* $g - h$ *and let* $y^* \in \partial h(x^*)$ *(i.e.,* $\partial h(x^*)$ *is nonempty and* $x^*$ *admits a neighbourhood* $U$ *such that* $g(x) - h(x) \geq g(x^*) - h(x^*), \quad \forall x \in U \cap dom\ g.$*) If*

$$y^* \in int\ (dom\ g^*) : \ and \quad \partial g^*(y^*) \subset U, \tag{11}$$

*then* $y^*$ *is a local minimum of* $h^* - g^*$ *((11) holds if* $g^*$ *is differentiable at* $y^*$*).*

## 2.1. Description of DCA for general d.c. programs.

For each fixed $x^* \in E$ we consider the problem

$$(S(x^*)) \qquad \inf\{h^*(y) - g^*(y) : y \in \partial h(x^*)\}$$

which is equivalent to the convex maximization one

$$\inf\{\langle x^*, y\rangle - g^*(y) : y \in \partial h(x^*)\}.$$

Similarly, for each fixed $y^* \in F$ we define the problem

$$(T(y^*)) \qquad \inf\{g(x) - h(x) : x \in \partial g^*(y^*)\}.$$

which is equivalent to

$$\inf\{\langle x, y^*\rangle - h(x) : x \in \partial g^*(y^*)\}.$$

Let $\mathcal{S}(x^*)$, $\mathcal{T}(y^*)$ denote the solution sets of Problems $(S(x^*))$ and $(T(y^*))$, respectively.

The complete form of DCA is based upon the duality of d.c. optimization defined by (P) and (D). It allows approximating a point $(x^*, y^*) \in \mathcal{P}_l \times \mathcal{D}_l$. From a point $x^{(0)} \in \operatorname{dom} g$ given in advance, the algorithm consists of constructing two sequences $\{x^{(k)}\}$ and $\{y^{(k)}\}$ defined by

$$y^{(k)} \in \mathcal{S}(x^{(k)}); \quad x^{(k+1)} \in \mathcal{T}(y^{(k)}). \tag{12}$$

The complete DCA can be viewed as a sort of decomposition approach of the primal and dual problems (P), (D). From a practical point of view, although problems $(S(x^{(k)}))$ and $(T(x^{(k)}))$ are simpler than (P), (D) (we work in $\partial h(x^{(k)})$ and $\partial g^*(y^{(k)})$ with convex maximization problems), they remain nonconvex programs and thus are still hard to solve. In practice the following simplified form of DCA is used:

- **Simplified form of DCA:**

The idea of the simplified DCA is quite simple: It constructs two sequences $\{x^{(k)}\}$ and $\{y^{(k)}\}$ (candidates to primal and dual solutions) which are easy to calculate and satisfy the following conditions:

(i) The sequences $\{g(x^{(k)}) - h(x^{(k)})\}$, and $\{h^*(y^{(k)}) - g^*(y^{(k)})\}$ are decreasing and

- If $g(x^{(k+1)}) - h(x^{(k+1)}) = g(x^{(k)}) - h(x^{(k)})$, then $\partial g(x^{(k)}) \cap \partial h(x^{(k)}) \neq \emptyset$,
- If $h^*(y^{(k+1)}) - g^*(y^{(k+1)}) = h^*(y^{(k)}) - g^*(y^{(k)})$, then $\partial g^*(y^{(k)}) \cap \partial h^*(y^{(k)}) \neq \emptyset$.

In such a case the DCA terminates at the $k^{th}$ iteration.

(ii) Every limit point $x^*$ (resp. $y^*$) of the sequence $\{x^{(k)}\}$ (resp. $\{y^{(k)}\}$) is a critical point of $g - h$ (resp. $h^* - g^*$).

Theorem 1 suggests quite naturally constructing two sequences $\{x^{(k)}\}$ and $\{y^{(k)}\}$, starting from a given point $x^o \in \operatorname{dom} g$ by setting

$$y^{(k)} \in \partial h(x^{(k)}); \quad x^{(k+1)} \in \partial g^*(y^{(k)}).$$

*Interpretation of the simplified DCA:*
At each iteration $k$ we do the following:

$$x^{(k)} \in \partial g^*(y^{(k-1)}) \rightarrow y^{(k)} \in \partial h(x^{(k)}) =$$

$$\text{argmin}\{h^*(y) - [g^*(y^{(k-1)}) + \langle x^{(k)}, y - y^{(k-1)}\rangle] : y \in F\}, \quad (D_k)$$

$$y^{(k)} \in \partial h(x^{(k)}) \to x^{(k+1)} \in \partial g^*(y^{(k)}) =$$

$$\text{argmin}\{g(x) - [h(x^{(k)}) + \langle x - x^{(k)}, y^{(k)}\rangle] : x \in E\}. \quad (P_k)$$

Problem $(P_k)$ is a convex program obtained from (P) by replacing $h$ with its affine minorization defined by $y^{(k)} \in \partial h(x^{(k)})$. Similarly, the convex problem $(D_k)$ is obtained from (D) by using the affine minorization of $g^*$ defined by $x^{(k)} \in \partial g^*(y^{(k-1)})$. Here we can see the complete symmetry between problems $(P_k)$ and $(D_k)$, and between the sequences $\{x^{(k)}\}$ and $\{y^{(k)}\}$ relative to the duality of d.c. optimization. The two forms of DCA are identical if $g^*$ and $h$ are essentially differentiable.

- **Well definiteness of DCA**:

DCA is well defined if one can construct two sequences $\{x^{(k)}\}$ and $\{y^{(k)}\}$ as above from an arbitrary initial point $x^o \in \text{dom}\, g$. We have $x^{(k+1)} \in \partial g^*(y^{(k)})$ and $y^{(k)} \in \partial h(x^{(k)})$, $\forall k \geq 0$. So $\{x^{(k)}\} \subset \text{range}\, \partial g^* = \text{dom}\, \partial g$ and $\{y^{(k)}\} \subset \text{range}\, \partial h = \text{dom}\, \partial h^*$. Then it is clear that

**Lemma 1** *The sequences $\{x^{(k)}\}$, $\{y^{(k)}\}$ in DCA are well defined if and only if*

$$dom\, \partial g \subset dom\, \partial h, \quad and\ dom\, \partial h^* \subset dom\, \partial g^*.$$

Since for $\theta \in \Gamma_o(E)$ we have ri (dom $\theta$) $\subset$ dom $\partial\theta \subset$ dom $\theta$ ([53]) we can say, under the essential assumption (8), that DCA is in general well defined.

**Remark 5** *A d.c. function $f$ has infinitely many d.c. decompositions. For example if $f = g - h$ then $f = (g + \theta) - (h + \theta)$ for every $\theta \in \Gamma_o(E)$ finite on the whole $E$. It is clear that the primal d.c. programs (P) corresponding to the two d.c. decompositions of the objective function $f$ are identical. But their dual programs are quite different and so is DCA relative to these d.c. decompositions. In other words, there are as many DCA as there are d.c. decompositions of the objective function $f$. It is therefore useful to find a suitable d.c. decomposition of $f$ since it may have an important influence on the efficiency of DCA for its solution. This question is intimately related to the regularization techniques in d.c. programming.*

## 2.2. Convergence of DCA for general d.c. programs.

Let $\rho_i$ and $\rho_i^*$, $(i = 1, 2)$ be real nonnegative numbers such that $0 \leq \rho_i < \rho(f_i)$ (resp. $0 \leq \rho_i^* < \rho(f_i^*)$) where $\rho_i = 0$ (resp. $\rho_i^* = 0$) if $\rho(f_i) = 0$ (resp. $\rho(f_i^*) = 0$ ) and $\rho_i$ (resp. $\rho_i^*$) may take the value $\rho(f_i)$ (resp. $\rho(f_i^*)$) if the upper bound is attained in (10). We next set $f_1 = g$ and $f_2 = h$.
Also let $dx^{(k)} := x^{(k+1)} - x^{(k)}$ and $dy^{(k)} := y^{(k+1)} - y^{(k)}$.

The basic convergence theorem of DCA for general d.c. programming is stated below. For the proof the reader is referred to [2], [50], [51].

**Theorem 2** *Suppose that the sequences $\{x^{(k)}\}$ and $\{y^{(k)}\}$ are defined by the simplified DCA. Then we have*

$$(i) \qquad (g-h)(x^{(k+1)}) \leq (h^*-g^*)(y^{(k)}) - \max\{\frac{\rho_2}{2}\|dx^{(k)}\|^2, \frac{\rho_2^*}{2}\|dy^{(k)}\|^2\}$$

$$\leq (g-h)(x^{(k)}) - \max\{\frac{\rho_1+\rho_2}{2}\|dx^{(k)}\|^2,$$

$$\frac{\rho_1^*}{2}\|dy^{(k-1)}\|^2 + \frac{\rho_2}{2}\|dx^{(k)}\|^2, \frac{\rho_1^*}{2}\|dy^{(k-1)}\|^2 + \frac{\rho_2^*}{2}\|dy^{(k)}\|^2\}.$$

*The equality $(g-h)(x^{(k+1)}) = (g-h)(x^{(k)})$ holds if and only if $x^{(k)} \in \partial g^*(y^{(k)})$, $y^{(k)} \in \partial h(x^{(k+1)})$and $(\rho_1+\rho_2)dx^{(k)} = \rho_1^* dy^{(k-1)} = \rho_2^* dy^{(k)} = 0$. In this case*

- *$(g-h)(x^{(k+1)}) = (h^*-g^*)(y^{(k)})$ and $x^{(k)}, x^{(k+1)}$ are the critical points of $g-h$ satisfying*
*$y^{(k)} \in (\partial g(x^{(k)}) \cap \partial h(x^{(k)}))$ and $y^{(k)} \in (\partial g(x^{(k+1)}) \cap \partial h(x^{(k+1)}))$,*
- *$y^{(k)}$ is a critical point of $h^*-g^*$ satisfying $[x^{(k)}, x^{(k+1)}] \subset ((\partial g^*(y^{(k)}) \cap \partial h^*(y^{(k)}))$,*
- *$x^{(k+1)} = x^{(k)}$ if $\rho(g)+\rho(h) > 0, y^{(k)} = y^{(k-1)}$ if $\rho(g^*) > 0$ and $y^{(k)} = y^{(k+1)}$ if $\rho(h^*) > 0$.*

*(ii) Similarly, for the dual problem we have*

$$(h^*-g^*)(y^{(k+1)}) \leq (g-h)(x^{(k+1)}) - \max\{\frac{\rho_1}{2}\|dx^{(k+1)}\|^2, \frac{\rho_1^*}{2}\|dy^{(k)}\|^2\}$$

$$\leq (h^*-g^*)(y^{(k)}) - \max\{\frac{\rho_1}{2}\|dx^{(k+1)}\|^2 + \frac{\rho_2}{2}\|dx^{(k)}\|^2,$$

$$\frac{\rho_1^*}{2}\|dy^{(k)}\|^2 + \frac{\rho_2}{2}\|dx^{(k)}\|^2, \frac{\rho_1^*+\rho_2^*}{2}\|dy^{(k)}\|^2\}.$$

*The equality* $(h^* - g^*)(y^{(k+1)}) = (h^* - g^*)(y^{(k)})$ *holds if and only if* $x^{(k+1)} \in \partial g^*(y^{(k+1)})$, $y^{(k)} \in \partial h(x^{(k+1)})$ *and* $(\rho_1^* + \rho_2^*)dy^{(k)} = \rho_2 dx^{(k)} = \rho_1 dx^{(k+1)} = 0$. *In this case*

- $(h^* - g^*)(y^{(k+1)}) = (g - h)(x^{(k+1)})$ *and* $y^{(k)}, y^{(k+1)}$ *are the critical points of* $h^* - g^*$ *satisfying* $x^{(k+1)} \in (\partial g^*(y^{(k)}) \cap \partial h^*(y^{(k)}))$ *and* $x^{(k+1)} \in (\partial g^*(y^{(k+1)}) \cap \partial h^*(y^{(k+1)}))$,
- $x^{(k+1)}$ *is a critical point of* $g - h$ *satisfying* $[y^{(k)}, y^{(k+1)}] \subset ((\partial g(x^{(k+1)}) \cap \partial h(x^{(k+1)}))$,
- $y^{(k+1)} = y^{(k)}$ *if* $\rho(g^*) + \rho(h^*) > 0, x^{(k+1)} = x^{(k)}$ *if* $\rho(h) > 0$ *and* $x^{(k+1)} = x^{(k+2)}$ *if* $\rho(g) > 0$.

*(iii) If* $\alpha$ *is finite, then*

- *the decreasing sequences* $\{(g - h)(x^{(k)})\}$ *and* $\{(h^* - g^*)(y^{(k)})\}$ *converge to the same limit* $\beta \geq \alpha$, *i.e.,* $\lim_{k \to +\infty}(g - h)(x^{(k)}) = \lim_{k \to +\infty}(h^* - g^*)(y^{(k)}) = \beta$.

- *If* $\rho(g) + \rho(h) > 0$ *(resp.* $\rho(g^*) + \rho(h^*) > 0$*), then the series* $\{\| x^{(i+1)} - x^{(i)} \|^2\}$ *(resp.* $\{\| y^{(i+1)} - y^{(i)} \|^2\}$*) converges. More precisely we have in this case*

$$\frac{\rho_1 + \rho_2}{2} \sum_{i=0}^{+\infty} \| x^{(i+1)} - x^{(i)} \| \leq (g - h)(x^{(0)}) - \alpha$$

$$(resp. \ \frac{\rho_1^* + \rho_2^*}{2} \sum_{i=0}^{+\infty} \| y^{(i+1)} - y^{(i)} \| \leq (h^* - g^*)y^{(0)}) - \alpha).$$

*(iv) If* $\alpha$ *is finite and the sequences* $\{x^{(k)}\}$ *and* $\{y^{(k)}\}$ *are bounded, then for every limit point* $x^*$ *of* $\{x^{(k)}\}$ *(resp.* $y^*$ *of* $\{y^{(k)}\}$*) there exists a limit point* $y^*$ *of* $\{y^{(k)}\}$ *(resp.* $x^*$ *of* $\{x^{(k)}\}$*) such that*

- $(x^*, y^*) \in [\partial g^*(y^*) \cap \partial h^*(y^*)] \times [\partial g(x^*) \cap \partial h(x^*)]$ *and* $(g - h)(x^*) = (h^* - g^*)(y^*) = \beta$,

- $\lim_{k \to +\infty}\{g(x^{(k)}) + g^*(y^{(k)})\} = \lim_{k \to +\infty}\langle x^{(k)}, y^{(k)} \rangle$.

**Comments on Theorem 2.**

(i) Properties (i) and (ii) prove that DCA is a descent method for both primal and dual programs. DCA provides critical points for (P) and (D) after finitely many operations if there is no strict decrease of the primal (or dual) objective function.

(ii) If $C$ and $D$ are convex sets such that $\{x^{(k)}\} \subset C$ and $\{y^{(k)}\} \subset D$, then Theorem 2 remains valid if we replace $\rho(f_i)$ by $\rho(f_i, C)$ and $\rho(f_i^*)$ by $\rho(f_i^*, D)$ for $i = 1, 2$ . By this way we may improve the results in the theorem.

(iii) In (ii) of Theorem 2, the convergence of the whole sequence $\{x^{(k)}\}$ (resp. $\{y^{(k)}\}$) can be ensured under the following conditions ([38], [42]):

- $\{x^{(k)}\}$ is bounded;
- The set of limit points of $\{x^{(k)}\}$ is finite;
- $\lim_{k\to+\infty} \|x^{(k+1)} - x^{(k)}\| = 0$.

(iv) The only difference between the simplified DCA and the complete DCA lies on the choice of $y^{(k)}$ in $\partial h(x^{(k)})$ and $x^{(k+1)}$ in $\partial g^*(y^{(k)})$. The convergence result of the complete DCA is thus improved: in Theorem 2, the nonemptiness of a subdifferential intersection is replaced by a subdifferential inclusion ([2], [47], [50]). In other words, the complete DCA permits to obtain a couple of elements $(x^*, y^*) \in \mathcal{P}_l \times \mathcal{D}_l$. In practice the simplified DCA usually yields a local minimizer which is also global ([2]).

(v) In general, the qualities (robustness, stability, rate of convergence and globality of sought solutions) of DCA, in both complete and simplified forms, depend upon the d.c. decomposition of the function $f$. Theorem 2 shows how strong convexity of d.c. components in primal and dual problems can influence on DCA. To make the d.c. components (of the primal objective function $f = g - h$) strongly convex we usually apply the following process

$$f = g - h = (g + \frac{\mu}{2}\|\cdot\|^2) - (h + \frac{\mu}{2}\|\cdot\|^2).$$

In this case the d.c. components in the dual problem will be differentiable.
In the same way, the inf-convolution of $g$ and $h$ with $\frac{\mu}{2}\|\cdot\|^2$ will make the d.c. components (in the dual problem) strongly convex and the d.c. components of the primal objective function differentiable. For a detailed study of regularization techniques in d.c. programming, see [2], [50].

(vi) It has been proved in [2], [4], [50] that DCA is finite for polyhedral d.c. programming.

### 2.3. How to restart simplified DCA for obtaining $x^*$ such that $\partial h(x^*) \subset \partial g(x^*)$.

As mentioned above, the complete DCA theoretically provides a $x^*$ such that $\partial h(x^*) \subset \partial g(x^*)$. In practice except the cases where the convex maximization problems $(S(x^{(k)})$ and $(T(y^{(k)}))$ are easy to solve, one generally uses the simplified DCA. It is worth noting that if the simplified DCA terminates at some point $x^*$ for which $\partial h(x^*)$ is not contained in $\partial g(x^*)$ then one can reduce the objective function value by restarting it from a new initial point $x^{(0)} = x^*$ with $y^{(0)} \in \partial h(x^{(0)})$ such that $y^{(0)} \notin \partial g(x^{(0)})$. In fact, since

$$g(x^{(1)}) + g^*y^{(0)}) = \langle x^{(1)}, y^{(0)}\rangle \leq h(x^{(1)}) - h(x^{(0)}) + \langle x^{(0)}, y^{(0)}\rangle.$$

and $\langle x^{(0)}, y^{(0)}\rangle < g(x^{(0)}) + g^*y^{(0)})$ because $y^{(0)} \notin \partial g(x^{(0)})$, we have

$$g(x^{(1)}) + g^*y^{(0)}) < h(x^{(1)}) - h(x^{(0)}) + g(x^{(0)}) + g^*y^{(0)}).$$

Hence,

$$g(x^{(1)}) - h(x^{(1)}) < g(x^{(0)}) - h(x^{(0)}).$$

## 3. Lagrangian duality with zero gap in d.c. optimization.

Problems ($S_{max}$) and ($S_1$) (stated in the Introduction) are d.c. programs which are formulated with the help of semi-norms. This feature can be exploited to yield a zero gap for the Lagrangian duality relative to these problems.

Now let $\psi \not\equiv 0$ and $\phi \not\equiv 0$ be two finite gauges on $E$ such that $M = \phi^{-1}(0)$ is a subspace contained in $\psi^{-1}(0)$. Consider the problem

$$(\mathrm{P_{max}}) \qquad S_{\psi\phi}(I) = \sup\{\psi(x): \quad \phi(x) \leq 1\}$$

and its reduced form

$$(\mathrm{PR_{max}}) \qquad \sup\{\psi(x): \quad x \in M^\perp,\ \phi(x) \leq 1\},$$

where $M^\perp$ denotes the orthogonal complement of $M$. Let $\mathcal{P}_{max}$ and $\mathcal{PR}_{max}$ be the solution set of ($P_{max}$) and ($PR_{max}$) respectively. We will first state the following result which is useful for the solution of ($\mathrm{P}_{\max}$) by DCA.

**Lemma 2** *$\mathcal{P}_{\max}$ is nonempty and $\mathcal{P}_{\max} = M + \mathcal{PR}_{\max}$.*

**Proof:** Since $E = M + M^{\perp}$, there is, for every $x \in M$, one and only one pair $(u, v) \in M \times M^{\perp}$ such that $x = u + v$. So

$$\psi(x) = \psi(u + v) \leq \psi(u) + \psi(v) = \psi(v),$$

and

$$\phi(x) = \phi(u + v) \leq \phi(u) + \phi(v) = \phi(v).$$

On the other hand we can write

$$\psi(v) = \psi(x - u) \leq \psi(x) + \psi(-u) = \psi(x),$$

$$\phi(v) = \phi(x - u) \leq \phi(x) + \phi(-u) = \phi(x).$$

Consequently $\phi(x) = \phi(u + v) = \phi(v)$ and $\psi(x) = \psi(u + v) = \psi(v)$. It then implies $\mathcal{P}_{\max} = M + \mathcal{PR}_{\max}$. The nonemptiness of $\mathcal{P}_{\max}$ is immediate from the compactness of $\{x \in M^{\perp} : \phi(x) \leq 1\}$. □

By writing the feasible domain as $\{x \in E : (1/2)\phi^2(x) \leq (1/2)\}$, ($P_{\max}$) has the form

$$(\mathrm{P}_{\max}) \qquad \zeta = -S_{\psi\phi}(I) = \inf\{-\psi(x) : \quad \frac{1}{2}\phi^2(x) \leq \frac{1}{2}\}.$$

The Lagrangian $L(x, \lambda)$ for ($\mathrm{P}_{\max}$) is then

$$L(x, \lambda) = \begin{cases} -\psi(x) + \frac{\lambda}{2}(\phi^2(x) - 1) & \text{if } \lambda \geq 0, \\ -\infty & \text{otherwise.} \end{cases}$$

Clearly, $-\psi(x) + \chi_C(x) = \sup\{L(x, \lambda) : \lambda \geq 0\}$. Thus ($\mathrm{P}_{\max}$) can be written as

$$\zeta = -S_{\psi\phi}(I) = \inf\{\sup\{L(x, \lambda) : \lambda \geq 0\} : x \in E\}.$$

For $\lambda \geq 0$ we have:

$$(P_\lambda) \qquad \gamma(\lambda) = \inf\{-\psi(x) + \frac{\lambda}{2}(\phi^2(x) - 1) : x \in E\}.$$

Obviously $\gamma$ is concave function and ($P_\lambda$) is a d.c. program.

The dual problem of ($P_{max}$) is

$$\text{(D)} \qquad \kappa = \sup\{\gamma(\lambda) : \lambda \geq 0\} = \sup\{\inf\{L(x,\lambda) : x \in E\} : \lambda \geq 0\}.$$

By definition of the Lagrangian we have

$$\zeta = \inf\{\sup\{L(x,\lambda) : \lambda \geq 0\} : x \in E\} \geq \tag{13}$$

$$\sup\{\inf\{L(x,\lambda) : x \in E\} : \lambda \geq 0\} = \kappa.$$

A point $(x^*, \lambda^*) \in E \times \mathbb{R}$ is said to be a saddle point of $L(x, \lambda)$ if

$$L(x^*, \lambda) \leq L(x^*, \lambda^*) \leq L(x, \lambda^*), \; \forall (x, \lambda) \in E \times \mathbb{R}.$$

Let us state now important results concerning the characterization of solutions of the dual problem and the stability of the Lagrangian duality.

**Theorem 3** *([2], [48]) (i) $\mathcal{P}_\lambda \neq \emptyset$ for every $\lambda > 0$ and dom $\gamma = ]0, +\infty[$.*
*(ii) $\mathcal{D} \neq \emptyset$ and $\gamma(\lambda) = \frac{-\lambda}{2} + \frac{K}{\lambda}$, where $K$ is a negative constant (dependent only on $\psi$ and $\phi$).*
*(iii) $\mathcal{D} = \{\lambda^*\} = \{\sqrt{-2K}\}, \;\; \gamma(\lambda^*) = -\lambda^*$.*
*(iv) $\zeta = \kappa$ and $\mathcal{P}_{max} = \mathcal{P}_{\lambda^*}$.*
*(v) $(x^*, \lambda^*) \in \mathcal{P}_{max} \times \mathcal{D}$ if and only if $(x^*, \lambda^*)$ is a saddle point of $L(x, \lambda)$.*

Consider now Problem ($P_\lambda$) with $\lambda = 1$:

$$(P_1) \qquad \gamma(1) + \frac{1}{2} = \inf\{\frac{1}{2}\phi^2(x) - \psi(x) : x \in E\}.$$

The following result can be used to determine a new d.c. optimization problem equivalent to ($P_{max}$).

**Theorem 4** *([2], [48])*

$$\mathcal{P}_{max} = \left\{\frac{x^*}{\phi(x^*)} : x^* \in \mathcal{P}_1\right\}. \tag{14}$$

$$\mathcal{P}_1 = \{\psi(x^*)x^* : x^* \in \mathcal{P}_{max}\}. \tag{15}$$

**Well-definiteness of the DCA applied to ($P_{max}$) and ($P_\lambda$).**

By using Lemma 1, it can be proved that DCA for ($P_{\max}$) and ($P_\lambda$) is well-defined ([2]). Also the following consequences of Theorems 3 and 4 are worth mentioning:

**Remark 6** *(i) Solving ($P_{\max}$) amounts to solving the d.c. program*

$$(P_\lambda) \qquad \gamma(\lambda) = \inf\{\frac{\lambda}{2}(\phi^2(x) - 1) - \psi(x) : x \in E\}.$$

*In fact one can solve ($P_{\max}$) by first solving ($P_\lambda$) with some $\lambda > 0$ for obtaining $K = \lambda(\frac{\lambda}{2} + \gamma(\lambda))$ and the solution $\lambda^*$ of dual problem (D). Then a solution of ($P_{\max}$) can be computed by solving ($P_{\lambda^*}$). We can also, by Theorem 4, solve ($P_{\max}$) by solving ($P_1$).*
*(ii) The form $\gamma(\lambda) = \frac{-\lambda}{2} + \frac{K}{\lambda}$ permits checking globality of solutions by comparing $\gamma(\lambda)$ with the approximate function $\tilde{\gamma}$ computed by DCA (see Section 6).*

## 4. Solving the Metric MDS Problem by DCA.

With the preceding background we now show how to solve the Euclidean metric MDS problem by DCA.

### 4.1. General Assumptions

As mentioned in Section 1, the dissimilarity matrix $\Delta$ may not be completely defined (i.e. some dissimilarities $\delta_{ij}$ may be unknown), but the two symmetric matrices $\Delta = (\delta_{ij})$ and $W = (w_{ij})$ must satisfy the following additional conditions:

(i) For $i \neq j$, $w_{ij} = 0$ if and only if $\delta_{ij}$ is unknown,

(ii) The weight matrix $W$ is irreducible, i.e., the associated graph of $W$, $G(N, \mathcal{S})$ with $N = \{1, ..., n\}$ and $\mathcal{S} = \{(i,j) \in N \times N : w_{ij} \neq 0\}$ is connected.

Under these assumptions we have:

- If $\delta_{ij} > 0$ for known $\delta_{ij}$ with $i \neq j$, and if we set $\delta_{ij} = 0$ otherwise, then $G(N, \mathcal{S})$ is the graph associated with $\Delta$ too.
- The normal case ($w_{ij} = 1$ for $i \neq j$) can occur if and only if the dissimilarity matrix $\Delta$ is completely defined.

Let $\mathcal{A}$ denote the set of matrices in $\mathcal{M}_{n,p}(\mathbb{R})$ whose rows are all identical, i.e.,

$$\mathcal{A} := \{X \in \mathcal{M}_{n,p}(\mathbb{R}) : \mathrm{X}_1 = \cdots = \mathrm{X}_\mathrm{n}\}.$$

Let $P_\mathcal{A}$ be the orthogonal projection on $\mathcal{A}$. Recall that the seminorms $\xi$ and $\eta$ are defined from (6).

**Lemma 3** *(i)* $\mathcal{A} = \{ev^T : \quad v \in \mathbb{R}^\mathrm{p}\}$ *is a* $p-$ *dimensional subspace of* $\mathcal{M}_{n,p}(\mathbb{R})$ *and*
$\mathcal{A}^\perp = \{Y \in \mathcal{M}_{n,p}(\mathbb{R}) : \sum_{\mathrm{i}=1}^{\mathrm{n}} \mathrm{Y}_\mathrm{i} = 0\}$.
*(ii)*$\mathcal{A} \subset \xi^{-1}(0); \quad \mathcal{A} \subset \eta^{-1}(0)$.
*(iii)*$P_\mathcal{A} = (1/n)ee^T; \quad P_{\mathcal{A}^\perp} = I - (1/n)ee^T$ *(e is the vector of ones in* $\mathbb{R}^\mathrm{n}$*).*
*(iv) If the weight matrix* $W$ *is irreducible (resp.* $W$ *is irreducible and* $w_{ij}\delta_{ij} > 0$ *whenever* $w_{ij} > 0$*), then* $\mathcal{A} = \eta^{-1}(0)$ *(resp.* $\mathcal{A} = \xi^{-1}(0)$*).*

**Proof.** (i) and (ii) are straightforward from the definition of $\mathcal{A}$. The proof of (iii) is easy.
Let $X \in \mathcal{M}_{n,p}(\mathbb{R})$ such that $\eta(X) = 0$ and $(i,j) \in \{1,\ldots,n\}^2$ with $i \neq j$. Since the matrix $W$ is irreducible there is a finite sequence $\{i_1,\ldots,i_r\} \subset \{1,\ldots,n\}$ verifying $w_{ii_1} > 0, w_{i_k i_{k+1}} > 0$ for $k = 1,\ldots,r-1$, and $w_{i_r j} > 0$. It follows that $X_i = X_{i_1} = \cdots = X_{i_r} = X_j$ and the proof is complete. □

The remaining of this section is devoted to the description of DCA applied to ($\mathrm{S}_1$) and ($\mathrm{S}_{\max}$) for solving the metric MDS problem with $\theta(t) = t^2$ and $\varphi$ being the Euclidean norm. Problems ($\mathrm{S}_1$ ) and ($\mathrm{S}_{\max}$) then take the forms ($\mathrm{P}_1$) and ($\mathrm{P}_{\max}$) where the seminorms $\xi$ and $\eta$ replace the gauges $\psi$ and $\phi$ respectively (Section 3).

According to the Lagrangian duality with zero gap stated in the last section (Theorem 4), solving Problem ($\mathrm{P}_1$) amounts to solving ($\mathrm{P}_{\max}$) which is a d.c. program too.

$$(\mathrm{P}_{\max}) \quad \max\{\xi(X) : \eta(X) \leq 1\} \Leftrightarrow \min\{\chi_C(X) - \xi(X) : X \in \mathcal{M}_{n,p}(\mathbb{R})\}, \tag{16}$$

where $C = \{X \in \mathcal{M}_{n,p}(\mathbb{R}) : (1/2)\eta^2(\mathrm{X}) \leq 1/2\}$.

Now we describe DCA (with and without regularization) for solving Problems ($\mathrm{P}_1$) and ($\mathrm{P}_{\max}$). First, observe that for both ($\mathrm{P}_1$) and ($\mathrm{P}_{\max}$) *we can restrict ourselves to* $\mathcal{A}^\perp$, where each of these problems has always a solution (Lemma 2). The scheme of DCA applied to ($\mathrm{P}_1$) and ($\mathrm{P}_{\max}$) can be described as follows:

Let $X^{(0)} \in \mathcal{A}^{\perp}\backslash\{0\}$ be arbitrary,

$$\text{DCAP}_1 \qquad X^{(k)} \longrightarrow Y^{(k)} \in \partial\xi(X^{(k)}) \longrightarrow X^{(k+1)} \in \partial\left(\frac{1}{2}\eta^2\right)^*(Y^{(k)})$$

$$\text{RDCAP}_1 \qquad X^{(k)} \longrightarrow Y^{(k)} \in \partial\left(\lambda\xi + \frac{\mu}{2}\|.\|^2\right)(X^{(k)}) \longrightarrow$$
$$X^{(k+1)} \in \partial\left(\frac{\lambda}{2}\eta^2 + \frac{\mu}{2}\|.\|^2\right)^*(Y^{(k)})$$

$$\text{DCAP}_{\max} \qquad X^{(k)} \longrightarrow Y^{(k)} \in \partial\xi(X^{(k)}) \longrightarrow X^{(k+1)} \in \partial\chi_C^*(Y^{(k)})$$

$$\text{RDCAP}_{\max} \qquad X^{(k)} \longrightarrow Y^{(k)} \in \partial\left(\lambda\xi + \frac{\mu}{2}\|.\|^2\right)(X^{(k)}) \longrightarrow$$
$$X^{(k+1)} \in \partial\left(\lambda\chi_C + \frac{\mu}{2}\|.\|^2\right)^*(Y^{(k)})$$

where *the sequences* $\{X^{(k)}\}$ *and* $\{Y^{(k)}\}$ *must be included in* $\mathcal{A}^{\perp}$. Performing this scheme thus reduces to calculating subdifferentials of the convex functions $\xi, ((1/2)\eta^2)^*$, $\chi_C^*, [(\lambda/2)\eta^2 + (\mu/2)\|.\|^2]^*$ and $[\lambda\chi_C + (\mu/2)\|.\|^2]^*$.

We will first calculate these subdifferentials and then specialize the results to the normal case ($w_{ij} = 1, \forall i \neq j$) with $\varphi$ being the Euclidean norm. These calculations are crucial (thus deserving a great attention) since they lead to the simplest expressions of the sequences $X^{(k)}$ and $Y^{(k)}$. As will be seen in Subsection 4.7, computing $X^{(k)}$ and $Y^{(k)}$ (in the normal Euclidean metric MDS) requires matrix-vector products only. It is important to note that all these results are due to a thorough study of (MDS) problem in its very convenient matrix formulation.

## 4.2. Calculation of $\partial\xi$.

By the definition

$$\xi(X) = \sum_{i<j} w_{ij}\delta_{ij}d_{ij}(X) \Rightarrow \partial\xi(X) = \sum_{i<j} w_{ij}\delta_{ij}\partial d_{ij}(X).$$

Using the row-representation of $\mathcal{M}_{n,p}(\mathbb{R})$ $d_{ij}$ can be expressed as:

$$d_{ij} = \varphi \circ Lij : (\mathbb{R}^p)^n \longrightarrow \mathbb{R}^p \longrightarrow \mathbb{R}$$

$$X \longmapsto L_{ij}(X) = X_i^T - X_j^T \longmapsto \varphi(X_i^T - X_j^T)$$

It follows that ([53])

$$\partial d_{ij}(X) = Lij^T\partial\varphi(L_{ij}(X)).$$

Hence

$$Y(i,j) \in \partial d_{ij}(X) \Leftrightarrow Y(i,j) = L_{ij}^T y, \quad y \in \partial\varphi(X_i^T - X_j^T)$$

which implies

$$Y(i,j)_k^T = 0 \text{ if } k \notin \{i,\, j\} \text{ and} Y(i,j)_i^T = -Y(i,j)_j^T \in \partial\varphi(X_i^T - X_j^T). \tag{17}$$

So $\xi$ is not differentiable on
$\Omega = \{X \in \mathcal{M}_{n,p}(\mathbb{R}) :$ X has two identical rows $\}$. Moreover if the norm $\varphi$ is differentiable on $\mathbb{R}^p \setminus \{0\}$, then $\xi$ is differentiable on the complement of $\Omega$ in $\mathcal{M}_{n,p}(\mathbb{R})$. If $\varphi$ is the Euclidean norm on $\mathbb{R}^p$, we can choose the subgradient $Y(i,j) \in \partial d_{ij}(X)$ defined by

$$Y(i,j)_i = -Y(i,j)_j = \begin{cases} \frac{X_i - X_j}{\|X_i^T - X_j^T\|} & \text{if } X_i \neq X_j, \\ 0 & \text{if } X_i = X_j. \end{cases} \tag{18}$$

In this case, each $Y \in \partial\xi(X)$ is called special subgradient of $\xi(X)$ and defined explicitly by

$$\begin{aligned} Y_k &= \sum_{i<j} w_{ij}\delta_{ij}Y(i,j)_k = \sum_{i<k} w_{ik}\delta_{ik}Y(i,k)_k + \sum_{j>k} w_{kj}\delta_{kj}Y(k,j)_k \\ &= \sum_{i<k} w_{ki}\delta_{ki}s_{ki}(X)(X_k - X_i) + \sum_{j>k} w_{kj}\delta_{kj}s_{kj}(X)(X_k - X_j) \\ &= \left[\sum_{i=1}^{n} w_{ki}\delta_{ki}s_{ki}(X)\right] X_k - \sum_{i=1}^{n} w_{ki}\delta_{ki}s_{ki}(X)X_i, \end{aligned}$$

where

$$s_{ij}(X) = \begin{cases} 1/(\|X_i^T - X_j^T\|) & \text{if } X_i \neq X_j \\ 0 & \text{orthewise.} \end{cases}$$

Let $B(X) = (b_{ij}(X))$ be the $n \times n$ matrix valued function defined by

$$b_{ij}(X) = \begin{cases} -w_{ij}\delta_{ij}s_{ij}(X) & \text{if } i \neq j \\ -\sum_{k=1,k\neq i}^{n} b_{ik} & \text{if } i = j. \end{cases} \tag{19}$$

It follows that $Y = B(X)X$. Throughout the sequel $\mathcal{N}(A)$ and Im$A$ denote the null space and the range of the matrix $A$ respectively.

**Proposition 1** *Let $\varphi$ be the Euclidean norm and $B(X)$ be the matrix defined by (19). Then*

*(i) $B(X)$ is symmetric positive semidefinite for all $X \in \mathcal{M}_{n,p}(\mathbb{R})$.*

*(ii) $\xi(X) = \langle X, B(X)X\rangle$.*

*(iii)* $\mathcal{N}(B(X)) \supset \mathcal{A}$, $\quad Im\,(B(X)) \subset \mathcal{A}^{\perp}$ *for all* $X \in \mathcal{M}_{n,p}(\mathrm{I\!R})$.

**Proof.** (i) $B(X)$ is symmetric, diagonally dominant whose diagonal entries are nonnegative. Thus it is positive semidefinite ([64]).
(ii) is the generalized Euler's relation for functions that are convex nondifferentiable and positively homogeneous of degree 1 ([53]).
(iii) is immediate from the fact that $\mathcal{A} = \{ev^T : v \in \mathrm{I\!R}^p\}$ and $B(X)e = 0$ for all $X \in \mathcal{M}_{n,p}(\mathrm{I\!R})$.

## 4.3. Calculation of $\partial\left((1/2)\eta^2\right)^*$.

We first state some results concerning the space $\mathcal{M}_{n,p}(\mathrm{I\!R})$ which will be needed for the calculation of $\partial((1/2)\eta^2)^*$.

**Proposition 2** *Let* $\Lambda$ *be an* $n \times n$ *real symmetric matrix,* $u \in \mathrm{I\!R}^n, u \neq 0$ *such that* $\Lambda u = 0$. *Let*

$$\mathcal{V} = \left\{\frac{1}{\|u\|^2} uu^T X : X \in \mathcal{M}_{n,p}(\mathrm{I\!R})\right\}, \mathcal{W} = \left\{(I - \frac{uu^T}{\|u\|^2})X : X \in \mathcal{M}_{n,p}(\mathrm{I\!R})\right\}.$$

*Then*

$$\mathcal{V} \subset \{X \in \mathcal{M}_{n,p}(\mathrm{I\!R}) : \Lambda X = 0\}, \tag{20}$$

$$\mathcal{V}^{\perp} = \mathcal{W}, \tag{21}$$

*and*

$$\{Y = \Lambda X : X \in \mathcal{M}_{n,p}(\mathrm{I\!R})\} = \{X \in \mathcal{M}_{n,p}(\mathrm{I\!R}) : \Lambda X = 0\}^{\perp} \subset \mathcal{W}, \tag{22}$$

$$\mathcal{V} = \{X \in \mathcal{M}_{n,p}(\mathrm{I\!R}) : \Lambda X = 0\} \text{ if and only if rank } \Lambda = n - 1. \tag{23}$$

**Proof.** • $\Lambda u = 0$ implies $\Lambda(\frac{1}{\|u\|^2} uu^T A) = 0, \quad \forall A \in \mathcal{M}_{n,p}(\mathrm{I\!R})$. Hence (20).
• Let $X \in \mathcal{V}, Y \in \mathcal{W} : X = \frac{1}{\|u\|^2} uu^T A,\ Y = (I - \frac{1}{\|u\|^2} uu^T)B,\ A, B \in \mathcal{M}_{n,p}(\mathrm{I\!R})$.
¿From $uu^T(I - \frac{1}{\|u\|^2} uu^T) = 0$ and the column-representation (2) follows

$$\langle X, Y\rangle = \sum_{k=1}^{p} (X^k)^T Y^k = \sum_{k=1}^{p} \langle \frac{uu^T}{\|u^2\|} A^k, (I - \frac{1}{\|u\|^2} uu^T)B^k\rangle = 0.$$

Thus the subspaces $\mathcal{V}$ and $\mathcal{W}$ are orthogonal. This is equivalent to (21) since $\mathcal{M}_{n,p}(\mathbb{R}) = \mathcal{V} + \mathcal{W}$.

• Denote by $\tilde{\Lambda}$ the symmetric matrix having $p$ diagonal blocks and each block is $\Lambda$. It is evident that $\mathcal{N}(\tilde{\Lambda})^\perp = Im(\tilde{\Lambda})$. Note that from column-identification (2) we have:

$$\tilde{\Lambda}\mathcal{X} = 0 \Leftrightarrow \Lambda X^k = 0\,, \quad k = 1, \ldots, p \quad \Leftrightarrow \Lambda X = 0, \tag{24}$$

$$\mathcal{Y} = \tilde{\Lambda}\mathcal{X} \Leftrightarrow Y^k = \Lambda X^k\,, \quad k = 1, \ldots, p \quad \Leftrightarrow Y = \Lambda X. \tag{25}$$

It follows that $\{Y = \Lambda X : X \in \mathcal{M}_{n,p}(\mathbb{R})\} = \{\mathrm{X} \in \mathcal{M}_{\mathrm{n,p}}(\mathbb{R}) : \Lambda\mathrm{X} = 0\}^\perp$. On the other hand, (20) and (21) imply

$$\{X \in \mathcal{M}_{n,p}(\mathbb{R}) : \Lambda \mathrm{X} = 0\}^\perp \subset \mathcal{V}^\perp = \mathcal{W}.$$

Hence (22) holds.

• If rank $\Lambda < n - 1$, there exists $v \notin \mathbb{R}u$ such that $\Lambda v = 0$. Let $X = v(e^p)^T$ ($e^p \in \mathbb{R}^{\mathrm{p}}$ whose every entry is 1). It is clear that $\Lambda X = 0$ and $X \notin \mathcal{V}$ (since, by definition of $\mathcal{V}$, $X \in \mathcal{V}$ implies $v \in \mathbb{R}u$). Conversely, if rank $\Lambda = n - 1$ then $\mathcal{N}(\Lambda) = \mathbb{R}u$ which implies $\Lambda X = 0 \Leftrightarrow \Lambda X^k = 0$, $k = 1, \ldots, p$, $\quad \Leftrightarrow X^k = \lambda_k u$, $(\lambda_k \in \mathbb{R}) \Leftrightarrow \mathrm{X} = \mathrm{u}\lambda^{\mathrm{T}}, \lambda = (\lambda_1, \ldots, \lambda_{\mathrm{p}})$. The proof is then complete according to

**Lemma 4** *Let $u$ be a nonnull vector in $\mathbb{R}^{\mathrm{n}}$. Then the subspace $\mathcal{V} = \{uu^T X : X \in \mathcal{M}_{n,p}(\mathbb{R})\}$ is equal to the $p-$ dimensional subspace $\{ux^T : x \in \mathbb{R}^{\mathrm{p}}\}$.*

**Proof.** Clearly the latter is a $p-$dimensional subspace which contains the former. So it remains to prove that for every $x \in \mathbb{R}^{\mathrm{p}}$ there exists $X \in \mathcal{M}_{n,p}(\mathbb{R})$ such that $x^T = u^T X, i.e., x^T = \sum_{i=1}^n u_i X_i$. Let $k$ be an index such that $u_k \neq 0$ and consider $X \in \mathcal{M}_{n,p}(\mathbb{R})$ defined by

$$X_k = \frac{1}{u_k}\left[x^T - \sum_{i=1,\, i\neq k}^{n} u_i X_i\right]$$

with $X_i$, for $i \neq k$, arbitrary. Clearly $X$ satisfies the required condition. □

Let $\Lambda^+$ be the pseudo-inverse of $\Lambda$. The following result is useful for a suitable computation of $\Lambda^+$.

**Proposition 3** *Let $\Lambda$ be an $n \times n$ symmetric matrix and $u \in \mathbb{R}^n, u \neq 0$ such that $\Lambda u = 0$. Then $\Lambda + \frac{1}{\|u\|^2}uu^T$ is nonsingular if and only if rank $\Lambda = n-1$.*
*In this case for every $y \in \mathbb{R}^n$ there exists $x \in Im\ \Lambda$ satisfying $\Lambda x = P_{Im\ \Lambda}y$ and*

$$(\Lambda+\frac{1}{\|u\|^2}uu^T)(x+\frac{u^Ty}{\|u\|^2}u) = y \quad i.e., \Lambda^+ = (\Lambda+\frac{1}{\|u\|^2}uu^T)^{-1}-\frac{1}{\|u\|^2}uu^T.$$

**Proof.** If $(\Lambda + \frac{1}{\|u\|^2}uu^T)$ is nonsingular then $\mathrm{Im}\left(\Lambda + \frac{1}{\|u\|^2}uu^T\right) = \mathbb{R}^n$. Thus

$$\mathrm{Im}\,\Lambda + \mathrm{Im}\left(\frac{1}{\|u\|^2)}uu^T\right) = \mathbb{R}^n = \mathrm{Im}\,\Lambda + \mathcal{N}(\Lambda).$$

This implies that $\mathcal{N}(\Lambda) = \mathrm{Im}\,(\frac{1}{\|u\|^2}uu^T) = \mathbb{R}u$. Thus rank $\Lambda$ = n-1. Conversely, if rank $\Lambda = n-1$, then $\mathcal{N}(\Lambda) = \mathbb{R}u$. Let $x \in \mathbb{R}^n$ such that

$$(\Lambda + \frac{1}{\|u\|^2}uu^T)x = 0, \text{i.e.,}\ \Lambda x = -\frac{1}{\|u\|^2}uu^Tx.$$

This and $\Lambda x \in \mathrm{Im}\,\Lambda$, $-\frac{1}{\|u\|^2}uu^Tx \in \mathcal{N}(\Lambda)$ imply that $\Lambda x = 0$ and $u^Tx = 0$. Hence $x = 0$. Thus we can deduce that $\mathcal{N}(\Lambda + \frac{1}{\|u\|^2}uu^T) = \{0\}$, i.e., $\Lambda + \frac{1}{\|u\|^2}uu^T$ is nonsingular. In this case the projection on $\mathcal{N}(\Lambda) = \mathbb{R}u$ is given by $P_{\mathcal{N}(\Lambda)} = (1/\|u\|^2)uu^T$ and

$$P_{\mathrm{Im}\ \Lambda} = I - \frac{1}{\|u\|^2}uu^T. \tag{26}$$

Let $y$ be an arbitrary vector in $\mathbb{R}^n$. Since $P_{\mathrm{Im}\ \Lambda}(y)$ is an element in $\mathrm{Im}\ \Lambda$, there exists $\bar{x} \in \mathrm{R}^n$ such that $\Lambda\bar{x} = P_{\mathrm{Im}\ \Lambda}(y)$. The decomposition $\mathbb{R}^n = \mathrm{Im}\ \Lambda + \mathcal{N}(\Lambda)$ ensures the existence of $x \in \mathrm{Im}\ \Lambda, x_1 \in \mathcal{N}(\Lambda)$ such that $\bar{x} = x + x_1$ and $\Lambda\bar{x} = \Lambda x = P_{\mathrm{Im}\ \Lambda}(y)$. Observing $\Lambda u = 0, u^Tx = 0$ (since $x \in \mathrm{Im}\ \Lambda$) we have from (26)

$$(\Lambda+\frac{1}{\|u\|^2}uu^T)(x+\frac{u^Ty}{\|u\|^2}u) = \Lambda x+\frac{1}{\|u\|^2}uu^T\frac{u^Ty}{\|u\|^2}u = P_{\mathrm{Im}\ \Lambda}y+\frac{u^Ty}{\|u\|^2}u = y. \tag{27}$$

This implies

$$x = (\Lambda + \frac{1}{\|u\|^2}uu^T)^{-1}y - \frac{u^Ty}{\|u\|^2}u.$$

Therefore

$$\Lambda^+ = (\Lambda + \frac{1}{\|u\|^2}uu^T)^{-1} - \frac{1}{\|u\|^2}uu^T. \quad \square \tag{28}$$

Now we state some fundamental properties of the function $(1/2)\eta^2$. We have

$$\frac{1}{2}\eta^2(X) = \frac{1}{2}\sum_{i<j} w_{ij}d_{ij}^2(X) \Rightarrow \partial(\frac{1}{2}\eta^2)(X) = \sum_{i<j} w_{ij}d_{ij}(X)\partial d_{ij}(X).$$

Thus

$$Y \in \partial\left(\frac{1}{2}\eta^2\right)(X) \Leftrightarrow Y = \sum_{i<j} w_{ij}d_{ij}(X)Y(i,j)$$

with $Y(i,j)$ defined by (17). It follows that $\eta^2$ is differentiable on $\mathcal{M}_{n,p}(\mathbb{R})$ if the norm $\varphi$ is differentiable on $\mathbb{R}^p \setminus \{0\}$. In particular, if $\varphi$ is the Euclidean norm on $\mathbb{R}^p$, then $Y = \nabla(\frac{1}{2}\eta^2)(X)$ is defined as

$$Y_k = \sum_{i<k} w_{ki}(X_k - X_i) + \sum_{j>k} w_{kj}(X_k - X_j) = (\sum_{i=1}^{n} w_{ki})X_k - \sum_{i=1}^{n} w_{ki}X_i.$$

Therefore $Y = VX$, with $V = (v_{ij})$ given by

$$v_{ij} = \begin{cases} -w_{ij} & \text{if } i \neq j, \\ \sum_{k=1}^{n} w_{ik} & \text{if } i = j. \end{cases} \tag{29}$$

Just as Proposition 1 holds for the function $\xi$, one has

**Proposition 4** *Let $\varphi$ be the Euclidean norm and $V$ be defined by (29). Then*

*(i) $V$ is positive semidefinite, $\nabla(\frac{1}{2})\eta^2(X) = VX$ and $\eta^2(X) = \langle X, VX\rangle$.*

*(ii) If the weight matrix $W$ is irreducible (as defined in Subsection 4.1), then $\mathcal{A} = \eta^{-1}(0) = \{X \in \mathcal{M}_{n,p}(\mathbb{R}) : \mathrm{VX} = 0\}$, rank $\mathrm{V} = \mathrm{n} - 1$ and*

$$\mathcal{A}^\perp = \{Y = VX : X \in \mathcal{M}_{n,p}(\mathbb{R})\} = \{\mathrm{Y} = \mathrm{V}^+\mathrm{X} : \mathrm{X} \in \mathcal{M}_{\mathrm{n,p}}(\mathbb{R})\}.$$

*Moreover the restriction of the seminorm $\eta$ to $\mathcal{A}^\perp$ is a quadratic norm. In the normal case $V = nI - ee^T$ and $V^+Y = Y/n$ for all $Y \in \mathcal{A}^\perp$. More precisely, $V^+ = V/n^2$ in this case.*

*(iii) $(\frac{1}{2}\eta^2)^*(Y) = \frac{1}{2}\langle Y, V^+Y\rangle$ if $Y \in \mathcal{A}^\perp$, $+\infty$ otherwise.*

**Proof.** (i) The positive semidefiniteness of $V$ is derived from [64] as in Proposition 1. Since $\nabla(\frac{1}{2}\eta^2)(X) = VX$ then from Euler's relation

follows $\eta^2(X) = \langle X, VX \rangle$.
(ii) In virtue of Proposition 3 we have, for $Y \in \mathcal{M}_{n,p}(\mathbb{R})$,

$$V^+Y = \left(V + \frac{1}{n}ee^T\right)^{-1} Y - \frac{1}{n}ee^T Y.$$

This implies for $Y \in \mathcal{A}^\perp$

$$X = V^+Y = (V + \frac{1}{n}ee^T)^{-1}Y, \tag{30}$$

i.e.,

$$(V + \frac{1}{n}ee^T)X = Y. \tag{31}$$

In particular, we have in the normal case ($V = nI - ee^T$): $nX = Y$, i.e., $X = Y/n$. The rest is immediate from Lemma 3 and Proposition 2 with $u = e$ and (24), (25).
(iii) It is well-known that if $A$ is an $n \times n$ real symmetric and positive semidefinite matrix, and $f(x) = \frac{1}{2}\langle x, Ax \rangle$, then

$$f^*(y) = \begin{cases} \frac{1}{2}\langle y, A^+y \rangle & \text{if } y \in \text{Im } A, \\ +\infty & \text{otherwise.} \end{cases}$$

On the other hand, by using the column-representation (2) it is clear from (24) and (25) that $V$ is positive semidefinite if and only if $\tilde{V}$ is positive semidefinite. Moreover the pseudo-inverse of $\tilde{V}$ is the transformed of $V^+$ by $\sim$: $(\tilde{V})^+ = \tilde{V}^+$. This completes the proof. □

It follows that $(\frac{1}{2}\eta^2)^*(Y) = \frac{1}{2}\langle V^+Y, Y \rangle + \chi_{\mathcal{A}^\perp}(Y)$ for $Y \in \mathcal{M}_{n,p}(\mathbb{R})$. Since $V^+$ is symmetric positive semidefinite, we have

$$\partial(\frac{1}{2}\eta^2)^*(Y) = V^+Y + \mathcal{A} \text{ for } Y \in \mathcal{A}^\perp.$$

## 4.4. Calculation of $\partial\chi_C^*$.

Recall that $C := \{X \in \mathcal{M}_{n,p}(\mathbb{R}) : \eta(\mathrm{X}) \leq 1\}$. Since the recession cone of $C$ is the subspace $\eta^{-1}(0) = \mathcal{A}$, the lineality space of $C$ is $\mathcal{A}$ too. Hence ([53]) $C = (C \cap \mathcal{A}^\perp) + \mathcal{A}$. We deduce that

$$\chi_C^*(Y) = \chi_{C\cap\mathcal{A}^\perp}^*(Y) \text{ if } Y \in \mathcal{A}^\perp, +\infty \text{ otherwise.}$$

But we have

$$\partial\chi_C^*(Y) = \{\overline{X} \in C : \sup_{X \in C}\langle X, Y \rangle = \langle \overline{X}, Y \rangle\} \tag{32}$$

and $C \cap \mathcal{A}^\perp$ is the unit ball of the quadratic norm $\eta$ on $\mathcal{A}^\perp$ (Lemma 3 and Proposition 4), then $\partial\chi_C^*(Y) \neq \emptyset$ if and only if $Y \in \mathcal{A}^\perp$. It is clear that $\partial\chi_C^*(Y) = C$ if $Y = 0$. Let us compute now $\partial\chi_C^*(Y)$ for $Y \in \mathcal{A}^\perp \setminus \{0\}$. It follows from (32) that

$$\partial\chi_C^*(Y) = \{\overline{X} + \mathcal{A} : \overline{X} \in \mathcal{A}^\perp, \eta(\overline{X}) = 1, \text{ and } Y \in \partial\chi_C(\overline{X})\}.$$

On the other hand there holds ([53])

$$Y \in \partial\chi_C(\overline{X}) \Leftrightarrow \lambda Y = V\overline{X} \text{ with } \lambda > 0$$

i.e., $\overline{X} = \lambda V^+ Y$ with $\lambda = \frac{1}{\langle Y, V^+Y\rangle^{1/2}}$. Since $VV^+Y = Y$ for $Y \in \mathcal{A}^\perp$, $\lambda$ is equal to $\frac{1}{\eta(V^+Y)}$ too. Finally $\partial\chi_C^*$ takes the form

$$\partial\chi_C^*(Y) = \begin{cases} \emptyset & \text{if } Y \notin \mathcal{A}^\perp, \\ V^+Y/\langle Y, V^+Y\rangle^{1/2} + \mathcal{A} & \text{if } Y \in \mathcal{A}^\perp \setminus \{0\}, \\ C & \text{if } Y = 0. \end{cases} \tag{33}$$

## 4.5. Calculation of $\partial\,(\lambda\chi_C + (\mu/2)\|.\|^2)^*$.

We have ([2], [50])

$$\partial[\lambda\chi_C + (\mu/2)\|.\|^2]^*(Y) = \nabla[\lambda\chi_C^*]_\mu(Y) = \left(I + \frac{\lambda}{\mu}\partial\chi_C\right)^{-1}\left(\frac{Y}{\mu}\right) = P_C\left(\frac{Y}{\mu}\right)$$

The orthogonal projection $P_C$ is easily computed by using

**Lemma 5** *Let $C$ be a closed convex set in $\mathbb{R}^n$ such that its recession cone $0^+C$ is a subspace $U$. Then*

$$P_C(y) = P_U(y) + P_{C\cap U^\perp}(P_{U^\perp}(y)) \quad \textit{for all } y \in \mathbb{R}^n.$$

**Proof.** Indeed, in this case the lineality space of $C$ is $U$: $0^+C \cap (-0^+C) = U$.
Hence, according to [53] we have $C = U + (C \cap U^\perp)$. Now let $y \in \mathbb{R}^n$ and $x \in C$ we can write $\|y - x\|^2 = \|y_U - x_U\|^2 + \|y_{U^\perp} - x_{U^\perp}\|^2$ where $y_U, y_{U^\perp}$ (resp. $x_U, x_{U^\perp}$) are the projections of $y$ (resp. $x$) on $U$ and on $U^\perp$. We have $x_{U^\perp} \in C \cap U^\perp$, so $z = P_C(y)$ is given by $z_U = y_U$ and $z_{U^\perp} = P_{C\cap U^\perp}(y_{U^\perp})$. □

In the Euclidean metric (MDS) problem where the closed convex set is $C = \{X \in \mathcal{M}_{n,p}(\mathbb{R}) : \eta(\mathrm{X}) \leq 1\}$, and $0^+C = \eta^{-1}(0) = \mathcal{A}$ we have, according to Lemma 5, $P_C(Y) = Y_{\mathcal{A}} + P_{C\cap\mathcal{A}^\perp}(Y_{\mathcal{A}^\perp}), \quad \forall Y \in \mathcal{M}_{n,p}(\mathbb{R})$. (Recall that $P_{\mathcal{A}} = \frac{1}{n}ee^T$ and $P_{\mathcal{A}^\perp} = I - \frac{1}{n}ee^T$ (Lemma 3)).

In the normal case with $\varphi$ being the Euclidean norm we have $V = nI - ee^T$, and $\eta^2(X) = \langle X, VX\rangle$. Hence

$$C\cap\mathcal{A}^\perp = \{X \in \mathcal{A}^\perp : \langle X, nX\rangle - \langle X, ee^TX\rangle \leq 1\} = \{X \in \mathcal{A}^\perp : \sqrt{n}\|X\| \leq 1\},$$

since $e^TX = 0$ for all $X \in \mathcal{A}^\perp$.
So, in this case, $P_C(Y)$ is explicitly given by

$$P_C(Y) = \begin{cases} Y & \text{if } \|Y_{\mathcal{A}^\perp}\| \leq 1/\sqrt{n}, \\ Y_{\mathcal{A}} + Y_{\mathcal{A}^\perp}/(\sqrt{n}\|Y_{\mathcal{A}^\perp}\|) & \text{ortherwise.} \end{cases}$$

In $\mathcal{A}^\perp$ this becomes

$$P_C(Y) = \begin{cases} Y & \text{if } \|Y\| \leq 1/\sqrt{n}, \\ Y/(\sqrt{n}\|Y\|) & \text{ortherwise.} \end{cases} \tag{34}$$

## 4.6. Calculation of $\partial\left((\lambda/2)\eta^2 + (\mu/2)\|.\|^2\right)^*$.

We have ([2], [50])

$$X \in \partial((\lambda/2)\eta^2 + (\mu/2)\|.\|^2)^*(Y) = \left[I + \frac{\lambda}{\mu}\partial\left(\frac{1}{2}\eta^2\right)\right]^{-1}\left(\frac{Y}{\mu}\right),$$

whence

$$X = \left[I + \frac{\lambda}{\mu}\partial\left(\frac{1}{2}\eta^2\right)\right]^{-1}\left(\frac{Y}{\mu}\right) \Leftrightarrow \frac{Y}{\mu} = \left[I + \frac{\lambda}{\mu}\partial\left(\frac{1}{2}\eta^2\right)\right](X).$$

If $\varphi$ is the Euclidean norm on $\mathbb{R}^p$, then using Proposition 4 we obtain

$$(\mu I + \lambda V)X = Y. \tag{35}$$

Indeed, Im $V \subset \mathcal{A}^\perp$ according to Lemma 3 and Proposition 4, it then follows that if $Y \in \mathcal{A}^\perp$, then the solution $X$ of (35) belongs to $\mathcal{A}^\perp$ too.

In the normal case (35) becomes

$$[\mu I + \lambda(nI - ee^T)]X = Y, \tag{36}$$

which implies

$$\mu e^TX + \lambda n e^TX - \lambda e^Tee^TX = e^TY = 0,$$

i.e. $e^T X = 0$. From (36) we have

$$X = (\mu + \lambda n)^{-1} Y. \tag{37}$$

To sum up, in the Euclidean case where the weight matrix $W = (w_{ij})$ is irreducible (verifying the assumptions in Subsection 4.1) we have the following properties:

(i) ($P_1$) is a d.c. optimization problem (Propositions 1, 3 and 4) of the form

$$(P_1) \qquad \min\left\{\frac{1}{2}\langle X, VX\rangle - \xi(X) : X \in \mathcal{M}_{n,p}(\mathbb{R})\right\}$$

with $\xi(X) = \langle X, B(X)X\rangle$. The quadratic form $\frac{1}{2}\langle X, VX\rangle$ is positive semidefinite on $\mathcal{M}_{n,p}(\mathbb{R})$ and positive definite on $\mathcal{A}^{\perp}$.

(ii) ($P_{\max}$) is the maximization of the seminorm $\xi$ on the unit ball of the quadratic seminorm $\eta$ which becomes a quadratic norm on $\mathcal{A}^{\perp}$.

Based upon the above calculations we are now in a position to describe an explicit form of DCA for solving Problems ($P_1$) and ($P_{\max}$) in the general case with $\varphi$ being the Euclidean norm of $\mathbb{R}^p$ (Subsection 4.1). In particular, in the normal case DCA takes the following simple form:

## 4.7. Algorithms.

Let $\varepsilon > 0$, $\lambda > 0$, $\mu > 0$ and $X^{(0)} \in \mathcal{A}^{\perp} \setminus \{0\}$ be given.

Let $k = 0, 1, \ldots$ until $\|X^{(k+1)} - X^{(k)}\| \leq \varepsilon$.

**DCAP$_1$:** *Take* $X^{(k+1)} = \frac{B(X^{(k)})X^{(k)}}{n}$.

**RDCAP$_1$:** *Take* $X^{(k+1)} = \frac{(\lambda B(X^{(k)}) + \mu I)X^{(k)}}{\mu + \lambda n}$.

**DCAP$_{\max}$:** *Take* $X^{(k+1)} = \frac{B(X^{(k)})X^{(k)}}{\eta(B(X^{(k)})X^{(k)})}$.

**RDCAP$_{\max}$:** *Take* $X^{(k+1)} = \begin{cases} \frac{(\lambda B(X^{(k)}) + \mu I)X^{(k)}}{\mu} & \text{if } \eta((\lambda B(X^{(k)}) + \mu I)X^{(k)}) \leq \mu, \\ \frac{(\lambda B(X^{(k)}) + \mu I)X^{(k)}}{\eta((\lambda B(X^{(k)}) + \mu I)X^{(k)})} & \text{otherwise.} \end{cases}$

Recall that in the normal case $\eta(Y) = \sqrt{n}\|Y\|$ in DCAP$_{\max}$ and RDCAP$_{\max}$.

**Remark 7** *(i) It should be noted that de Leeuw's algorithm for solving the Euclidean metric MDS problem is a special case of DCA applied to the d.c. program ($P_1$). Indeed DCAP$_1$ is given by (for a given $X^{(0)} \in \mathcal{A}^\perp \setminus \{0\}$)*

$$X^{(k+1)} = V^+ B(X^{(k)})X^{(k)} \in \mathcal{A}^\perp,$$

*where the computation of $X = V^+Y$ with $Y \in \mathcal{A}^\perp$ amounts to solving the nonsingular linear system $(V + (1/n)ee^T)X = Y$, according to Proposition 3 and (30). We thus recover de Leeuw's algorithm [12], [13]. Remark also that the seminorm $\xi$ is differentiable (and so is the loss function $\sigma$) on the set of local minimizers of ($P_1$), according to Theorem 1 (see also de Leeuw [12]). Moreover in virtue of Proposition 4 we have for $W$ irreducible $\rho(\eta^2) = 0 < \rho(\eta^2, \mathcal{A}^\perp)$. In particular $\rho(\eta^2, \mathcal{A}^\perp) = n$ in the normal case. It follows from Theorem 2 and its comments that $\lim_{k\to+\infty} \|X^{(k+1)} - X^{(k)}\| = 0$. This property is crucial for the majorization method by de Leeuw. The d.c. programming provides thus a new insight into the majorization method which becomes very easy to understand.*

*Note also that the proximal regularization applied to DCAP$_{\max}$ ensures the convergence of $\|X^{(k+1)} - X^{(k)}\|$ to zero (as k tends to infinity) in RCDAP$_{\max}$ (Theorem 2 and Section 3). Finally convergence results of these algorithms are ensured by Theorem 2 and regularization techniques.*

*(ii) In the general case (when some of dissimilarity data $\delta_{ij}$ are missing but the dissimilarity matrix $\Delta = (\delta_{ij})$ is irreducible as defined in Subsection 4.1), computing $X^{(k+1)}$ in DCAP$_1$ (resp. RDCAP$_1$) amounts to solving the linear system (31) (resp. (35)) whose matrix is symmetric positive definite. In our numerical experiments, DCA requires matrix-vector products and only one Cholesky factorization. Adapted existing methods are currently being investigated to efficiently solve these systems. On the other hand, the computation of $X^{(k+1)}$ in DCAP$_{\max}$ requires as in DCAP$_1$ the solution of the linear system (31), but its regularized version RDCAP$_{\max}$ is no longer explicit as in (35). We instead have to compute in $\mathcal{A}^\perp$ the projection over the unit ball $C \cap \mathcal{A}^\perp$ of the quadratic norm $\eta$ on $\mathcal{A}^\perp$:*

$$C \cap \mathcal{A}^\perp = \{X \in \mathcal{A}^\perp : \eta(X) = \langle X, VX\rangle \leq 1\}.$$

*(iii) In the general case of the Euclidean metric MDS problem, DCAP$_1$ requires solution of (31) with the matrix $V + \frac{1}{n}ee^T$ which might destroy the sparsity of $V$ while the matrix $\mu I + \lambda V$ of the linear system (35) whose solution is required by RDCAP$_1$ has the same sparsity as V. This interesting effect, among others, of regularization techniques must be*

*taken into account in the large scale setting with sparse weight matrices W (see Subsection 4.1).*

## 5. Normal Euclidean metric MDS as parametric trust region problem and solution by the parametric DCA.

Let us rewrite below the normal Euclidean metric MDS problem ($P_1$)

$$(\mathrm{P}_1) \quad \min\left\{\frac{1}{2}\eta^2(X) - \xi(X) : X \in \mathcal{M}_{n,p}(\mathbb{R})\right\}$$

and the equivalent form ($P_{\max}$):

$$(\mathrm{P}_{\max}) \quad \min\{-\xi(X) : \eta(X) \leq 1\}$$

where

- $\eta^2(X) = \langle VX, X\rangle$ with $V = nI - ee^T$, Im $V = \mathcal{A}^\perp$ and $VX = nX$ for all $X \in \mathcal{A}$,
- $\xi(X) = \langle B(X)X, X\rangle$ with $B(X)$ being positive semidefinite for all $X \in \mathcal{M}_{n,p}(\mathbb{R})$ and Im $B(X) \subset \mathcal{A}^\perp$.

As has been said before (Lemma 2), for solving ($P_{\max}$) we can restrict ourselves to $\mathcal{A}^\perp$ and then content ourselves with solving the simpler problem

$$(\mathrm{PR}_{\max}) \quad \min\left\{-\langle B(X)X, X\rangle : X \in \mathcal{A}^\perp, \|X\| \leq \frac{1}{\sqrt{n}}\right\}.$$

Since Im $B(X) \subset \mathcal{A}^\perp$, the constraint "$X \in \mathcal{A}^\perp$" does not cause any trouble to numerical solution of ($PR_{\max}$). Its presence is only to ensure that we work in $\mathcal{A}^\perp$.

Now let $X^{(k)} \in \mathcal{A}^\perp$ be an approximate solution to ($PR_{\max}$ ). Consider the parametric trust region problem

$$(\mathrm{TRP}_k) \quad \min\left\{-\frac{1}{2}\langle B(X^{(k)})X, X\rangle : X \in \mathcal{A}^\perp, \|X\| \leq \frac{1}{\sqrt{n}}\right\}.$$

DCA is known to be robust and efficient for solving the trust region problem, especially for the large scale setting ([2], [51]) because it requires only matrix-vector products and allows exploiting sparsity. Theoretically, DCA (with at most $2m + 2$ simple restarting procedures, $m$

being the number of negative distinct eigenvalues of the matrix in the quadratic form) converges to a global solution of the trust region problem ([51]). However in practice we do not need to use the restarting procedure because DCA almost always converges to a global solution. We shall describe below the parametric DCA for solving ($PR_{\max}$):

$X^{(0)}$ is chosen in advance in $\mathcal{A}^{\perp} \setminus \{0\}$

$$X^{(k+1)} = \begin{cases} \frac{1}{\sqrt{n}} \frac{(B(X^{(k)})+\rho_k I)X^{(k)}}{(\|B(X^{(k)})+\rho_k I)X^{(k)}\|} & \text{if } \|B(X^{(k)}) + \rho_k I)X^{(k)}\| > \frac{\rho_k}{\sqrt{n}}, \\ \frac{(B(X^{(k)})+\rho_k I)X^{(k)}}{\rho_k} & \text{otherwise.} \end{cases} \tag{38}$$

where the sequence of positive numbers $\rho_k$ must only verify $\rho_k \geq \rho > 0$ for every $k > 0$ since the matrix $B(X)$ is positive semidefinite for all $X$.

Let us give some indications about the construction of $\{X^{(k)}\}$. Consider the following d.c. decomposition for ($TRP_k$):

$$\min \left\{ \chi_S(X) + \frac{\rho_k}{2}\|X\|^2 - \left[\frac{1}{2}\langle B(X^{(k)}) + \rho_k I)X, X\rangle\right] : X \in \mathcal{M}_{n,p}(\mathbb{R}) \right\} \tag{39}$$

with

$$G_k(X) = \chi_S(X) + \frac{\rho_k}{2}\|X\|^2, \; H_k(X) = \frac{1}{2}\langle B(X^{(k)}) + \rho_k I)X, X\rangle, \tag{40}$$

$$S = \left\{ X \in \mathcal{A}^{\perp} : \|X\| \leq \frac{1}{\sqrt{n}} \right\}.$$

Instead of completely solving (39) we use only one step of DCA (relative to the above d.c. decomposition) applied to (39) with $X^{(k)}$ as the initial vector ([51]). The formulation (38) is obtained by following the same computations as in $RDCAP_{\max}$.
It is clear that we can replace in (40) $H_k$ by

$$\tilde{H}_k(X) = \frac{1}{2}\langle(\lambda B(X^{(k)}) + \rho_k I)X, X\rangle, \quad \lambda > 0$$

without altering the equivalence of the corresponding d.c. programs. In this case we obtain an algorithm which is nothing but $RDCAP_{\max}$ when $\rho_k$ is constantly equal to $\rho$. In other words, we can say that the parametric DCA and $RDCAP_{\max}$ are two different interpretations of the same algorithm.
Another version of the parametric DCA consists in setting $X^{(k+1)}$ as a solution of ($TRP_k$) and is called the complete parametric DCA. This algorithm is evidently more expensive than the latter one.

**Remark 8** *In the general case aforementioned in Subsection 4.1, the parametric trust region problem* $(TRP_k)$ *will be defined by*

$$(TRP_k) \qquad \min\left\{-\frac{1}{2}\langle B(X^{(k)})X, X\rangle : X \in \mathcal{A}^{\perp}, \langle VX, X\rangle\} \leq 1\right\}.$$

*According to Proposition 4,* $\eta(X) = \langle VX, X\rangle$ *is a quadratic norm on* $\mathcal{A}^{\perp}$*, so the above results for the normal case can be extended to the general case.*

## 6. Computational experiments.

We have implemented our algorithms in FORTRAN (with double precision) and run on a SGI Origin 2000 multiprocessor with IRIX system. We consider the case where the dissimilarity matrix $\Delta$ is complete, say all $\delta_{ij}$ are known. For Problem $(\text{EDP})_1$ with missing data, our algorithm is composed of two phases. In phase 1 (finding a suitable initial point for DCA applied to this problem) we first compute the estimations $\tilde{\delta}_{ij}$ of $\delta_{ij}$, for $(i,j) \notin S$ as being a kind of shortest path between nodes i and j which preserve the triangle inequality, i.e. $\tilde{\delta}_{ij} \leq \tilde{\delta}_{ik} + \tilde{\delta}_{kj}$ for all pairs $(i,j)$. We then apply DCA to the new problem (which is rather a (MDS) problem than a distance geometry problem) where all pairwise "distances" are known. In phase 2, we solve the original problem $(\text{EDP})_1$ with missing data by applying DCA from the point provided by phase 1. To avoid lengthening the paper, this part is not included herein and we present only computational results for the case of complete dissimilarity data. It is worth mentioning that our algorithm globally solved such distance geometry problems with up to 4096 x 3 = 12288 variables in less than 30 minutes. Numerical experiments concerning Problem (EDP)1 with missing data will be presented elsewhere. Our experiments were carried out for two cases:

- **Case 1:** Dissimilarities are the distances between the objects (in this case the optimal value is 0).
- **Case 2**: Dissimilarities are positive numbers (in this case the optimal value is not known a priori).

To generate the data, in the case 1, we first have considered $n$ objects $X_i^T \in \mathbb{R}^p$ whose coordinates are random numbers in the interval $(0, 10)$, and then took $\delta_{ij} = \|X_i^T - X_j^T\|$, $i, j = 1, \ldots, n$. In the case 2, $\delta_{ij}$ are random numbers in $(0, 100)$.

In Table 1 we present the performance of four algorithms: number of iterations (abbr. *iter*) and CPU time in seconds (abbr. *CPU*) for 75 problems (3 problems of each dimension) with the data in the case 1. For $\text{RDCAP}_1$ and $\text{RDCAP}_{\max}$ we take $\mu = 0.001, \lambda = 1$.

In view of the relation between $\mathcal{P}_1$ and $\mathcal{P}_{\max}$ (c.f. Theorem 4), we took the initial point for $\text{DCAP}_1$ and $\text{RDCAP}_1$ as

$$X^{01} = P_{A^\perp}(\tilde{X})$$

and then $\text{DCAP}_{\max}$ and $\text{RDCAP}_{\max}$ were started from the point

$$X^{02} = X^{01}/\eta(X^{01}),$$

where $\tilde{X}$ was randomly generated in $(0, 1)$.

All algorithms stopped when $\sigma(X^{(k+1)}) \leq 10^{-6}$.

Likewise, Table 2 indicates the performance of four algorithms for 75 problems (3 problems of each dimension) with the data in the case 2. The stopping criterion for all algorithms was $|\sigma(X^{(k+1)}) - \sigma(X^{(k)})| \leq 10^{-3}\sigma(X^{(k+1)})$ while the initial point was chosen by a similar way as in the case 1.

In Table 3 we compare the performance of two regularized algorithms with different choices of parameter $\mu$ ($\lambda$ is always equal to one) with the data in the case 1.

**Comments.** For the case 1:

(i) The solutions obtained by all algorithms in the case 1 are global since the optimal value is $\approx 10^{-7}$.

(ii) $\text{DCAP}_{\max}$ is more efficient than $\text{DCAP}_1$. We observe from many examples that the sequences $\{\|X^{(k+1)} - X^{(k)}\|\}$ and $\{\sigma(X^{(k)})\}$ in the former decrease faster than in the latter.

(iii) The regularized algorithms ($\text{RDCAP}_1$ and $\text{RDACP}_{\max}$) with a "good" choice of the parameter $\mu$ and $\lambda$ are more efficient than the nonregularized ones. However the optimal choice of parameters $\lambda$ and $\mu$ remains an open problem. It is not simple but worth of further developments. For instance, the results presented in Table 3 show that $\text{RDCAP}_1$ with $\mu < 1$ is more efficient than that when $\mu \geq 1$ while $\text{RDCAP}_{\max}$ has the same behavior with $\mu \in (0.001, 10)$.

| $n \times p$ | $DCAP_1$ | | $RDCAP_1$ | | $DCAP_{max}$ | | $RDCAP_{max}$ | |
|---|---|---|---|---|---|---|---|---|
| | iter | CPU | iter | CPU | iter | CPU | iter | CPU |
| 100 × 2 | 92 | 0.412 | 92 | 0.408 | 92 | 0.342 | 92 | 0.342 |
| 200 × 2 | 84.67 | 1.629 | 84.67 | 1.625 | 84.67 | 1.359 | 84.67 | 1.352 |
| 300 × 2 | 84.33 | 4.832 | 84.67 | 4.848 | 84.33 | 4.149 | 84.33 | 4.149 |
| 400 × 2 | 108 | 12.422 | 108 | 12.420 | 108 | 10.987 | 108 | 10.982 |
| 500 × 2 | 93 | 18.460 | 93 | 18.459 | 93 | 16.227 | 93 | 16.222 |
| 600 × 2 | 99.33 | 30.503 | 99.33 | 30.448 | 99.33 | 26.640 | 99.33 | 26.640 |
| 700 × 2 | 92.67 | 40.128 | 92.67 | 40.127 | 92.67 | 35.501 | 92.67 | 35.001 |
| 800 × 2 | 96.67 | 56.851 | 96.67 | 56.752 | 97.67 | 49.812 | 97.67 | 49.712 |
| 900 × 2 | 103.33 | 76.562 | 103.33 | 76.611 | 103.33 | 68.548 | 103.33 | 68.042 |
| 1000 × 2 | 98.33 | 92.308 | 98.33 | 92.289 | 98.66 | 81.586 | 98.33 | 80.546 |
| 1500 × 2 | 94.67 | 361.505 | 94.67 | 361.022 | 92 | 323.832 | 92 | 323.732 |
| 2000 × 2 | 102 | 711.176 | 102 | 711.001 | 95 | 605.074 | 92 | 604.684 |
| 2500 × 2 | 106.33 | 1152.532 | 106.33 | 1152.232 | 93.67 | 941.214 | 93.67 | 940.534 |
| 100 × 3 | 118 | 0.586 | 118 | 0.585 | 118 | 0.479 | 118 | 0.479 |
| 200 × 3 | 125.67 | 2.67 | 125.67 | 2.66 | 125.67 | 2.21 | 125.67 | 2.21 |
| 300 × 3 | 116.33 | 7.711 | 116.33 | 7.713 | 116.33 | 6.625 | 116.33 | 6.623 |
| 400 × 3 | 132.33 | 17.666 | 132.33 | 17.664 | 132.33 | 15.541 | 132.33 | 15.540 |
| 500 × 3 | 123 | 28.091 | 123 | 28.082 | 123 | 24.819 | 123 | 24.818 |
| 600 × 3 | 125.33 | 43.930 | 125.33 | 43.886 | 125.33 | 38.861 | 125.33 | 38.857 |
| 700 × 3 | 138.67 | 68.461 | 138.67 | 68.444 | 138.67 | 61.062 | 138.67 | 61.000 |
| 800 × 3 | 133 | 89.111 | 133 | 89.086 | 133 | 78.884 | 133 | 78.423 |
| 900 × 3 | 131.67 | 112.817 | 131.67 | 112.809 | 131.67 | 100.095 | 131.67 | 100.001 |
| 1000 × 3 | 130.67 | 140.559 | 130.67 | 140.552 | 130.67 | 123.860 | 130.67 | 130.01 |
| 1500 × 3 | 132 | 592.564 | 132 | 592.491 | 127 | 527.525 | 127 | 527.002 |
| 2000 × 3 | 138.67 | 1132.736 | 138.67 | 1132.980 | 123 | 925.639 | 123 | 925.012 |
| 2500 × 3 | 133 | 1719.769 | 133 | 1718.447 | 116 | 1372.106 | 116 | 1371.254 |

*Table 11.1.* The performance of four algorithms in the case 1.

| $n \times p$ | $DCAP_1$ | | $RDCAP_1$ | | $DCAP_{max}$ | | $RDCAP_{max}$ | |
|---|---|---|---|---|---|---|---|---|
| | iter | CPU | iter | CPU | iter | CPU | iter | CPU |
| 100 × 2 | 4 | 0.015 | 4 | 0.015 | 4 | 0.011 | 4 | 0.011 |
| 200 × 2 | 4 | 0.059 | 4 | 0.060 | 4 | 0.046 | 4 | 0.046 |
| 300 × 2 | 4 | 0.181 | 4 | 0.213 | 4 | 0.164 | 4 | 0.163 |
| 400 × 2 | 4 | 0.377 | 4 | 0.378 | 4 | 0.359 | 4 | 0.359 |
| 500 × 2 | 4 | 0.650 | 4 | 0.741 | 4 | 0.631 | 4 | 0.631 |
| 600 × 2 | 4 | 1.004 | 4 | 1.038 | 4 | 0.986 | 4 | 0.985 |
| 800 × 2 | 4 | 1.933 | 4 | 1.973 | 4 | 1.935 | 4 | 1.933 |
| 900 × 2 | 4 | 2.509 | 4 | 2.582 | 4 | 2.505 | 4 | 2.505 |
| 1000 × 2 | 4 | 3.123 | 4 | 3.145 | 4 | 3.152 | 4 | 3.152 |
| 1500 × 2 | 4 | 13.758 | 4 | 14.444 | 4 | 14.287 | 4 | 14.225 |
| 2000 × 2 | 4 | 24.861 | 4 | 25.552 | 4 | 25.592 | 4 | 25.452 |
| 2500 × 2 | 4 | 39.175 | 4 | 39.185 | 4 | 41.541 | 4 | 41.421 |
| 100 × 3 | 35 | 0.102 | 35 | 0.103 | 16 | 0.048 | 16 | 0.048 |
| 200 × 3 | 31 | 0.325 | 31 | 0.335 | 11.33 | 0.125 | 11.33 | 0.124 |
| 300 × 3 | 18.33 | 0.744 | 18.33 | 0.746 | 10.67 | 0.446 | 10.67 | 0.446 |
| 400 × 3 | 21.33 | 1.854 | 21.33 | 1.837 | 9.66 | 0.897 | 7.66 | 0.895 |
| 500 × 3 | 31.67 | 4.729 | 31.67 | 4.739 | 12.33 | 1.960 | 12.33 | 0.960 |
| 600 × 3 | 10.67 | 2.702 | 10.67 | 2.728 | 7.66 | 2.033 | 7.66 | 2.032 |
| 700 × 3 | 28.33 | 9.628 | 28.33 | 9.684 | 11.33 | 4.128 | 11.33 | 4.028 |
| 800 × 2 | 25.67 | 11.884 | 25.67 | 11.089 | 13.33 | 6.488 | 13.33 | 6.298 |
| 900 × 3 | 20.67 | 12.510 | 20.67 | 12.550 | 8.33 | 5.533 | 8.33 | 5.213 |
| 1000 × 3 | 16.67 | 12.795 | 16.67 | 12.158 | 9.67 | 7.896 | 9.67 | 7.894 |
| 1500 × 3 | 15.67 | 56.418 | 15.67 | 56.412 | 8.67 | 34.014 | 8,67 | 34.002 |
| 2000 × 3 | 10.33 | 68.981 | 10.33 | 69.901 | 7 | 51.189 | 7 | 51.002 |
| 2500 × 3 | 8 | 86.086 | 8 | 86.857 | 7 | 80.839 | 7 | 80.326 |

*Table 11.2.* The performance of four algorithms in the case 2.

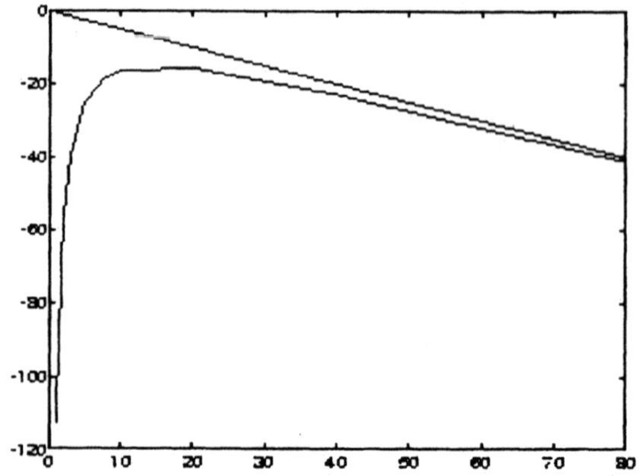


*Figure 11.1.* The graphs of the functions $\tilde{\gamma}(\lambda)$ and $-\lambda/2$

For the case 2:

According to Theorem 3, the dual objective function $\gamma(\lambda)$ is of the form $\gamma(\lambda) = \frac{-\lambda}{2} + \frac{K}{\lambda}$ for $\lambda > 0$ where $K$ is a negative constant (depending only on the dissimilarity matrix $\Delta = (\delta_{ij})$). The graph of $\gamma(\lambda)$ is quite simple: it looks like a hyperbola with two asymptotes $\lambda = 0$ and $\mu(\lambda) = -\lambda/2$ and admits a unique maximum at $\lambda^* = \sqrt{-2K}$ such that $\gamma(\lambda^*) = -\lambda^*$ (Theorem 3). Clearly if the computed solutions by DCA applying to $(P_\lambda)$ are global then we should have: ($\tilde{\gamma}(\lambda)$ denoting the approximate dual objective function computed by DCA)

$\lambda[\frac{\lambda}{2} + \tilde{\gamma}(\lambda)] = \tilde{K}$ for every $\lambda > 0$ and the curve of $\tilde{\gamma}(\lambda)$ must have the above-displayed characteristics.

Numerical results seemed to suggest that $\tilde{\gamma}$ verifies the above conditions (see Fig.1).

**Conclusion.**

We have presented in this paper the d.c. programming approach and solution algorithm (DCA) to the Euclidean metric MDS problem, and in particular to the Euclidean distance geometry problem. These tools seem to be quite suitable to the problems being considered. DCA actually requires matrix-vector products (and additionally only one Cholesky factorization in the solution of the Euclidean distance geometry problem with incomplete data), and allows exploiting sparsity in the large-scale setting. Robustness, efficiency and nonexpensiveness of DCA with respect to known methods should make it possible to solve this problem in the very large-scale setting. It might be expected that the d.c. approach could be also successful in solving the general distance geometry problem (5). We plan to address this issue in future work.

*Acknowledgments.* The authors are grateful to the referees and Professor Hoang Tuy for their constructive comments and helpful discussions, which improved the quality of this paper.

*Table 11.3.* The performance of Regularized algorithms with different choices of $\mu$.

| $n \times p$ | Algorithm | $\mu = 10$ | | $\mu = 1$ | | $\mu = 0.1$ | | $\mu = 0.01$ | | $\mu = 0.001$ | |
|---|---|---|---|---|---|---|---|---|---|---|---|
| | | iter | CPU | iter | CPU | iter | CPU | iter | CPU | iter | CPU |
| $50 \times 2$ | $RDCAP_1$ | 114 | 0.028 | 71 | 0.018 | 69 | 0.017 | 69 | 0.018 | 69 | 0.018 |
| | $RDCAP_{max}$ | 69 | 0.017 | 69 | 0.017 | 69 | 0.017 | 69 | 0.017 | 69 | 0.016 |
| $100 \times 2$ | $RDCAP_1$ | 67 | 0.27 | 62 | 0.192 | 61 | 0.182 | 61 | 0.180 | 61 | 0.152 |
| | $RDCAP_{max}$ | 61 | 0.174 | 61 | 0.172 | 61 | 0.174 | 61 | 0.172 | 61 | 0.168 |
| $200 \times 2$ | $RDCAP_1$ | 76 | 1.622 | 72 | 1.574 | 72 | 1.574 | 72 | 1.572 | 72 | 1.572 |
| | $RDCAP_{max}$ | 72 | 1.572 | 72 | 1.572 | 72 | 1.570 | 72 | 1.570 | 72 | 1.570 |
| $300 \times 2$ | $RDCAP_1$ | 119 | 7.555 | 115 | 7.353 | 115 | 7.355 | 115 | 7.352 | 115 | 7.352 |
| | $RDCAP_{max}$ | 115 | 7.265 | 115 | 7.262 | 115 | 7.262 | 115 | 7.262 | 115 | 7.265 |
| $500 \times 2$ | $RDCAP_1$ | 96 | 23.582 | 95 | 23.252 | 94 | 22.867 | 94 | 22.876 | 94 | 22.762 |
| | $RDCAP_{max}$ | 94 | 21.842 | 94 | 21.642 | 94 | 21.902 | 94 | 21.702 | 94 | 21.102 |
| $1000 \times 2$ | $RDCAP_1$ | 116 | 159.882 | 115 | 129.885 | 115 | 130.162 | 115 | 130.082 | 115 | 129.851 |
| | $RDCAP_{max}$ | 115 | 157.452 | 115 | 158.082 | 115 | 157.874 | 115 | 158.982 | 115 | 157.462 |

# References

[1] ABDO Y. ALFAKIH, A. KHANDANI & H. WOLKOWICZ, *An interior-point method for the Euclidean distance matrix completion problem*, Research Report CORR 97-9, University of Waterloo, Waterloo, Ontario N2L 3G1, Canada.

[2] LE THI HOAI AN, *Contribution à l'optimisation non convexe et l'optimisation globale. Théorie, Algorithms et Applications*, Habilitation à Diriger des Recherches, Université de Rouen, Juin 1997.

[3] LE THI HOAI AN, PHAM DINH TAO and L.D. MUU (1996), *Numerical solution for Optimization over the efficient set by D.c. Optimization Algorithm*, Operations Research Letters, 19, pp. 117-128.

[4] LE THI HOAI AN and PHAM DINH TAO (1997), Solving a class of linearly constrained indefinite quadratic problems by D.c. algorithms, *Journal of Global Optimization,* **11** pp. 253-285.

[5] AUSLENDER, *Optimisation, méthodes numériques*, Ed. Masson. Paris, 1976.

[6] R. BEALS, D.H. KRANTZ & A. TVERSKY, *Foundation of multidimensional scaling*, Psychol. Rev. Vol 75, pp. 12–142, 1968.

[7] W. BICK, H. BAUER, P.J. MUELLER & O. GIESEKE, *Multidimensional scaling and clustering techniques (theory and applications in the social science)*, Institut fur angewandte sozialforchung, Universitat zu koln, 1977.

[8] H. BREZIS, *Opérateurs maximaux monotones*, Mathematics Studies 5, North Holland, 1973.

[9] G.M. CRIPPEN, *A novel approach to calculation of conformation: distance geometry*, J. Computational physics, Vol 24 (1977), pp. 96-107.

[10] G.M. CRIPPEN, *Rapid calculation of coordinates from distance measures*, J. Computational physics, Vol 26 (1978), pp. 449-452.

[11] G.M. CRIPPEN & T.F. HAVEL, *Distance Geometry and Molecular Conformation*, Research Studies Press, Taunton, Somerset, UK, 1988.

[12] J. DE LEEUW, *Applications of convex analysis to multidimensional scaling*, Recent developments in statistics, J.R. Barra *et al.*, editors, North-Holland Publishing Company, pp. 133-145, 1977.

[13] J. DE LEEUW, *Convergence of the majorization method for multidimansional scaling*, Journal of classification, Vol 5 (1988), pp. 163-180.

[14] R.O. DUDA and P.E. HART, *Pattern Classification and Scene Analysis*, Wiley, New York, 1973.

[15] W. GLUNT, T.L. HAYDEN, & M. RAYDAN, Molecular conformation from distance matrices, J. Comp. Chem., 14 (1993), pp. 114-120.

[16] L. GUTTMAN, *A general nonmetric technique for finding the smallest coordinate space for a configuration of points*, Psychometrika, Vol 33 (1968), pp. 469-506.

[17] T.F. HAVEL, *An evaluation of computational strategies for use in the determination of protein structure from distance geometry constraints obtained by nuclear magnetic resonance*, Prog. Biophys. Mol. Biol., 56 (1991), pp. 43-78.

[18] B. A. HENDRICKSON, *The Molecule Problem: Determining Conformation from Pairwise Distances*, Ph.D. Thesis, Cornell University, Ithaca, New York, 1991.

[19] B. A. HENDRICKSON, *The Molecule Problem: Exploiting structure in global optimization*, SIAM J. Optim., 5 (1995), pp. 835-857.

[20] J.B. HIRIART-URRUTY, *From convex optimization to non convex optimization. Part I: Necessary and sufficent conditions for global optimality*, Nonsmooth Optimization and Related Topics, Ettore Majorana International Sciences, Series 43, Plenum Press, 1988.

[21] R. HORST & H. TUY, *Global Optimization (Deterministic Approaches)*, second edition, Springer-Verlag, Berlin New York, 1993.

[22] H. KONNO, *Maximization of a convex quadratic function under linear constraints*, Mathematical Programming, 11 (1976), pp. 117-127.

[23] H. KONNO, P.T. THACH and H. TUY, *Optimization on Low Rank Nonconvex Structures*, Kluwer, Dordrecht-Boston-London, 1997.

[24] J.B. KRUSKAL, *Multidimensional scaling by optimizing goodness-of-fit to a non metric hypothesis*, Psychometrika, Vol 29, pp. 1-28, 1964.

[25] J.B. KRUSKAL, *Nonmetric multidimensional scaling: a numerical method*, Psychometrika, Vol 29 (1964), pp. 115-229.

[26] J.B. KRUSKAL & M. WISH, *Multidimensional Scaling*, Newbury Park, CA. Sage, 1978.

[27] P.J. LAURENT, *Approximation et Optimisation*, Hermann, Paris, 1972.

[28] M. LAURENT, *Cuts, matrix completions and a graph rigidity*, Mathematical Programming, Vol. 79 (1997), Nos 1-3, pp. 255-283.

[29] W.J.M. LEVET, J.P. VAN DE GEER & R. PLOMPl, *Tridiac comparaisons of musical intervals*, Bristish. J. Math. Statist. Psychol. Vol 19 (1966), pp. 163-179.

[30] J.C. LINGOES & E.E. ROSKAM, *A mathematical and emprical analysis of two multidimensional scaling algorithms*, Psychometrika, Vol 38 (1973), monograph supplement.

[31] P.L. LIONS & B. MERCIER, *Splitting algorithms for the sum of two nonlinear operators*, SIAM J. Numer. Anal. 16,6 (1979), pp. 964-979.

[32] P. MAHEY and PHAM DINH TAO, *Partial regularization of the sum of two maximal monotone operators,* Math. Modell. Numer. Anal. ($M^2AN$), Vol. 27 (1993), pp. 375-395.

[33] P. MAHEY and PHAM DINH TAO, *Proximal decomposition of the graph of maximal monotone operator*, SIAM J. Optim. 5(1995), pp. 454-468.

[34] B. MIRKIN, *Mathematical Classification and Clustering*, in Book Series "Nonconvex Optimization and its Applications", P.M. Pardalos and R. Horst eds., Kluwer Academic Publishers.

[35] J.J. MORE & ZHIJUN WU, *Global continuation for distance geometry problems*, SIAM J. Optim., Vol 7, No 3 (1997), pp. 814-836.

[36] J.J. MORE & Z. WU, *Issues in large-scale global molecular optimization*, preprint MCS-P539-1095, Argonne National Laboratory, Argonne, Illinois 60439.

[37] G.L. NEMHAUSER & L.A. WOLSEY, *Integer and Combinatorial Optimization*, John Wiley & Sons, 1988.

[38] A.M. OSTROWSKI, *Solutions of equations and systems of equations*, Academic Press, New York, 1966.

[39] P.M. PARDALOS & J.B. ROSEN, *Constrained Global Optimization: Algorithms and applications*, Lecture Notes in Computer Science, 268, Springer-Verlag, Berlin, 1987.

[40] PHAM DINH TAO, *Eléments homoduaux d'une matrice A relatif à un couple des normes* $(\phi, \psi)$. *Applications au calcul de* $S_{\phi\psi}(A)$, Séminaire d'analyse numérique, Grenoble, $n^o$236, 1975.

[41] PHAM DINH TAO, *Calcul du maximum d'une forme quadratique définie positive sur la boule unité de* $\psi_\infty$, Séminaire d'analyse numérique, Grenoble, $n^o$247, 1976.

[42] PHAM DINH TAO, *Contribution à la théorie de normes et ses applications à l'analyse numérique*, Thèse de Doctorat d'Etat Es Science, Université Joseph Fourier- Grenoble, 1981.

[43] PHAM DINH TAO, Convergence of subgradient method for computing the bound norm of matrices, Linear Alg. and Its Appl, Vol 62 (1984), pp. 163-182.

[44] PHAM DINH TAO, *Algorithmes de calcul d'une forme quadratique sur la boule unité de la norme maximum*, Numer. Math., Vol 45 (1984), pp. 377-401.

[45] PHAM DINH TAO, *Algorithms for solving a class of non convex optimization problems. Methods of subgradients,* Fermat days 85. Mathematics for Optimization, Elsevier Science Publishers B.V. North-Holland, 1986.

[46] PHAM DINH TAO, *Iterative behaviour, Fixed point of a class of monotone operators. Application to non symmetric threshold function,* Discrete Mathematics 70 (1988) ,pp. 85-105.

[47] PHAM DINH TAO, *Duality in d.c. (difference of convex functions) optimization. Subgradient methods*, Trends in Mathematical Optimization, International Series of Numer Math. Vol 84 (1988), Birkhauser, pp. 277-293.

[48] PHAM DINH TAO et LE THI HOAI AN, *Stabilité de la dualité lagrangienne en optimisation d.c. (différence de deux fonctions convexes)*, C.R. Acad. Paris, t.318, Série I (1994), pp. 379-384.

[49] PHAM DINH TAO and LE THI HOAI AN, *Lagrangian stability and global optimality on nonconvex quadratic minimization over Euclidean balls and spheres*, Journal of Convex Analysis, 2(1995), pp. 263-276.

[50] PHAM DINH TAO and LE THI HOAI AN (1997), Convex analysis approach to d.c. programming: Theory, Algorithm and Applications (dedicated to Professor Hoang Tuy on the occasion of his 70th birthday), *Acta Mathematica Vietnamica,* Vol. 22, $N^o 1$, 1997, pp 289-355.

[51] PHAM DINH TAO and LE THI HOAI AN (1998), *D.c. optimization algorithms for trust region problem.* SIAM J. Optimization, vol 8, No 2, pp. 476-505.

[52] B. POLYAK, *Introduction to Optimization. Optimization Software,* Inc., Publication Division, New York, 1987.

[53] R.T ROCKAFELLAR, *Convex Analysis*, Princeton University, Princeton, 1970.

[54] B.D. RIPLEY, *Pattern Recognition and Neural Networks*, Cambridge University Press, 1996.

[55] R.T ROCKAFELLAR, *Monotone operators and the proximal point algorithm*, SIAM J. Control and Optimization, Vol.14, $N^o 5$ (1976).

[56] J. B. SAXE, *Embeddability of Weighted Graphs in k-space is Strongly NP-hard*, Proc. 17 Allerton Conference in Communications, Control and Computing, 1979, pp. 480-489.

[57] R.N SHEPARD, *Representation of structure in similarity data; problem and prospects*, Psychometrika, Vol 39 (1974), pp. 373-421.

[58] J.E. SPINGARN, *Partial inverse of a monotone operator*, Appl. Math. Optim. 10 (1983), pp. 247-265.

[59] Y. TAKANE, F.W. YOUNG & J. DE LEEUW *Nonmetric individual differences multidimensional scaling : an alternating least squares method with optimal scaling features*, Psychometrika, Vol 42 (1977), pp. 7-67.

[60] J.F. TOLAND, *On subdifferential calculus and duality in nonconvex optimization*, Bull. Soc. Math. France, Mémoire 60 (1979), pp. 177-183.

[61] H. TUY, *A general deterministic approach to global optimization via d.c. programming*, Fermat Days 1985: Mathematics for Optimization, North-Holland, Amsterdam, (1986), pp. 137-162.

[62] H. TUY, *Introduction to Global Optimization*, Les Cahiers du GERAD, Groupe d'Etudes et de Recherche en Analyse des Décision, Montréal, Québec, 1994.

[63] H. TUY, *Convex Analysis and Global Optimization*, Kluwer 1998.

[64] R. VARGA *Matrix iterative analysis*, Prentice Hall, 1962.

[65] A.R. WEBB, *Multidimensional scaling by iterative majorization using radial basis functions*, Pattern Recognition, 28(5) (1995), pp. 753-759.

[66] Z. ZOU, RICHARD.H. BIRD, & ROBERT B. SCHNABEL, *A Stochastic/Pertubation Global Optimization Algorithm for Distance Geometry Problems*, J. of Global Optimization, 11(1997), pp. 91-105.

Chapter 12

# ON MINIMIZATION OF SUMS OF HETEROGENEOUS QUADRATIC FUNCTIONS ON STIEFEL MANIFOLDS.

T. Rapcsák[1]

*Senior Researcher, Computer and Automation Institute, Hungarian Academy of Sciences, Budapest, Hungary.*

[1]This research was supported in part by the Hungarian National Research Foundation, Grant Nos. OTKA-T016413 and OTKA-T017241

**Abstract** The minimization of functions $\sum_{i=1}^{k} \frac{1}{2}\mathbf{x}_i^T \mathbf{A}_i \mathbf{x}_i$ is studied under the constraint that vectors $\mathbf{x}_1, \mathbf{x}_2, \dots, \mathbf{x}_k \in R^n$ form an orthonormal system and $\mathbf{A}_1, \dots, \mathbf{A}_k$, $(k \leq n)$ are given symmetric $n \times n$ matrices. The set of feasible points determines a differentiable manifold introduced by Stiefel in 1935. The optimality conditions are obtained by the global Lagrange multiplier rule, and variable metric methods along geodesics are suggested as solving methods for which a global convergence theorem is proved. Such problems arise in various situations in multivariate statistical analysis.

**Keywords:** Quadratic optimization, quadratic equality constraints, Stiefel manifolds.

## 1. Introduction

In 1935, Stiefel introduced a differentiable manifold consisting of all the orthonormal vector systems $\mathbf{x_1}, \mathbf{x_2}, \dots, \mathbf{x_k} \in R^n$, where $R^n$ is the $n$-dimensional Euclidean space and $k \leq n$. James (1976) dealt with



*A. Migdalas et al. (eds.), From Local to Global Optimization,* 277–290.

the topology of Stiefel manifolds. Bolla, Michaletzky, Tusnády and Ziermann (1998) analysed the maximization of sums of heterogeneous quadratic functions on Stiefel manifolds assuming the positive definiteness of the quadratic forms. They formulated the optimality conditions, studied some structural properties of the problem, proposed an iterative algorithm and proved its convergence to a critical point. Statistical applications of these global optimization models (e.g., dynamical factor analysis) are quoted. Their results are mainly based on the matrix theory and the given representation of Stiefel manifolds.

Here, the aim is to reformulate the above optimization problem as a smooth nonlinear optimization one to obtain the first-order and second-order necessary and sufficient optimality conditions based on classical and new results in optimization theory and to show how it is possible to use a generalization of the classical nonlinear optimization methods for solving them. A new theoretical approach elaborated for smooth nonlinear optimization is suggested based on the Riemannian geometry and the global Lagrange multiplier rule [7,9].

## 2. Optimization problem

Consider the following optimization problem:

$$\begin{aligned} &\min \sum_{i=1}^{k} \frac{1}{2}\mathbf{x}_i^T \mathbf{A}_i \mathbf{x}_i \\ &\mathbf{x}_i^T \mathbf{x}_j = \delta_{ij}, \ 1 \le i, j \le k \le n, \\ &\mathbf{x}_i \in R^n, \ i = 1, \dots, k, \ n \ge 2, \end{aligned} \tag{2.1}$$

where $\mathbf{A}_i$, $i = 1, \dots, k$, are given symmetric $n \times n$ matrices and $\delta_{ij}$ is the Kronecker's delta. Since the constraint set is compact and the objective function is continuous, optimization problem (2.1) has a global minimum point. Obviously, the lower bound of the global minimum is $\sum_{i=1}^{k} \lambda_{i,min}$, where $\lambda_{i,min}$ denote the minimal eigenvalues of the matrices $A_i$, $i = 1, \dots, k$, but in general, this value is not attained because the minimal eigenvectors of the matrices $A_i$, $i = 1, \dots, k$, are pairwise orthogonal in some cases only. The constraints of problem (2.1) can be written as

$$\mathbf{x}_i^T \mathbf{x}_i = 1, \qquad i = 1, \dots, k, \tag{2.2a}$$

$$\mathbf{x}_i^T \mathbf{x}_j = 0, \qquad i, j = 1, \dots, k, \ i \ne j, \tag{2.2b}$$

$$\mathbf{x}_i \in R^n, \qquad i = 1, \dots, k, \quad n \ge 2.$$

It follows from orthogonality that $n \geq k \geq 2$. Equalities (2.2a) and (2.2b), and equalities (2.2b) determine a compact and a noncompact differentiable manifold, respectively. So, optimization problem (2.1) consists of the minimization of the sums of heterogeneous quadratic functions over a special differentiable manifold, a Stiefel manifold.

First, optimization problem (2.1) is reformulated as a smooth nonlinear optimization one in order to use its machinery. Let us introduce the following notations:

$$\mathbf{x} = (\mathbf{x}_1, \ldots, \mathbf{x}_k) \in R^{kn}, \qquad J = \{(i,j) |\ i,j = 1, \ldots, k,\ i \neq j\},$$

$$\mathbf{A} = \begin{pmatrix} \mathbf{A}_1 & \cdots & 0 \\ \vdots & \ddots & \vdots \\ 0 & \cdots & \mathbf{A}_k \end{pmatrix}, \quad \mathbf{C}_l = \begin{pmatrix} 0 & \cdots & 0 & \cdots & 0 \\ \vdots & \ddots & \vdots & \vdots & \vdots \\ 0 & \cdots & I_n & \cdots & 0 \\ \vdots & \vdots & \vdots & \ddots & \vdots \\ 0 & \cdots & 0 & \cdots & 0 \end{pmatrix}, \quad l = 1, \ldots, k,$$

$$\mathbf{C}_{ij} = \begin{pmatrix} 0 & \cdots & 0 & \cdots & 0 & \cdots \\ \vdots & \ddots & \vdots & \ddots & \vdots & \vdots \\ 0 & \cdots & 0 & \cdots & I_n & \cdots \\ \vdots & \ddots & \vdots & \ddots & \vdots & \vdots \\ 0 & \cdots & I_n & \cdots & 0 & \cdots \\ 0 & \cdots & 0 & \cdots & 0 & \cdots \end{pmatrix}, \quad (i,j) \in J,$$

where $\mathbf{A}$, $\mathbf{C}_l$, $\mathbf{C}_{ij}$ are $kn \times kn$ blockdiagonal matrices, $I_n$ is the identity matrix in $R^n$, $\mathbf{C}_l$ and $\mathbf{C}_{ij}$ contain $I_n$ in the $l$-th diagonal block and in the $(i,j)$ as well as $(j,i)$ blocks, respectively. The $kn \times kn$ symmetric matrices $\mathbf{C}_{ij}$ are defined for all the pairs of different indices belonging to $J$, given by the $k(k-1)/2$ combinations of the indices $1, \ldots, k$.

In the case of a compact Stiefel manifold, problem (2.1) is equivalent to

$$\begin{aligned} &\min \frac{1}{2}\mathbf{x}^T \mathbf{A}\mathbf{x} \\ h_l(\mathbf{x}) = \frac{1}{2}\mathbf{x}^T \mathbf{C}_l \mathbf{x} - \frac{1}{2} = 0, \qquad &l = 1, \ldots, k, \qquad (2.3) \\ h_{ij}(\mathbf{x}) = \frac{1}{2}\mathbf{x}^T \mathbf{C}_{ij} \mathbf{x} = 0, \qquad &(i,j) \in J, \\ \mathbf{x} \in R^{kn}, \qquad &n \geq 2. \end{aligned}$$

Problem (2.3) is of the basic equality constrained problems in smooth optimization which type of constraints is studied in most of the classical literature (e.g., Luenberger, 1973). The difficulty in the solution of problems (2.3) originated mostly from the intersections of the quadratic equality constraints, which results in the fact that the feasible region is a nonconvex and possibly disconnected subset of the hypersphere $\sum_{i=1}^{k}\sum_{j=1}^{n} x_{ij}^2 = k$ in $R^{kn}$. Adams (1962) gave necessary and sufficient conditions depending on $n$ and $k$ for ensuring that every point of the whole $(n-1)$-dimensional hypersphere should admit an orthonormal $k$-frame.

The next step is the investigation of the structure of the constraint set. Let

$$
\begin{aligned}
f(\mathbf{x}) &= \frac{1}{2}\mathbf{x}^T\mathbf{A}\mathbf{x}, \qquad \mathbf{x} \in R^{kn}, \\
M_{n,k} &= \{\mathbf{x} \in R^{kn} |\ h_l(\mathbf{x}) = 0,\ l = 1,\ldots,k; \quad h_{ij}(\mathbf{x}) = 0, \quad (i,j) \in J\}, \qquad (2.4) \\
TM_{n,k}(\mathbf{x}) &= \{\mathbf{v} \in R^{kn} |\ \nabla h_l(\mathbf{x})\mathbf{v} = 0,\ l = 1,\ldots,k; \\
&\qquad \nabla h_{ij}(\mathbf{x})\mathbf{v} = 0, \quad (i,j) \in J\}, \quad \mathbf{x} \in M_{n,k},
\end{aligned}
$$

where the symbol $\nabla$ denotes the gradient vector of a function. Though the next statement is known in the literature, the optimization theoretical reformulation of the original problem makes a simple proof possible.

**Theorem 2.1** *The set $M_{n,k}$ is a $C^\infty$ differentiable manifold with the dimension $kn - \frac{k(k+1)}{2}$.*

**Proof** It is sufficient to show that the gradients of the equality constraints given by the row vectors

$$
\begin{aligned}
\mathbf{x}^T\mathbf{C}_l &= (\ldots,\mathbf{x}_i,\ldots), \qquad l = 1,\ldots,k, \\
\mathbf{x}^T\mathbf{C}_{ij} &= (\ldots,\mathbf{x}_i,\ldots,\mathbf{x}_j,\ldots), \qquad (i,j) \in J,
\end{aligned}
\qquad (2.5)
$$

are linearly independent on $M_{n,k}$. In problem (2.3), the number of equalities is equal to $k + k(k-1)/2 = \frac{k(k+1)}{2} < kn$. As

$$
\begin{aligned}
\mathbf{x}^T\mathbf{C}_l\mathbf{C}_l\mathbf{x} &= 1, \qquad l = 1,\ldots,k, \qquad \mathbf{x} \in M_{n,k}, \\
\mathbf{x}^T\mathbf{C}_{ij}\mathbf{C}_{ij}\mathbf{x} &= 2, \qquad (i,j) \in J, \qquad \mathbf{x} \in M_{n,k}, \\
\mathbf{x}^T\mathbf{C}_i\mathbf{C}_j\mathbf{x} &= 0, \qquad (i,j) \in J, \qquad \mathbf{x} \in M_{n,k}, \qquad (2.6) \\
\mathbf{x}^T\mathbf{C}_l\mathbf{C}_{ij}\mathbf{x} &= 0, \qquad l = 1,\ldots,k; \quad (i,j) \in J, \qquad \mathbf{x} \in M_{n,k}, \\
\mathbf{x}^T\mathbf{C}_{ij}\mathbf{C}_{kl}\mathbf{x} &= 0, \qquad (i,j),(k,l) \in J,\ (i,j) \neq (k,l), \qquad \mathbf{x} \in M_{n,k},
\end{aligned}
$$

these relations mean that the gradients are different from zero, and any two of them are orthogonal, from which the statement follows. ∎

# 3. Optimality conditions

In this part, the first-order and second-order necessary and sufficient optimality conditions of problem (2.3) are stated. The gradients of the equality constraints are linearly independent, the feasible set is a differentiable manifold, moreover, the feasible set $M_{n,k}$ is a Riemannian smooth $(kn - \frac{k(k+1)}{2})$-manifold in $R^{kn}$ where the Riemannian metric on $M_{n,k}$ is induced by the Euclidean one of $R^{kn}$. Now, the global Lagrange multiplier rule developed for the case of equality constraints (Rapcsák, 1991, 1997) is used to obtain the first-order and second-order optimality conditions of problem (2.3). The geometric background of the rule can be found in these references as well.

Let the Lagrangian function be

$$L(\mathbf{x}, \mu(\mathbf{x}), \eta(\mathbf{x})) =$$

$$f(\mathbf{x}) - \sum_{l=1}^{k} \mu_l(\mathbf{x}) h_l(\mathbf{x}) - \sum_{(i,j)\in J} \eta_{ij}(\mathbf{x}) h_{ij}(\mathbf{x}), \qquad \mathbf{x} \in M_{n,k}, \tag{3.1}$$

the geodesic gradient vector of the Lagrangian function,

$$\nabla^g L(\mathbf{x}, \mu(\mathbf{x}), \eta(\mathbf{x})) =$$

$$\nabla f(\mathbf{x}) - \sum_{l=1}^{k} \mu_l(\mathbf{x}) \nabla h_l(\mathbf{x}) - \sum_{(i,j)\in J} \eta_{ij}(\mathbf{x}) \nabla h_{ij}(\mathbf{x}), \qquad \mathbf{x} \in M_{n,k}, \tag{3.2}$$

and the geodesic Hessian matrix of the Lagrangian function

$$H^g L(\mathbf{x}, \mu(\mathbf{x}), \eta(\mathbf{x})) =$$

$$Hf(\mathbf{x}) - \sum_{l=1}^{k} \mu_l(\mathbf{x}) Hh_l(\mathbf{x}) - \sum_{(i,j)\in J} \eta_{ij}(\mathbf{x}) Hh_{ij}(\mathbf{x}), \qquad \mathbf{x} \in M_{n,k}, \tag{3.3}$$

where the symbol $H$ denotes the Hessian matrix of a function, and

$$\mu(\mathbf{x})^T = \nabla f(\mathbf{x}) J\mathbf{h}^T(\mathbf{x})[J\mathbf{h}(\mathbf{x}) J\mathbf{h}(\mathbf{x})^T]^{-1}, \qquad \mathbf{x} \in M_{n,k}, \tag{3.4}$$

$$J\mathbf{h}(\mathbf{x}) = \begin{pmatrix} \nabla h_1(\mathbf{x}) \\ \vdots \\ \nabla h_l(\mathbf{x}) \end{pmatrix}, \qquad \mathbf{x} \in M_{n,k},$$

$$\eta(\mathbf{x})^T = \nabla f(\mathbf{x}) J\hat{\mathbf{h}}^T(\mathbf{x})[J\hat{\mathbf{h}}(\mathbf{x}) J\hat{\mathbf{h}}(\mathbf{x})^T]^{-1}, \qquad \mathbf{x} \in M_{n,k}, \tag{3.5}$$

$$J\hat{\mathbf{h}}(\mathbf{x}) = \begin{pmatrix} \nabla h_{12}(\mathbf{x}) \\ \vdots \\ \nabla h_{ij}(\mathbf{x}) \\ \vdots \end{pmatrix}, \qquad \mathbf{x} \in M_{n,k}.$$

The global Lagrange multiplier rule based on Theorem 15.2.1 and Corollary 15.2.1 in [9, p.276] is formulated for the case of problem (2.3) in the following statement. First, a definition of geodesic convex sets is recalled where the geodesic is used in the classical meaning. If $M$ is a Riemannian $C^2$ manifold, then a set $\mathcal{C} \subseteq \mathcal{M}$ is geodesic convex if any two points of $\mathcal{C}$ are joined by a geodesic belonging to $\mathcal{C}$, moreover, a singleton is geodesic convex.

**Theorem 3.1** *If the point $\mathbf{x}_0 \in M_{n,k}$ is a (strict) local minimum of problem (2.3), then*

$$\begin{aligned} \nabla^g L(\mathbf{x}_0, \mu(\mathbf{x}_0), \eta(\mathbf{x}_0)) &= 0, \\ \mathbf{v}^T H^g L(\mathbf{x}_0, \mu(\mathbf{x}_0), \eta(\mathbf{x}_0))\mathbf{v} \geq (>)0, &\qquad \mathbf{v} \in TM_{n,k}(\mathbf{x}_0). \end{aligned} \tag{3.6}$$

*If $\mathcal{C} \subseteq M_{n,k}$ is an open geodesic convex set and there exists a point $\mathbf{x}_0 \in \mathcal{C}$ such that*

$$\begin{aligned} \nabla^g L(\mathbf{x}_0, \mu(\mathbf{x}_0), \eta(\mathbf{x}_0)) &= 0, \\ \mathbf{v}^T H^g L(\mathbf{x}, \mu(\mathbf{x}), \eta(\mathbf{x}))\mathbf{v} \geq (>)0, \qquad \mathbf{v} \in TM_{n,k}(\mathbf{x}), &\qquad \mathbf{x} \in \mathcal{C}, \end{aligned} \tag{3.7}$$

*then the point $\mathbf{x}_0$ is a (strict) global minimum of the function $f$ on $\mathcal{C}$.*

Let

$$S(\mathbf{x}) = \tag{3.8}$$

$$\begin{pmatrix} \mathbf{x}_1^T \mathbf{A}_1 \mathbf{x}_1 I_n & \frac{1}{2}(\mathbf{x}_2^T \mathbf{A}_2 \mathbf{x}_1 + \mathbf{x}_1^T \mathbf{A}_1 \mathbf{x}_2) I_n & \cdots & \frac{1}{2}(\mathbf{x}_k^T \mathbf{A}_k \mathbf{x}_1 + \mathbf{x}_1^T \mathbf{A}_1 \mathbf{x}_k) I_n \\ \frac{1}{2}(\mathbf{x}_1^T \mathbf{A}_1 \mathbf{x}_2 + \mathbf{x}_2^T \mathbf{A}_2 \mathbf{x}_1) I_n & \mathbf{x}_2^T \mathbf{A}_2 \mathbf{x}_2 I_n & \cdots & \frac{1}{2}(\mathbf{x}_k^T \mathbf{A}_k \mathbf{x}_2 + \mathbf{x}_2^T \mathbf{A}_2 \mathbf{x}_k) I_n \\ \vdots & \vdots & \ddots & \vdots \\ \frac{1}{2}(\mathbf{x}_1^T \mathbf{A}_1 \mathbf{x}_k + \mathbf{x}_k^T \mathbf{A}_k \mathbf{x}_1) I_n & \frac{1}{2}(\mathbf{x}_2^T \mathbf{A}_2 \mathbf{x}_k + \mathbf{x}_k^T \mathbf{A}_k \mathbf{x}_2) I_n & \cdots & \mathbf{x}_k^T \mathbf{A}_k \mathbf{x}_k I_n \end{pmatrix}$$

where $\mathbf{x} = (\mathbf{x}_1, \dots, \mathbf{x}_k) \in M_{n,k}$.

By Theorem 3.1, the first-order optimality condition of problem (2.3) at a point $\mathbf{x}_0$ is

$$\mathbf{A}\mathbf{x}_0 - \sum_{l=1}^{k} (\mathbf{x}_0^T \mathbf{A} \mathbf{C}_l \mathbf{x}_0) \mathbf{C}_l \mathbf{x}_0 - \sum_{(i,j) \in J} \frac{1}{2} (\mathbf{x}_0^T \mathbf{A} \mathbf{C}_{ij} \mathbf{x}_0) C_{ij} \mathbf{x}_0 = 0, \qquad \text{i.e.,}$$

$$\mathbf{A}\mathbf{x}_0 = S(\mathbf{x}_0)\mathbf{x}_0, \tag{3.9}$$

and the second-order necessary condition at a point $\mathbf{x}_0$ is equivalent to the positive semidefiniteness of the matrix

$$\left(\mathbf{A} - \sum_{l=1}^{k}(\mathbf{x}_0^T\mathbf{A}\mathbf{C}_l\mathbf{x}_0)\mathbf{C}_l - \sum_{(i,j)\in J}\frac{1}{2}(\mathbf{x}_0^T\mathbf{A}\mathbf{C}_{ij}\mathbf{x}_0)\mathbf{C}_{ij}\right)_{|TM_{\mathbf{x}_0}} = (\mathbf{A} - S(\mathbf{x}_0))_{|TM_{\mathbf{x}_0}}, \tag{3.10}$$

where the symbol $|TM_{\mathbf{x}}$ denotes the restriction to the tangent space $TM_{n,k}(\mathbf{x})$ given in the form of

$$\begin{aligned} TM_{n,k}(\mathbf{x}) = \{\mathbf{v} \in R^{kn} |\ &\mathbf{x}_l^T\mathbf{v}_l = 0, \quad l = 1,\ldots,k; \\ &\mathbf{x}_i^T\mathbf{v}_j + \mathbf{x}_j^T\mathbf{v}_i = 0, \quad (i,j) \in J\}, \mathbf{x} \in M_{n,k}. \end{aligned} \tag{3.11}$$

It is obvious that the tangent spaces $TM_{n,k}(\mathbf{x})$, $\mathbf{x} \in M_{n,k}$, take part of the corresponding tangent spaces of the hypersphere $\sum_{i=1}^{k}\sum_{j=1}^{n} x_{ij}^2 = k$ in $R^{kn}$. As problem (2.3) has at least one stationary point and $\det(\mathbf{A} - S(\mathbf{x}_0)) = 0$, thus several stationary points can be expected.

**Corollary 3.1***If we consider the maximization problem*

$$\begin{gathered} \max \sum_{i=1}^{k}\frac{1}{2}\mathbf{x}_i^T\mathbf{A}_i\mathbf{x}_i \\ \mathbf{x}_i^T\mathbf{x}_j = \delta_{ij}, \qquad 1 \le i,j \le k \le n, \\ \mathbf{x}_i \in R^n, \quad i = 1,\ldots,k, \qquad n \ge 2, \end{gathered} \tag{3.12}$$

*then the first-order and second-order necessary optimality conditions at a point* $\mathbf{x}_0$ *are as follows:*

$$\begin{aligned} \mathbf{A}\mathbf{x}_0 &= S(\mathbf{x}_0)\mathbf{x}_0, \\ (\mathbf{A} - S(\mathbf{x}_0))_{|TM_{\mathbf{x}_0}} &\quad \textit{is negative semidefinite}, \end{aligned} \tag{3.13}$$

Bolla, Michaletzky, Tusnády and Ziermann (1997) studied problem (3.12) assuming the positive definiteness of the matrices $\mathbf{A}_1,\ldots,\mathbf{A}_k$ and proved that the first-order and second-order optimality conditions at a point $\mathbf{x}_0$ are

$$\begin{aligned} \mathbf{A}\mathbf{x}_0 &= \hat{S}(\mathbf{x}_0)\mathbf{x}_0, \\ \hat{S}(\mathbf{x}_0) &\quad \text{is symmetric}, \\ (\mathbf{A} - \hat{S}(\mathbf{x}_0))_{|TM_{\mathbf{x}_0}} &\quad \text{is negative semidefinite}, \end{aligned} \tag{3.14}$$

$$\hat{S}(\mathbf{x}) = \begin{pmatrix} \mathbf{x}_1^T\mathbf{A}_1\mathbf{x}_1 I_n & \mathbf{x}_1^T\mathbf{A}_1\mathbf{x}_2 I_n & \cdots & \mathbf{x}_1^T\mathbf{A}_1\mathbf{x}_k I_n \\ \mathbf{x}_2^T\mathbf{A}_2\mathbf{x}_1 I_n & \mathbf{x}_2^T\mathbf{A}_2\mathbf{x}_2 I_n & \cdots & \mathbf{x}_2^T\mathbf{A}_2\mathbf{x}_k I_n \\ \vdots & \vdots & \ddots & \vdots \\ \mathbf{x}_k^T\mathbf{A}_k\mathbf{x}_1 I_n & \mathbf{x}_k^T\mathbf{A}_k\mathbf{x}_2 I_n & \cdots & \mathbf{x}_k^T A_k\mathbf{x}_k I_n \end{pmatrix}, \qquad (3.15)$$

where $\mathbf{x} = (\mathbf{x}_1, \ldots, \mathbf{x}_k) \in M_{n,k}$.

The main difference between the first-order and second-order optimality conditions (3.13) and (3.14) is that $\hat{S}(\mathbf{x})$, $\mathbf{x} \in M_{n,k}$, are, in general, nonsymmetric matrices, which is unusual in optimization theory, while $S(\mathbf{x})$, $\mathbf{x} \in M_{n,k}$, are symmetric.

**Lemma 3.1** *Optimality conditions (3.13) and (3.14) are equivalent.*

**Proof** If all the $k$ blocks consisting of one $n$-vector each are multiplied by $\mathbf{x}_j$, $j = 1, \ldots, k$, in the first-order optimality condition (3.13), then equalities

$$\mathbf{x}_j A_i \mathbf{x}_i = \mathbf{x}_j A_j \mathbf{x}_i, \qquad (i,j) \in J, \qquad (3.16)$$

are obtained at the point $\mathbf{x}_0$, from which the statement follows. ∎

**Example 3.1** Let $n = 2$, $k = 2$ and

$$\mathbf{A}_1 = \begin{pmatrix} 2 & 0 \\ 0 & 1 \end{pmatrix}, \qquad \mathbf{A}_2 = \begin{pmatrix} 2 & 0 \\ 0 & 3 \end{pmatrix},$$

$$\mathbf{x} = (1,0,0,1), \qquad \mathbf{y} = (0,1,1,0),$$

then $f(\mathbf{x}) = 5$, $f(\mathbf{y}) = \mathbf{3}$,

$$S(\mathbf{x}) = \begin{pmatrix} 2 & 0 & 0 & 0 \\ 0 & 2 & 0 & 0 \\ 0 & 0 & 3 & 0 \\ 0 & 0 & 0 & 3 \end{pmatrix}, \qquad S(\mathbf{y}) = \begin{pmatrix} 1 & 0 & 0 & 0 \\ 0 & 1 & 0 & 0 \\ 0 & 0 & 2 & 0 \\ 0 & 0 & 0 & 2 \end{pmatrix},$$

$$\mathbf{A} - S(\mathbf{x}) = \begin{pmatrix} 0 & 0 & 0 & 0 \\ 0 & -1 & 0 & 0 \\ 0 & 0 & -1 & 0 \\ 0 & 0 & 0 & 0 \end{pmatrix}, \qquad \mathbf{A} - S(\mathbf{y}) = \begin{pmatrix} 1 & 0 & 0 & 0 \\ 0 & 0 & 0 & 0 \\ 0 & 0 & 0 & 0 \\ 0 & 0 & 0 & 1 \end{pmatrix},$$

$$TM_{n,k}(\mathbf{x}) \subseteq \{\mathbf{v} \in R^4 |\ v_1 = 0,\ v_4 = 0\},$$
$$TM_{n,k}(\mathbf{y}) \subseteq \{\mathbf{v} \in \mathbf{R}^4 |\ \mathbf{v}_2 = \mathbf{0},\ \mathbf{v}_3 = \mathbf{0}\}.$$

It follows that $(\mathbf{A} - S(\mathbf{x}))\mathbf{x} = 0$, $\mathbf{A} - S(\mathbf{x})$ is negative definite on $TM_{n,k}(\mathbf{x})$, i.e., $\mathbf{x}$ is a strict local maximum point, $(\mathbf{A} - S(\mathbf{y}))\mathbf{y} = 0$, $\mathbf{A} - S(\mathbf{y})$ is positive definite on $TM_{n,k}(\mathbf{y})$, i.e., $\mathbf{y}$ is a strict local minimum point.

# 4. Variable metric methods along geodesics

A general framework for variable metric methods along geodesics was developed to find a stationary point or a local optimum of smooth nonlinear optimization problems defined on a Riemannian manifold (Gabay, 1982; Rapcsák and Thang, 1995; Rapcsák, 1997, 1998). These algorithms represent a generalization of unconstrained optimization methods and constrained optimization methods under equalities and proceed in the case of problem (2.3) as follows.

Let us consider problem (2.3). Starting from an initial feasible solution $\mathbf{x_0} \in M_{n,k} \subseteq R^{kn}$, let $\mathbf{x_i} \in M_{n,k} \subseteq R^{kn}$ be the feasible solution of the $i$th iteration step, $D_i$ a $kn \times kn$ symmetric matrix defining a linear map $D_i : TM_{n,k}(\mathbf{x_i}) \to TM_{n,k}(\mathbf{x_i})$, and a positive definite quadratic form on $TM_{n,k}(\mathbf{x_i})$ with a uniform lower bound at every iteration point (i.e., $\mathbf{v}_i D_i \mathbf{v}_i \geq \varepsilon |\mathbf{v}_i|^2$, $\mathbf{v}_i \in TM_{n,k}(\mathbf{x_i})$, $\varepsilon > 0$ for all $i$) and $G$ the induced Riemannian metric of $M_{n,k}$.

1. Compute the direction $\mathbf{p_i} \in R^{kn}$ as follows:

$$\mathbf{p_i} = -D_i(\mathbf{A} - S(\mathbf{x}_i))\mathbf{x_i}. \qquad (4.1)$$

2. Let

$$\mathbf{x_{i+1}} = \gamma_{\mathbf{x_i}}(t_i, \mathbf{p}_i), \qquad (4.2)$$

where $\gamma_{\mathbf{x_i}}(t_i, \mathbf{p}_i)$ means the arc of the geodesic starting from $\mathbf{x_i}$ with tangent $\mathbf{p_i}$ and the stepsize $t_i$ is determined by an exact geodesic search to find the first local minimum along the above geodesic, i.e.,

$$t_i = \arg\min\{f(\gamma_{\mathbf{x_i}}(t, \mathbf{p}_i)) | \ t \in R_+\} \qquad (4.3)$$

or chosen according to the Armijo principle (e.g., Ortega and Rheinboldt, 1970), i.e., given $\alpha \in (0, \frac{1}{2})$, let $t_i = 2^{-l_i}$, with $l_i$ the smallest integer such that

$$f(\gamma_{\mathbf{x}_i}(2^{-l_i}, \mathbf{p}_i)) \leq f(\mathbf{x}_i) - \alpha 2^{-l_i} |\mathbf{x}_i^T(\mathbf{A} - S(\mathbf{x}_i))D_i(\mathbf{A} - S(\mathbf{x}_i))\mathbf{x}_i|. \qquad (4.4)$$

Note that $(\mathbf{A} - S(\mathbf{x}))\mathbf{x}$, $\mathbf{x} \in M_{n,k}$, are the orthogonal projections of the gradient vectors $\nabla f$ with respect to the Euclidean metric of $R^{kn}$ to the tangent spaces of the Stiefel manifold. So, this general framework contains several well-known nonlinear optimization algorithms: gradient projection method and reduced gradient method by choosing $D_i = I$ for all $i$; Newton-type methods and Newton-type methods along geodesics by choosing $D_i$ based on the Hessian matrix $HL$ of the Lagrangian function and the geodesic Hessian matrix $H^gL$ at the $i$th iteration point, respectively. Thus, numerical techniques of smooth nonlinear optimization can be applied in these cases as well.

An algorithm is said to be globally convergent if the algorithm is guaranteed to generate a sequence of points converging to a critical point for arbitrary starting point. The global convergence theorems proved for variable metric methods along geodesics (Rapcsák and Thang, 1995; Rapcsák, 1997, 1998) can be adapted to this case. An advantage of variable metric methods along geodesics is that global convergence can be proved for a class of methods without assuming the positive or negative definiteness of the matrices $\mathbf{A}_1, \ldots, \mathbf{A}_k$.

**Theorem 4.1** *If the sequence $\{\mathbf{x}_i\}$ is generated by (4.1), (4.2) and (4.3) or (4.1), (4.2) and (4.4), and if a mapping is defined by $D_i(\mathbf{A} - S(\mathbf{x}_i))\mathbf{x}_i \in TM_{n,k}(\mathbf{x}_i)$, for all $i$ at the iteration points and $D_i$ are positive definite on $TM_{n,k}(\mathbf{x}_i)$ with a uniform lower bound for all $i$, then the sequence $\{\mathbf{x}_i\}$ is either finite terminating at a critical point, or infinite, and every accumulation point is critical. If the critical values of $f$ are distinct, the whole sequence $\{\mathbf{x}_i\}$ converges to a critical point.*

If $D_i = I$ for all $i$ at the iteration points, then the statement is as follows.

**Corollary 4.1** *If the sequence $\{\mathbf{x}_i\}$ is generated by (4.1), (4.2) and (4.3) or (4.1), (4.2) and (4.4), then the sequence $\{\mathbf{x}_i\}$ is either finite terminating at a critical point, or infinite, and every accumulation point is critical. If the critical values of $f$ are distinct, the whole sequence $\{\mathbf{x}_i\}$ converges to a critical point.*

## 5. Proof of Theorem 4.1

Let $W_i$ denote the connected component containing the $\mathbf{x}_i$ of the level set

$$\{\mathbf{x} \in M \subseteq \mathcal{M} |\ f(\mathbf{x}) \leq f(\mathbf{x}_i)\},$$

where $\mathcal{M}$ is equal to $R^n$ endowed with a Riemannian metric $\mathcal{G}(\mathbf{x})$, $\mathbf{x} \in \mathcal{M}$, and $M$ is a Riemannian submanifold with the induced metric $G$,

$$\mathbf{p}_i = -D_i^2 \nabla^G f(\mathbf{x}_i)^T, \qquad \nabla^G f(\mathbf{x}_i)^T \in R^n, \tag{5.1}$$

where $\nabla^G f^T$ is the Riemannian gradient vector (e.g., Rapcsák, 1997) and

$$f(\gamma_{\mathbf{x}_i}(2^{-l_i}, \mathbf{p}_i)) \leq f(\mathbf{x}_i) - \alpha 2^{-l_i} |\nabla^G f(\mathbf{x}_i) \mathcal{G}(\mathbf{x}_i) D_i^2 \nabla^G f(\mathbf{x}_i)^T|. \tag{5.2}$$

The following statement is a combination of two global convergence theorems found in Rapcsák and Thang, 1995 and Rapcsák, 1997, 1998, respectively.

**Theorem 5.1** *If $f$ is continuously differentiable, $W_0$ is a compact set in a complete Riemannian submanifold $M \subseteq \mathcal{M}$, the sequence $\{\mathbf{x}_i\}$ is generated by (5.1), (4.2) and (4.3) or (5.1), (4.2) and (5.2), and if a mapping is defined by $D_i \nabla^G f(\mathbf{x}_i)^T \in TM(\mathbf{x}_i)$, for all i at the iteration points and satisfies there the Lipschitz condition, $D_i$ are positive definite on $TM(\mathbf{x}_i)$ with a uniform lower bound for all i, and $D_i$ and $\mathcal{G}$ commute on $TM(\mathbf{x}_i)$ at every iteration point, then the sequence $\{\mathbf{x}_i\}$ is either finite terminating at a critical point, or infinite, and every accumulation point is critical. If the critical values of $f$ are distinct, the whole sequence $\{\mathbf{x}_i\}$ converges to a critical point.*

By Theorem 11.1.1 in [9], if $R^{kn}$ is endowed with a Riemannian metric $\mathcal{G}$ in problem (2.3), then the Riemannian gradient vector is equal to

$$\nabla^G f^T = (I - \mathcal{G}^{-1}[J\mathbf{h}^T, J\hat{\mathbf{h}}^T]([J\mathbf{h}, J\hat{\mathbf{h}}]\mathcal{G}^{-1}[J\mathbf{h}^T, J\hat{\mathbf{h}}^T])^{-1}[J\mathbf{h}^T, J\hat{\mathbf{h}}])\mathcal{G}^{-1}\nabla f^T. \tag{5.3}$$

In the case of problem (2.3), the space $R^{kn}$ is endowed with the Euclidean metric, i.e., $\mathcal{G}(\mathbf{x}) = I_{kn}$, $\mathbf{x} \in R^{kn}$, the Riemannian metrics of the Stiefel manifolds are induced by the Euclidean metric, the objective function is, obviously, continuously differentiable, the set $W_0$ is compact, the Stiefel manifolds are complete, the matrices $D_i^2$ can be replaced by $D_i$ for all $i$, because $\mathbf{p}_i$ given by (4.1) is a descent direction for every i, thus the statement of Theorem 4.1 is a consequence of Theorem 5.1. ∎

**Acknowledgement.** This is to express my acknowledgement to Gy. Michaletzky whose precious ideas greatly contributed to this paper.

# References


[1] Adams, J. F., Vector fields on spheres, Annales of Mathematics 75 (1962) 603-632.

[2] Bolla, M., Michaletzky, Gy., Tusnády, G. and Ziermann, M., Extrema of sums of heterogeneous quadratic forms, Linear Algebra and Applications 269 (1998) 331-365.

[3] Gabay, D., Minimizing a differentiable function over a differentiable manifold, Journal of Optimization Theory and Applications 37 (1982) 177-219.

[4] James, J. M., The topology of Stiefel manifolds, London Mathematical Society, Lecture Notes Series 24, Cambridge University Press, Cambridge, 1976.

[5] Luenberger, D. G., Introduction to linear and nonlinear programming, Addison-Wesley Publishing Company, Reading, 1973.

[6] Ortega, J.M. and Rheinboldt, N.C., Iterative solution of nonlinear equations in several variables, Academic Press, New York, New York, 1970.

[7] Rapcsák, T., Geodesic convexity in nonlinear optimization, Journal of Optimization Theory and Applications 69 (1991) 169-183.

[8] Rapcsák, T. and Thang, T. T., Nonlinear coordinate representations of smooth optimization problems, Journal of Optimization Theory and Applications 86 (1995) 459-489.

[9] Rapcsák, T., Smooth nonlinear optimization in $R^n$, Kluwer Academic Publishers, Boston, London, Dordrecht, 1997.

[10] Rapcsák, T., Variable metric methods along geodesics, in: New trends in mathematical programming, F. Giannessi, S. Komlósi and T. Rapcsák (eds.), Kluwer Academic Publishers (1998) 257-275.

[11] Stiefel, E., Richtungsfelder und Fernparallelismus in $n$-dimensionalen Mannigfaltigkeiten, Commentarii Mathematici Helvetici 8 (1935/6) 305-353.

Chapter 13

# A NONLINEAR LAGRANGIAN FUNCTION FOR DISCRETE OPTIMIZATION PROBLEMS

X.Q. Yang[1] and C.J. Goh

[1] yangx@maths.uwa.edu.au

**Abstract** It is known that global optimality conditions for nonconvex optimization problems are, in general, difficult to obtain. Moreover, zero duality gap property does not, in general, hold for a nonconvex optimization problem and there are examples where a nonzero duality gap exists for a discrete linear program when the conventional (linear) Lagrangian dual formulation is applied. In this paper, a nonlinear Lagrangian function is introduced and applied to establish zero duality gaps between discrete optimization problems and their nonlinear Lagrangian dual problems. An example of a discrete linear program is given to illustrate the results. An application of the results is also presented to a discrete network optimization problem.

**Keywords**: Discrete optimization, nonlinear Lagrangian function, network optimization, zero duality gap.

## 1. Introduction

Global optimization has many applications in finance and engineering. The theory of global optimization and extensive solution methods are given in e.g., Horst and Tuy (1991) and the references cited therein. In the dual program approach for nonconvex optimization problems, the zero duality gap property plays an important role. In the conventional Lagrangian theory for constrained convex optimization problems, zero



*A. Migdalas et al. (eds.), From Local to Global Optimization,* 291–304.

duality gap always exists, see Luenberger (1984) and Rockafellar (1974). The zero duality gap property is established for some nonconvex optimization using generalized convexity conditions, see Jeyakumar and Wolkowicz (1990) and Thach (1993).

Recently in Ben-Tal and Teboulle (1996) the optimization problem of minimizing a concave quadratic function subject to finitely many convex quadratic constraints is reformulated as an equivalent minimax convex problem. In Li (1995), a $p$-th power transformation is introduced to guarantee a zero duality gap in the transformed space for a class of nonconvex twice differentiable optimization problems. In contrast to the *linear* combination of the cost and constraint functions in the conventional (linear) Lagrangian theory for inequality constrained optimization problems, a *nonlinear* Lagrangian function is introduced which has a minimax structure in Goh and Yang (1996a, 1996b). This nonlinear Lagrangian function can be explained as the weighted Tchebyshev norm of the cost and constraint functions. The zero duality gap property between the primal problem and its nonlinear dual problem is equivalent to fact that there is a shifted supporting cone to the epigraph of the value function for the perturbed optimization problem. Without any convexity assumption, zero duality gaps for continuous (nonconvex) optimization problems are established in Goh and Yang (1996b).

The study is continued on the nonlinear Lagrangian function. The aim of this paper is to consider the nonlinear Lagrangian function for optimization problems which have discrete variables. Discrete optimization problems have many applications in network flows and production scheduling, see Ahuja et al (1993). As shown in Bazaraa and Shetty (1979), for a discrete optimization problem, a zero duality gap may exist even though the cost and the constraints are linear functions. See also Example 1 in section 4. Properties of the nonlinear Lagrangain function will be investigated. Zero duality gaps will be established for discrete optimization problems and their nonlinear Lagrangian dual problems.

The outline of the paper is as follows. In next section, a nonlinear Lagrangian function for a discrete optimization problem is introduced. It is observed that the value function of a discrete optimization problem is a step function. In section 3, various zero duality gaps are provided for the cases of finite elements or infinite elements in the discrete set. In section 4, some applications are given e.g. to network optimization problems.

## 2. A Nonlinear Lagrangian Function

Consider the following discrete inequality constrained optimization problem P:

$$\begin{array}{lll} \inf & f_0(x) & \\ \text{subject to} & f_i(x) \leq 0, & i = 1, \cdots, m, \\ & x \in X, & \end{array}$$

where $X \subset R^n$ is a discrete set of finite elements or infinite elements, $f_i : X \to R$, $i = 0, 1, 2, \cdots, m$ are real-valued functions. Since $X$ may have an infinite number of elements in one case, 'inf' is used, instead of 'min'.

Let the set of feasible solutions of problem P be

$$X_0 = \{x \in R^n : x \in X, f_i(x) \leq 0, \quad i = 1, 2, \cdots, m\}$$

and

$$A = \{e = (1, e_1, \cdots, e_m)^\top : e_i > 0, i = 1, \cdots, m\}.$$

The set $A$ will be the feasible set for the dual optimization problem to be constructed in Definition 3.

**Definition 1** The nonlinear Lagrangian function of P is defined as

$$N(u, e) = \max_{0 \leq i \leq m} \left\{ \frac{f_i(u)}{e_i} \right\},$$

where $e \in A, u \in X$.

Note that the function $N(u, e)$ can be interpreted as a weighted Tchebyshev metric of the vector $(f_0(u), f_1(u), \cdots, f_m(u))^\top$.

**Definition 2** The nonlinear Lagrangian dual function for P is defined as follows

$$\phi(e) = \inf_{u \in X} N(u, e),$$

where $e \in A$.

**Proposition 1** The dual function $\phi(e)$ is nonincreasing, i.e., if $e_1, e_2 \in A$, $e_1 - e_2 \in R_+^{m+1}$, then

$$\phi(e_1) \leq \phi(e_2).$$

**Proof.** Follows directly from the definition of $\phi$.

It is clear that computing the nonlinear dual function $\phi(e)$ is equivalent to solving a minimax discrete optimization problem or a convex composite discrete optimization problem which is not an easy task. Thus the difficulty in the following dual optimization problem is to evaluate the dual function $\phi(e)$. However, the constraint has a simple structure.

**Definition 3** The dual optimization problem DP of problem P is defined as

$$\begin{array}{ll} \sup & \phi(e) \\ \text{subject to} & e \in A. \end{array}$$

**Theorem 1** *(Weak duality) Assume that $f_0(x) \geq 0, \forall x \in X_0$. Let $x \in X_0$ and $e \in A$. Then*

$$f_0(x) \geq \phi(e). \tag{1}$$

**Proof:** It follows from the assumption and $x \in X_0$ that

$$f_0(x) \geq \max_{0 \leq i \leq m} \left\{ \frac{f_i(x)}{e_i} \right\} = N(x, e).$$

Thus

$$f_0(x) \geq N(x, e) \geq \inf_{u \in X} N(u, e) = \phi(e).$$

Thus, (1) holds.

It is also not difficult to see that an equality constraint, if any, can be easily converted into two inequality constraints. Thus our results are applicable to optimization problems with both inequality and equality constrained optimization problems, see Example 1.

- **Geometric Interpretation**

The family of perturbed problems (Problem $\text{P}_y$ associated with problem P is defined by

$$\begin{array}{ll} \inf & f_0(x) \\ \text{subject to} & f_i(x) \leq y_i, \quad i = 1, \cdots, m, \\ & x \in X, \end{array}$$

where the vector $y = (y_1, \cdots, y_m)^\top$ is a perturbation to the original problem P. When $y = 0$, the perturbed problem reduces to the original problem P. Let the value function $w : R^m \to R$ be defined by

$$w(y) = \inf\{f_0(x) \mid f_i(x) \le y_i, i = 1, \cdots, m\ x \in X\}, \tag{2}$$

which has an effective domain $\text{dom}(w) = \{y \mid \exists x \in X \text{ s.t. } f_i(x) \le y_i, i = 1, \cdots, m\}$. Define the epigraph of $w(y)$ as:

$$epi(w) = \left\{ \begin{pmatrix} y \\ y_0 \end{pmatrix} \mid y_0 \ge w(y) \right\} \subset R^{m+1}.$$

Since $X$ is a discrete set, the feasible set for Problem $\text{P}_y$ may not change as $y$ changes. Thus $w(y)$ is a step function. Clearly $w(y)$ is a monotone non-increasing step function of $y$. The epigraph $epi(w)$ is always non-convex. In general, a supporting hyperplane does not exist at $y = 0$. However, there is a shifted supporting cone $a - R_+^{m+1}$ at any boundary point $a$ of $epi(w)$. This shifted cone is measured by the weighted Tchebyshev metric as in Definition 1.

## 3. Zero Duality Gaps

In this section, we establish zero duality gaps for discrete optimization problems and their nonlinear Lagrangian dual problems. We consider the cases where $X$ is a set of finite number of elements (Theorem 2), or a set of infinite number of elements (Theorem 3). In next section, we show an application of the result in a network optimization problem.

**Theorem 2** *(Zero duality gap) Assume that $X$ is a discrete set with a finite number of elements and $X_0 \ne \emptyset$. If $f_0(x) \ge 0, \forall x \in X_0$, then*

$$\min_{x \in X_0} f_0(x) = \sup_{e \in A} \phi(e), \tag{3}$$

*Note that here 'min' is used since $X$ is a discrete set of finite elements.*

**Proof:** If the equality does not hold, then by weak duality (1), there exists $\epsilon > 0$ such that

$$\min_{x \in X_0} f_0(x) - \epsilon \ge \phi(e), \quad \forall e \in A. \tag{4}$$

For $\epsilon/4 > 0$, there exists $\hat{x} \in X_0$ such that

$$f_0(\hat{x}) \ge \min_{x \in X_0} f_0(x) \ge f_0(\hat{x}) - \frac{\epsilon}{4}. \tag{5}$$

It is clear from (3) and (4) that

$$f_0(\hat{x}) - \epsilon \geq \min_{x \in X_0} f_0(x) - \epsilon \geq \phi(e), \forall e \in A. \tag{6}$$

Then, by the assumption of the theorem and (5),

$$\begin{aligned} \phi(e) &= \min\left\{ \min_{x \in X_0} \max_{0 \leq i \leq m} \left\{ \frac{f_i(x)}{e_i} \right\}, \min_{x \in X \setminus X_0} \max_{0 \leq i \leq m} \left\{ \frac{f_i(x)}{e_i} \right\} \right\} \\ &= \min\left\{ \min_{x \in X_0} f_0(x), \min_{x \in X \setminus X_0} \max_{0 \leq i \leq m} \left\{ \frac{f_i(x)}{e_i} \right\} \right\} \\ &\geq \min\left\{ f_0(\hat{x}) - \frac{\epsilon}{4}, \min_{x \in X \setminus X_0} \max_{0 \leq i \leq m} \left\{ \frac{f_i(x)}{e_i} \right\} \right\}. \end{aligned} \tag{7}$$

From (6) and (7), we have

$$\phi(e) \geq \min_{x \in X \setminus X_0} \max_{0 \leq i \leq m} \left\{ \frac{f_i(x)}{e_i} \right\}, \forall e \in A. \tag{8}$$

Let $i$ $(i = 1, \cdots, m)$ be fixed and any $\gamma > 0$ be given. One of the the two cases may happen.

Case I. There exists $x \in X \setminus X_0$ such that $f_i(x) > 0$. Since $X \setminus X_0$ has only a finite number of elements, there exists $\tau_i > 0$ such that $f_i(x) > \tau_i$, for any $x \in X \setminus X_0$ and $f_i(x)$ being positive.

In this case, let $e_i$ be such that

$$\frac{\tau_i}{\gamma} \geq e_i.$$

Then

$$\frac{f_i(x)}{e_i} \geq \gamma, \quad \forall x \in X \setminus X_0 \text{ and } f_i(x) \text{ being positive.}$$

Case II. For all $x \in X \setminus X_0$, $f_i(x) \leq 0$. Let $e_i = 1$.

Let $x \in X \setminus X_0$. There always exists $i$ such that Case I happens. Thus

$$\max_{0 \leq i \leq m} \left\{ \frac{f_i(x)}{e_i} \right\} \geq \gamma, \quad \forall x \in X \setminus X_0.$$

So

$$\min_{x\in X\setminus X_0}\max_{0\le i\le m}\left\{\frac{f_i(x)}{e_i}\right\}\ge\gamma.$$

Notice that $\gamma$ is given arbitrarily, so

$$\min_{x\in X\setminus X_0}\max_{0\le i\le m}\left\{\frac{f_i(x)}{e_i}\right\}\to+\infty,\qquad \text{as } e\to+0 \text{ in } A,$$

a contradiction to (8).

Next we consider the case when $X$ is a discrete set but has an infinite number of elements. The proof of next result is similar to that of Theorem 4.2 in Goh and Yang (1996b) with some modification to allow for the case that $X$ is now a discrete set but has an infinite number of elements. In Goh and Yang (1996b) the case that $X$ is a connected set was considered.

**Theorem 3** *(Zero duality gap) Assume that $f_0(x)\ge 0,\forall x\in X_0$ and $X$ is a discrete set, but has an infinite number of elements. Let $f_0$ be uniformly continuous and $f_i(i=1,\cdots,m)$ be continuous functions on an open set containing $X$. Then*

$$\inf_{x\in X_0} f_0(x)=\sup_{e\in A}\phi(e). \tag{9}$$

**Proof:** If (9) does not hold, then by weak duality (1) as in the first part of the proof for Theorem 2, there exist $\epsilon>0$ and $\hat{x}\in X_0$ such that

$$f_0(\hat{x})\ge\inf_{x\in X_0} f_0(x)\ge f_0(\hat{x})-\frac{\epsilon}{4}. \tag{10}$$

$$f_0(\hat{x})-\epsilon\ge\inf_{x\in X_0} f_0(x)-\epsilon\ge\phi(e),\forall e\in A, \tag{11}$$

and

$$\phi(e)\ge\inf_{x\in X\setminus X_0}\max_{0\le i\le m}\left\{\frac{f_i(x)}{e_i}\right\},\forall e\in A. \tag{12}$$

It follows from the uniform continuity of $f_0$ and the boundedness of $X$ that there exists $\delta>0$ such that if $x\in X\setminus X_0$ and $d(x,X_0)\le\delta$, then

$$-\frac{\epsilon}{2}<f_0(x)-f_0(x_2)<\frac{\epsilon}{2},\ \text{for some } x_2\in X_0.$$

Note that $f_0(x_2) \geq \inf_{x \in X_0} f_0(x) \geq f_0(\hat{x}) - \epsilon/4$. Then

$$f_0(\hat{x}) - \frac{3\epsilon}{4} < f_0(x). \tag{13}$$

From (10)-(12), we have

$$\phi(e) \geq \inf_{\substack{x \in X \setminus X_0, \\ d(x, X_0) > \delta}} \max_{0 \leq i \leq m} \left\{ \frac{f_i(x)}{e_i} \right\}, \forall e \in A. \tag{14}$$

It is clear that there exists $\beta > 0$ such that for any $x \in X \setminus X_0$, satisfying $d(x, X_0) > \delta$, $f_i(x) > \beta$, for some $i$. Define

$$B_i = \{x \in X \setminus X_0 : d(x, X_0) > \delta, f_i(x) > \beta\}.$$

Let

$$\hat{e}_i = \begin{cases} \min_{x \in B_i} \left\{ \frac{f_i(x)}{f_0(\hat{x}) - \frac{\epsilon}{2}} \right\}, & \text{if } B_i \neq \emptyset, \\ 1, & \text{if } B_i = \emptyset. \end{cases}$$

It is clear that $\hat{e} = (1, \hat{e}_1, \cdots, \hat{e}_m) \in A$. Moreover,

$$\begin{aligned} \phi(\hat{e}) &\geq \inf_{\substack{x \in X \setminus X_0, \\ d(x, X_0) > \delta}} \max_{0 \leq i \leq m} \left\{ \tfrac{f_i(x)}{\hat{e}_i} \right\} \\ &\geq \max_{0 \leq i \leq m} \tfrac{f_i(w)}{\hat{e}_i} - \tfrac{\epsilon}{3}, \end{aligned}$$

where for some $w \in X \setminus X_0, d(w, X_0) > \delta$ from the definition of '$inf$'. Now there exists $k_w$ such that $w \in B_{k_w}$ and

$$\max_{0 \leq i \leq m} \frac{f_i(w)}{\hat{e}_i} \geq \frac{f_{k_w}(w)}{\hat{e}_{k_w}}.$$

So

$$\phi(\hat{e}) \geq \frac{f_{k_w}(w)}{\hat{e}_{k_w}} - \frac{\epsilon}{3},$$

and hence

$$\phi(\hat{e}) \geq f_{k_w}(w) \left( \min_{v \in B_{k_w}} \frac{f_{k_w}(v)}{f_0(\hat{x}) - \frac{\epsilon}{2}} \right)^{-1} - \frac{\epsilon}{3}.$$

Since $w \in B_{k_w}$, we have

$$\phi(\hat{e}) \geq f_0(\hat{x}) - \frac{5\epsilon}{6} > f_0(\hat{x}) - \epsilon.$$

This contradicts (11) and the proof is complete.

**Corollary 1** (Zero duality gap) Assume that $f_0(x) \geq 0, \forall x \in X_0$ and $X$ is a discrete and bounded set, but has an infinite number of elements. Let $f_i(i = 0, 1, \cdots, m)$ be continuous functions on an open set containing $X$. Then

$$\inf_{x \in X_0} f_0(x) = \sup_{e \in A} \phi(e). \tag{15}$$

**Proof:** Clearly $f_0$ is uniformly continuous on an open set containing $X$. The result follows from Theorem 3.

## 4. Applications

### • An Example

**Example 1** Consider the example given in page 181, Bazaraa and Shetty (1979):

$$\begin{array}{ll} \min & f_0(x) \\ \text{subject to} & g_1(x) = 0, \\ & x = (x_1, x_2) \in X, \end{array}$$

where $X = \{(0,0), (0,4), (4,4), (4,0), (1,2), (2,1)\}$, $f_0(x) = -2x_1 + x_2 + 4, g_1(x) = x_1 + x_2 - 3$. It is clear that $\min_{X_0} f_0(x) = 1$. If the linear Lagrangian dual is applied, then there is a nonzero duality gap, see Bazaraa and Shetty (1979) since

$$\max_{\lambda} \min_{X} \{f_0(x) + \lambda g_1(x)\} = -2.$$

Next we apply the nonlinear Lagrangian dual formulation to the problem. We calculate the dual cost. To do this, we rewrite the optimization problem as

$$\begin{array}{ll} \min & f_0(x) \\ \text{subject to} & f_1(x) \leq 0, \\ & f_2(x) \leq 0, \\ & x = (x_1, x_2) \in X, \end{array}$$

where $f_1(x) = x_1 + x_2 - 3$, $f_2(x) = -x_1 - x_2 + 3$.

Then the dual function is

$$\phi(e) = \min_{u \in X} \max_{0 \leq i \leq 2} \left\{ \frac{f_i(u)}{e_i} \right\},$$

where $e_0 = 1, e_i > 0, i = 1, 2$. It is not difficult to verify that

$$\max_{e_i > 0} \phi(e) = 1,$$

where $e_0 = 1$. Hence there exists a zero duality gap when the nonlinear Lagrangian dual formulation is applied.

- **Network optimization problems**

Consider the multi-commodity network flow problem NP with integer variables: given a network $\mathcal{G} = (\mathcal{N}, \mathcal{A})$,

$$\begin{array}{ll} \min & f(x_1, \cdots, x_k) \\ \text{subject to} & \sum_{l=1}^{k} x_l^j \leq d_j, \quad j = 1, \cdots, n, \\ & Ex_l = b_l, \quad l = 1, \cdots, k, \\ & 0 \leq x_l \leq v_l, \quad x_l^j \text{ integer}, \ l = 1, \cdots, k, \end{array}$$

where $m$ is the number of nodes in $\mathcal{N}$, $n$ is the number of arcs in $\mathcal{A}$, $d_j$ is the shared capacity on arc $j$, $f$ can be either linear or convex or even nonconvex and $E$ is the $m \times n$ node-arc incident matrix. The model of a nonconvex cost function $f$ can be found in Dem'yanov et al (1996).

Let $X_0'$ be the feasible set of NP.

The nonlinear dual function $\phi'(e)$ is defined to be the optimal value of the following optimization problem

$$\begin{array}{ll} \min\limits_{u} & N'(u, e) \\ \text{subject to} & Eu_l = b_l, \ l = 1, \cdots, k, \\ & 0 \leq u_l \leq v_l, \quad u_l^j \text{ integer}, \ l = 1, \cdots, k, \end{array}$$

where $e \in A$ and

$$N'(u, e) = \max \left\{ f(u_1, \cdots, u_k), \frac{\sum_{l=1}^{k} u_l^1 - d_1}{e_1}, \cdots, \frac{\sum_{l=1}^{k} u_l^n - d_n}{e_n} \right\}$$

is the nonlinear dual function. Consider the dual optimization problem DNP of NP,

$$\begin{array}{ll} \sup & \phi'(e) \\ \text{subject to} & e \in E. \end{array}$$

Then we have

**Theorem 4** *(Zero duality gap) Assume that $f(x) \geq 0, \forall x \in X_0'$ and $X_0'$ has a finite number of elements (or $X_0'$ has an infinite number of elements and is bounded and $f$ is continuous on an open set containing $X$). Then*

$$\inf_{x \in X_0'} f(x_1, \cdots, x_k) = \sup_{e \in A} \phi'(e). \qquad (16)$$

**Proof.** Follows from Theorems 3.1 and 3.2.

When evaluating the nonlinear dual function $\phi'(e)$, smoothing methods proposed in Bertsekas (1976) and Yang (1994) can be applied. Note that the dual function $\phi'(e)$ has a convex composite structure. Another approach is to consider the first-order approximation of $N'(u, e)$ at $u_0$:

$$N'(u, e) = N'(u_0, e) + \gamma^\top (u - u_0) + o(\|u - u_0\|),$$

where $\gamma \in \partial_u N'(u_0, e)$ - the Clarke subgradient. Thus the problem of evaluating $\phi'(e)$ is reduced to a set of linear cost network optimization problems with integer variables:

$$\begin{array}{lcc} \min\limits_{u} & & \gamma^\top (u - u_0) \\ \text{subject to} & & Eu_l = b_l, \ l = 1, \cdots, k, \\ & 0 \leq u_l \leq v_l, & u_l^j \text{ integer}, \ l = 1, \cdots, k, \end{array}$$

Each of the above linear problems can be solved efficiently using the network optimization method, for example, Lagrangian relaxation method in Ahuja et al (1993). In particular, if $x_l^j$ is a continuous variable, each subproblem is a linear cost network optimization problem, hence can be solved using network simplex method.

On the other hand, it is worth noting that in last ten years, extensive research has been done to the development of heuristic methods, such as simulated annealing and tabu search, see Glover et al (1993) and Reeves

(1993). As in these heuristic methods, only function values are computed. Thus there is an advantage to combine the nonlinear Lagrangian theory developed in this paper and heuristic methods.

Moreover, other problems that can be formulated in this way include the network optimization problem with side constraints, see Ahuja et al (1993).

## 5. Conclusions

In this paper a nonlinear Lagrangian method for discrete optimization problems was developed. The nonlinear Lagrangian method for continuous optimization problems was presented in Goh and Yang (1996b). This opens a way to tackle the difficult, but important, nonconvex/discrete optimization problems. The class of nonlinear Lagrangian methods provides zero duality gaps under very weak conditions. However, these methods and theories are of a preliminary study and more refinements are needed. Computational experiments for large scale nonconvex (discrete) optimization problems are needed to test the efficiency of the proposed method and to compare with other existing methods for solving these problems.

# References


[1] Ahuja, R.K., Magnanti, T.L. and Orlin, J.B.(1993), *Network flows*, Prentice Hall, Englwood Cliffs, New Jersey.

[2] Bazaraa, M.S. and Shetty, C.M.(1979), *Nonlinear Programming*, John Wiley & Sons, New York, Singapore.

[3] Ben-Tal, A. and Teboulle, M. (1996) Hidden convexity in some non-convex quadratically constrained quadratic programming, *Mathematical Programming*, Vol. 72, pp. 51-63.

[4] Bertsekas, D.P.(1976), Nondifferentiable optimization via approximation, *Math. Prog.* Vol. 3, 1-25.

[5] Dem'yanov, V.F., Stavroulakis, G.E., Polyakova, L.N. and Panagiotopoulos, P.D. (1996) *Quasidifferentiability and Nonsmooth Modelling in Mechanics, Engineering and Economics*, Kluwer Academic Publishers, Dordrecht / Boston / London.

[6] Glover, F, Taillard, E. and de Werra, D. (1993), A user's guide to tabu search, *Annals of Operations Research* Vol. 41, 3-28.

[7] Goh, C.J. and Yang, X.Q. (1996a) A sufficient and necessary condition for nonconvex constrained optimization, *Applied Mathematics Letters* (to appear).

[8] Goh, C.J. and Yang, X.Q. (1996b) A nonlinear Lagrangian theory for nonconvex optimization problems, Preprint, the University of Western Australia, Australia.

[9] Horst, R. and Tuy, H. (1991) *Global Optimization – Deterministic Approaches*, Springer-Verlag, Berlin.

[10] Jeyakumar, V. and Wolkowicz, H.(1990), Zero duality gaps in infinite-dimensional programming, *Journal of Optimization Theory and Applications*, Vol. 67, 87-108.

[11] Li, D.(1995), Zero duality gap for a class of nonconvex optimization problem, *Journal of Optimization Theory and Applications*, Vol. 85(2), pp. 309-324.

[12] Luenberger, D.G.(1984), *Linear and Nonlinear Programming, second ed.*, Addison Wesley, Reading, MA.

[13] Reeves, C.R.(eds)(1993), Modern Heuristic Techniques for Combinatorial Problems, John Wiley & Sons, INC, New York and Toronto.

[14] Rockafellar, R.T.(1974), Conjugate Duality and Optimization, CBMS Lecture Notes Series, Vol. 162, SIAM Publications, Philadephia, Pennsylvania.

[15] Thach, R.T.(1993), Global optimality criterion and a duality with zero gap in nonconvex optimization, *SIAM J. Math. Anal.* Vol. 24, pp. 1537-1556.

[16] Yang, X.Q.(1994), Smoothing approximations to nonsmooth optimization problems, *J. Austral. Math. Soc.*, B, Vol. 36, pp. 274-285.

Chapter 14

# LOCAL AND GLOBAL BEHAVIOR OF MOVING POLYTOPE ALGORITHMS

Beidi Soumana Hamma

*Parallel Algorithms Team,*
*CERFACS, European Centre for Research*
*and Advanced Training in Scientific Computing*
*42 av. Coriolis, F-31057 Toulouse-Cedex,*
*France.*

**Abstract**

This paper deals with algorithms based on the Moving Polytope Method which solves nonlinear optimization problems without using gradients. We first review two algorithms based on the Moving Polytope Method: the original algorithm by Nelder and Mead [2], and a parallel algorithm by Dennis and Torczon [1]. We then introduce a new algorithm that we call Adaptive Direct Search (ADS) which retains the advantages of the two previous algorithms and attempts to find an improved local minimum.

**Keywords:** Direct Search Methods, Numerical Optimization, Parallel Computing, Improvement of Local Minimum

## 1. Introduction

We are concerned with algorithms of the Moving Polytope Method for solving the following optimization problem



*A. Migdalas et al. (eds.), From Local to Global Optimization,* 305–325.

$$\text{minimize } f(x) \text{ where } x \in \mathbb{R}^n$$

without using gradients.

The original method, also known as Downhill Simplex, Nonlinear Simplex, or simply Nelder-Mead Simplex (Method) is due to J.A. Nelder and R. Mead (1965) [2]. Our interest in this method is based on the following observations:
- first, the method requires only function evaluations, not derivatives;
- second, it is used by many people, for instance who wish to have something working quickly and easy to implement . As illustration we can read in [1] "*...the 1989 Science Citation Index lists more than 215 citations for the classic Nelder-Mead paper...*", and, "*... Nelder-Mead simplex algorithm appears in most available commercial software libraries.*";
- third, it is suitable for parallel computation.

Finally, we are also thinking about how to derive a "global" optimization moving polytope algorithm.

After reviewing the original algorithm [2], and a parallel modified algorithm by Dennis and Torczon, [1], we introduce a new algorithm, which we call Adaptive Direct Search (ADS). The ADS algorithm has the following interesting characteristics:
- it retains all the benefits of the algorithms of [1] and [2]
- it is parallelizable
- it is more stable than the algorithm of [2]
- finally, it sometimes behaves like a global optimizer by escaping from local minimum states.

The three algorithms ( [1], [2] and ADS), start from a Simplex (we prefer the term Polytope) which is a geometrical figure of N+1 vertices (in dimension N) and their interconnecting lines. They then perform some geometrical operations (reflection, expansion, contraction) on some candidate vertices in order to move the Polytope in descent "directions". That is, the candidates are first reflected; if this makes an improvement (i.e., it gives a better function value), an expansion will be attempted. Otherwise the polytope is reduced. By those geometrical operations the polytope changes in volume and in form, thus the name of Moving Polytope.

In the strategy of the original algorithm, the choices made are the following:
- the candidate vertex which will be (first) affected is $X_h$ the one with the highest function value. That is, the (only) search direction is $< X_h, X_0 >$
- the centre of the geometrical operations is $X_0$ the centroid of all vertices except $X_L$ (the lowest function value vertex).

In the strategy of the Dennis-Torczon algorithm, the choices are the following:
- all vertices are candidates at each operation. There are $N$ search directions $< X_j, X_L >$ for $j = 1, 2, ..., N$ and $j \neq L$
- the centre of operations is the vertex $X_L$ with the lowest function value.

In our approach, we allow the polytope to attempt an escape from the local minima by adding an adaptive search. That is, when a search fails (the reflection doesn't improve) in we do not immediately contract the polytope as is done in the other two algorithms. Instead, we try first to improve the situation by changing the candidates for the geometrical operations and/or changing the centre. Thus, contractions are made only after this adaptive search fails.

The ADS implementation is based on the following strategies:

- The Nelder-Mead strategy plus additional search directions before contractions.

- The Dennis-Torczon Strategy, with supplementary searches along the direction $< X_h, X_0 >$ (which is not a search direction for Dennis-Torczon) as well as along the directions $< X_{h-1}, X_0 >$ , $< X_{h-2}, X_0 >$, ...

Finally we present some numerical experiments on the minimization of standard functions (some being constrained) using ADS.

# 2. Description of the algorithms

## 2.1. The Sequential Moving Polytope of Nelder and Mead

We first present the original algorithm (which is sequential) as introduced by its authors, Nelder and Mead [1]. It can be seen as an extension of the simplex method of Spendley, Hext and Himsworth [3].

**Definition 1** *A set of* $(N+1)$ *mutually equidistant points in N-dimensional space is known as a* regular simplex (or polytope).

In this paper we will call a polytope any set of $(N+1)$ points in N-dimensional space. The points will be called the vertices of the polytope.

Thus, in two dimensions a polytope is a triangle and in three dimensions it is a tetrahedron.

**Notation:** We will denote a polytope in this form $\mathbf{S} = \{X_1, X_2, ..., X_{N+1}\}$.

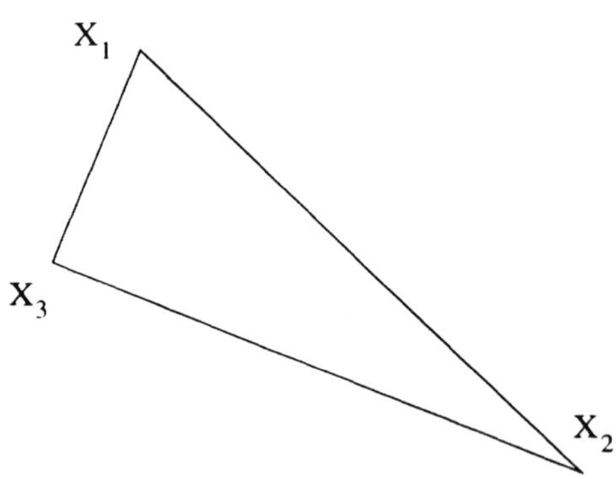


*Figure 14.1.* A polytope $\mathbf{S} = \{X_1, X_2, X_3\}$ in $\mathbb{R}^2$.

In Spendley and Himsworth [3], the polytope remains regular during the entire process. Nelder and Mead proposed several modifications to the method which allow the polytopes to become non-regular. This results in a direct search method which gives good results for small dimensional problems .

**Underlying idea.** The principle is the following. We start with a polytope composed of $N + 1$ vertices (for an N-dimensional optimization problem). Then, by applying three basic geometrical operations *reflection, expansion* and *contraction* we try to move away from $X_h$ (the vertex with the highest objective function value) in order to attempt to replace it with a new "better" vertex. We may also replace $X_{h-1}$, the second highest function value vertex. This procedure will move the whole polytope in the descent directions. The polytope may eventually change, the sizes of the edges becoming irregular, and converge towards a small polytope around a local minimum.

The procedure can be summarized in the following five basic steps:

Step 1: Choose an initial polytope S (i.e. $N + 1$ vertices) such that the edges joining the vertices span $\mathbb{R}^n$.

Step 2: Take as a search direction $[X_h, X_0]$, the direction joining $X_h$ and the centroid $X_0$ of all vertices expected $X_h$.

Step 3: If this direction behaves as a descent direction, we replace $X_h$ by a chosen point in this direction. That point may be chosen closer (reflection) or the furthest possible (expansion) according to the results of the basic geometrical operations applied on $X_h$.

Step 4: The polytope is eventually updated with the new vertices according to step 3, aiming to "move" it in the descent directions. It may also be reduced.

Step 5: Stop if a stopping condition is satisfied otherwise return to step 2.

At the end, the last polytope should be a small one, with vertices very close together, and so hopefully near a local minimum.

**Definition 2 Reflection, Expansion, Contraction and Polytope Reduction**
For $X \in \mathbb{R}^n$ we give the following definitions:

1. Image of $X$ by **Reflection** in $X_0$.
*The image $X_r$ of $X$ by reflection in $X_0$ with reflection factor $\alpha > 0$ is*

*given by:*

$$X_r - X_0 = \alpha(X_0 - X) ,$$

*i.e.* $X_r = (1+\alpha)X_0 - \alpha X$.

2. Image of $X$ by **Expansion** in $X_0$ (assuming $X_r$ is its reflected image). *The image $X_e$ of $X$ by expansion in $X_0$ with expansion factor $\gamma > 0$ is given by:*

$$X_e - X_0 = \gamma(X_r - X_0) ,$$

*i.e.* $X_e = \gamma X_r + (1-\gamma)X_0$.

3. Image of $X$ by **Contraction** with respect to $X_0$. *The image $X_c$ of $X$ by contraction with respect to $X_0$ with contraction factor $0 < \beta < 1$ is given by:*

$$X_c - X_0 = \beta(X - X_0) ,$$

*i.e.* $X_c = \beta X + (1-\beta)X_0$.

**Definition 3** *A reflection is said to be a* success *if and only if* $f(X_r) \leq f(X_L)$. *An expansion is said to be a* success *if and only if* $f(X_e) \leq f(X_L)$.

**Sequential Moving Polytope algorithm**

Initializations of:
  $S0 = \{X_1, X_2, ..X_{N+1}\}$;
  $\alpha, \gamma, \beta$ /* geometrical factors */
  Compute $f(X_1), f(X_2), ..., f(X_{N+1})$;
**While** ( Stopping-Criteria = false ) **do**
  Find $X_L, X_{h-1}, X_h, f_L, f_{h-1}, f_h$;
  Find $X_0$ the centroid of all points expect $X_h$;
  Reflect $X_h$;
  **If** (the-reflection-is-a-success) **then**
    Expand $X_h$;
    **If** (the-expansion-is-a-success) **then**
      Replace $X_h$ by $X_e$;
    **else**
      Replace $X_h$ by $X_r$;
    **end if**;
  **else**
  Contract $X_r$;

```
    If (this-contraction-is-a-success) then
        Replace X_h by X_c;
    else
        Reduce the polytope; /* by halving all distances from X_0 */
    end if;
end if;
End While;
```

**Definition 4** *A partial contraction, i.e. a contraction of* $X_h$ *in* $X_r$ *(assuming* $X_r$ *is the reflected image of* $X_h$*), is said to be a* success *if and only* $f(X_c) \leq f(X_h)$.

**Description of the algorithm.** Recall that the aim of this procedure is to move away from $X_h$, the highest function value vertex. We present in more detail the different phases of the procedure. From phase A to phase G we can follow how the procedure works:

- **PHASE (A)**
  We start with the initial polytope $\mathbf{S} = \{X_1, X_2, ..X_{N+1}\}$ (i.e. with $(N+1)$ points) and find their function values $f_1, f_2, ..., f_{N+1}$ .

- **PHASE (B)**
  We find the highest function value $f_h$, the next highest function value $f_{h-1}$ and $f_L$ the lowest function value and the corresponding points $X_h, X_{h-1}$ and $X_L$.

- **PHASE (C)**
  We find the centroid of all points except $X_h$. Let this be denoted $X_0$ and evaluate $f_0 = f(X_0)$

$$X_0 = \frac{1}{N} \sum_{i \neq h} X_i.$$

- **PHASE (D)**
  It would seem reasonable to try to move away from $X_h$. We reflect $X_h$ in $X_0$ to find $X_r$.

- **PHASE (E)**: We now compare $f_r$ with $f_L$

  - **SUB-PHASE (E.1)**: $f_r < f_L$
    The reflection is a success, this means we have obtained the

lowest function value so far. The direction $< X_0, X_r >$ appears to be a good one to move along. We therefore make an **expansion** in this direction and find $X_e$

  * **SUB-SUB-PHASE (E.1.1)**: $f_e < f_L$
    The expansion is a success, we then **replace** $X_h$ by $X_e$. Test convergence for the updated polytope. If it has not converged return to step (B).

  * **SUB-SUB-PHASE (E.1.2)**: $f_e \not< f_L$
    In this case the expansion has failed. It seems that we have moved too far. We replace $X_h$ by $X_r$ and test convergence for the updated polytope. If it has not converged return to step (B).

- **SUB-PHASE (E.2)**: $f_r \not< f_L$ but $f_r < f_{h-1}$
  The reflection has failed ($f_r > f_L$). Nevertheless, since $f_r < f_{h-1}$, we can consider that there is an improvement compared to a worst point ($X_{h-1}$): replace $X_h$ by $X_r$. Test convergence for the updated polytope. If it has not converged return to step B.

- **SUB-PHASE (E.3)**: $f_r \not< f_L$ and $f_r \not< f_{h-1}$
  The reflection has failed ( $f_r > f_L$), and we have $f_r > f_{h-1}$, that is, no improvement at all. Proceed to step F.

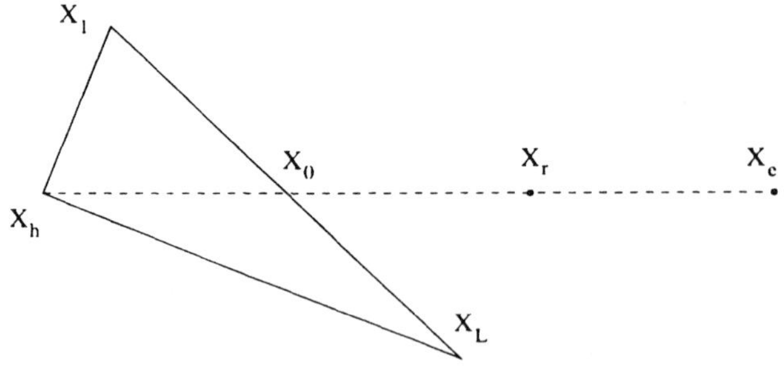


*Figure 14.2.* The polytope $\mathbf{S} = \{X_L, X_2, X_h\}$ with the reflected and expanded points.

- **PHASE (F)**: We now compare $f_r$ with $f_h$. Recall that $f_r \not< f_L$ and $f_r \not< f_{h-1}$

- **SUB-PHASE (F.1)**: $f_r < f_h$
  Replace $X_h$ by $X_r$, then make a contraction of $X_r$ with respect to $X_0$ and find $X_c$, then go to step G.

- **SUB-PHASE (F.2)**: $f_r \not< f_h$
  Make a contraction of $X_h$ with respect to $X_0$ and find $X_c$ (in this case we also call it $X_c$ *Bis*), then go to step G.

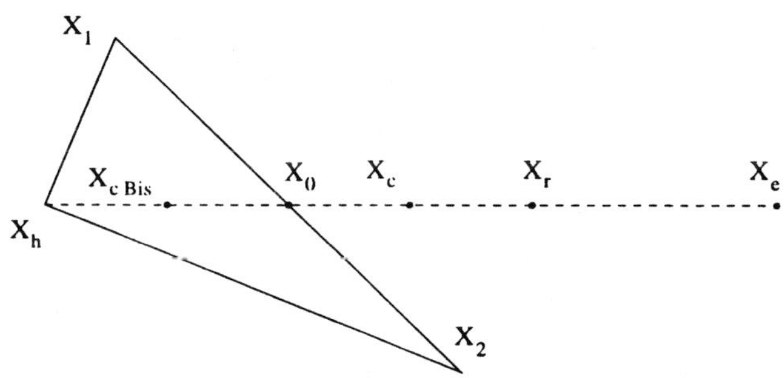


*Figure 14.3.* The polytope **S** with the reflected, expanded and contracted points.

- **PHASE (G)**: Evaluate $f(X_c) = f_c$.

  - **SUB-PHASE (G.1)**:$f_c < f_h$
    Replace $X_h$ by $X_c$. Test convergence for the updated polytope. If it has not converged return to step B.

  - **SUB-PHASE (G.2)**: $f_c \not< f_h$
    This means that all the efforts to find a better value than $f_h$ have failed. We then reduce the size of our polytope by halving the distance of each vertex of the polytope from $X_L$, i.e.,

$$X_i = (X_i + X_L)/2 \qquad i = 1, 2, ..., N + 1$$

    Test convergence for the updated polytope. If it has not converged return to step B.

**Properties of Nelder and Mead Moving Polytope.**

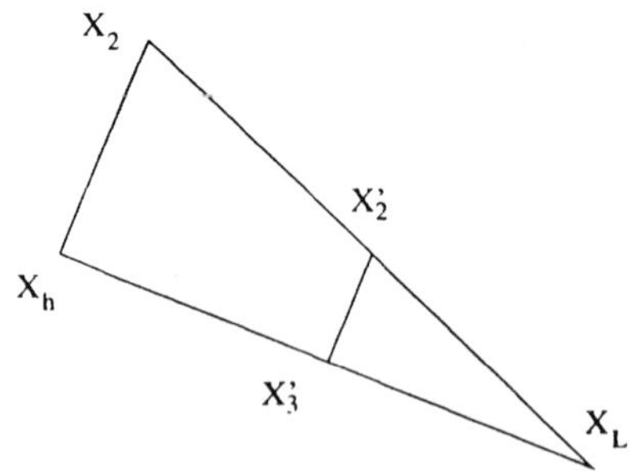


*Figure 14.4.* The polytope and its reduction.

At each iteration we have the following properties:

(i) The vertex $X_h$ is always replaced.

(ii) $X_L$ is never replaced.

(iii) $X_{h-1}$ may be replaced.

(iv) At each iteration we evaluate the function, at least, for two new points.

(v) $X_{h-2}$ , $X_{h-3}$ ... are replaced only when the polytope is reduced. This happens when the procedure fails totally ( SUB-PHASE G.2)

(vi) In $\mathbb{R}^2$, for instance, the polytope $\mathbf{S}= \{X_L, X_2, X_h\}$, after one iteration becomes one of the following polytopes (see figure 5): $\{X_L, X_2, X_e\}$ or $\{X_L, X_2, X_r\}$ or $\{X_L, X_2, X_c\}$ or $\{X_L, X_2, X_c'\}$ or the reduced polytope $\{X_L, X_2', X_3'\}$.

## 2.2. The Parallel Moving Polytope of Dennis and Torczon

In the sequential algorithm described in the previous section, the search direction is $[X_h, X_0]$. The idea of Dennis and Torczon is to use $N$ directions instead of one. Furthermore, while the sequential algorithm does all the geometrical operations with respect to $X_0$ (the centroid of all vertices expect $X_h$), they suggest to use $X_L$ (the lowest function value vertex) as basis for all operations. That is a reflection, in their case, will consist in reflecting all the other $N$ vertices in $X_L$. Expansion and contraction are also made for the $N$ vertices with respect to $X_L$. Another thing they have added is the notion of success for an expansion

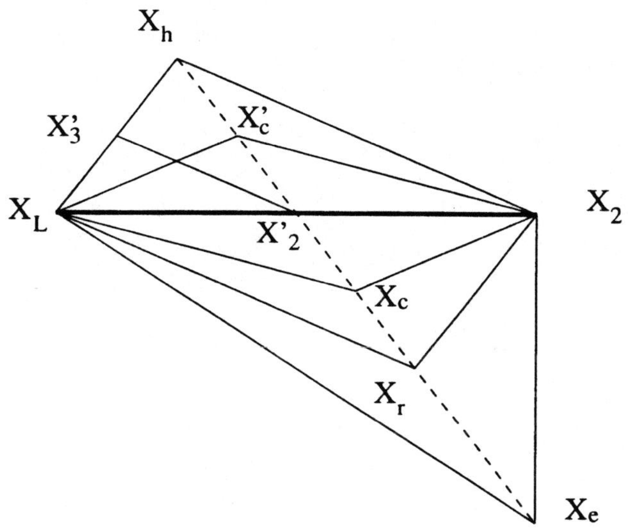


*Figure 14.5.* The polytope $\mathbf{S} = \{X_L, X_2, X_h\}$ with all its subsequent possible forms.

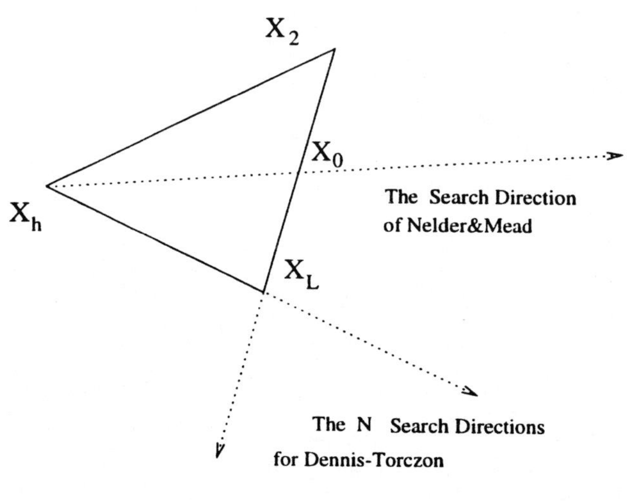


*Figure 14.6.* Search directions.

**Definition 5** *For the Dennis-Torczon algorithm, an expansion will be said a* success *if and only if the expansion ameliorates the reflection.*

Recall that in the sequential case only one amelioration compared to the worst point was required.

### parallel moving polytope algorithm

Choose coefficients:
  $\alpha,\ \beta,\ \gamma$ ;   /* for reflection, expansion and contraction */
Initializations:
  $S_0 = \{X_1^0, X_2^0, ..., X_{N+1}^0\}$;
Evaluate $f(X_1^0)$ , $f(X_2^0)$ , ..., $f(X_{N+1}^0)$;
Find $X_{min}^0$ = Argmin $\{f(X_i^0)\ i = 1, ..., N+1\}$;
Swap $X_{min}^0$ with $X_1^0$; /* so that $X_1^k$ is always the lowest function value vertex */
Set $f_{min}^0 = f(X_1^0)$;
Set $k = 1$;
**While** (Stopping-Criteria = false) **do**
  $X_{ref} = X_1^{k-1}$ ;   /* centre of geometrical operations */
  Reflect all vertices in $X_{ref}$;   /* this gives $Xr_i^k$ for each $i$ */
  Find $f_{min}^k = \{f(Xr_i^k)\ i = 1, ..., N+1\}$;
  **if** ( $f_{min}^k < f_{min}^{k-1}$) **then**
    Expand all vertices with respect to $X_{ref}$;   /* this gives $Xe_i^k$ */
    Find $fe_{min}^k = \{f(Xe_i^k)\ i = 1, ..., N+1\}$;
    **if** ($fe_{min}^k < f_{min}^k$) **then**
      Replace $X_i^k$ by $Xe_i^k$   $i = 2, ..., N+1$;
      Rename the polytope as $\mathbf{S_{k+1}} = \{X_1^{k+1}, X_2^{k+1}, ..., X_{N+1}^{k+1}\}$;
    **else**
      Replace $X_i^k$ by $Xr_i^k$   $i = 2, ..., N+1$;
      Rename the polytope as $\mathbf{S_{k+1}} = \{X_1^{k+1}, X_2^{k+1}, ..., X_{N+1}^{k+1}\}$;
    **endif**;
  **else**
    Contract all vertices with respect to $X_{ref}$;  /* this gives $Xc_i^k$ */
    Replace $X_i^k$ by $Xc_i^k$ ;   $i = 2, ..., N+1$
    Rename the polytope as $\mathbf{S_{k+1}} = \{X_1^{k+1}, X_2^{k+1}, ..., X_{N+1}^{k+1}\}$;
  **endif**
  Find $X_{min}^{k+1}$ = Argmin $\{ f(X_i^{k+1})\ i = 1, ..., N+1 \}$;
  **if** ( $f(X_{min}^{k+1}) < f(X_1^{k+1})$ )
    Swap $X_{min}^{k+1}$ and $X_1^{k+1}$;

**endif**;
$k = k + 1$;
**End while**;

**Properties of Dennis-Torczon algorithm.** At each iteration we have the following properties:

(i) $X_L$ is not replaced.

(ii) All other vertices are replaced.

(iii) At each iteration we evaluate the function, at least, at $2N$ new points (2 points for Nelder-Mead).

## 2.3. The Adaptive Direct Search (ADS) algorithm

### Motivation

First, in the two previous algorithms, the search directions are fixed in advance and they do not take into account the progress of the algorithm. That is, during the first iterations as well as during the last, the search directions remain always the same. There is no way to know whether, even after the algorithm fails in a direction, a contraction (or a reduction) is justified by the fact that the polytope is moving towards convergence.

Second, we try to answer the following questions:

During the early iterations, is a contraction "welcome", since it reduces the polytope and thus reduces (prematurely) the region of interest?

During the latter iterations, why keep only one search direction, especially when an attempt is made to improve the current region of attraction?

For all these reasons, we introduce an adaptive search (ADS) procedure which is supposed to select some supplementary search directions adapted to the state of the algorithm. This procedure is designed to prevent unprofitable contractions by trying other search directions in addition to the basic ones. It will also attempt to escape from the attraction regions of local minima.

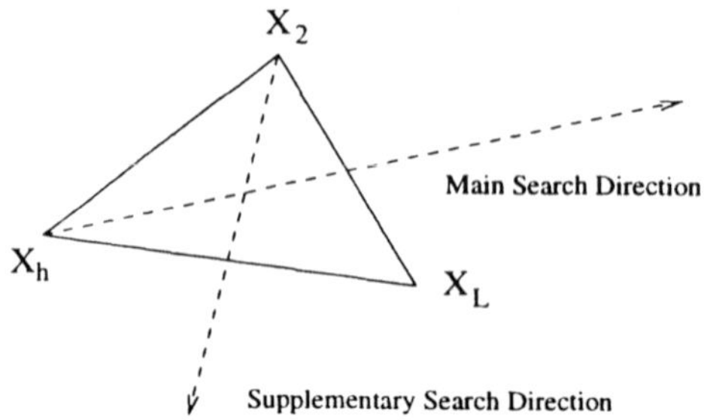


*Figure 14.7.* Example of ADS supplementary search directions for the sequential algorithm.

The main search direction still is $[X_h, X_0]$. The supplementary search directions are $[X_h, X_0]$, $[X_{h-1}, X_0]$, ..., $[X_2, X_0]$. That is, with ADS we have a maximum of $N$ possible search directions compared to the one of Nelder-Mead.

In the case of Dennis-Torczon + ADS, the main directions are (still) the $N$ $[X_2, X_L]$, $[X_3, X_L]$, ... $[X_{N+1}, X_L]$. The supplementary search directions are again , in order of importance: $[X_h, X_0]$, $[X_{h-1}, X_0]$, ..., $[X_2, X_0]$. That is, with ADS we have a maximum of $2N$ possible search directions compared to the $N$ of Dennis-Torczon.

## ADS sequential

Initializations of:
  $S0 = \{X_1, X_2, ..X_{N+1}\}$;
  $\alpha, \gamma, \beta$ /* geometrical factors */ ;
Compute $f(X_1), f(X_2), ..f(X_{N+1})$;
**While** ( Stopping-Criteria = false ) **do**
  Find $X_L, X_{h-1}, X_h, f_L, f_{h-1}, f_h$;
  Find $X_0$ the centroid of all points except $X_h$;
  Reflect $X_h$;
  **If** (the-reflection-is-a-success) **then**
    Expand $X_h$;
    **If** (the-expansion-is-a-success) **then**
      Replace $X_h$ by $X_e$;
    **else**

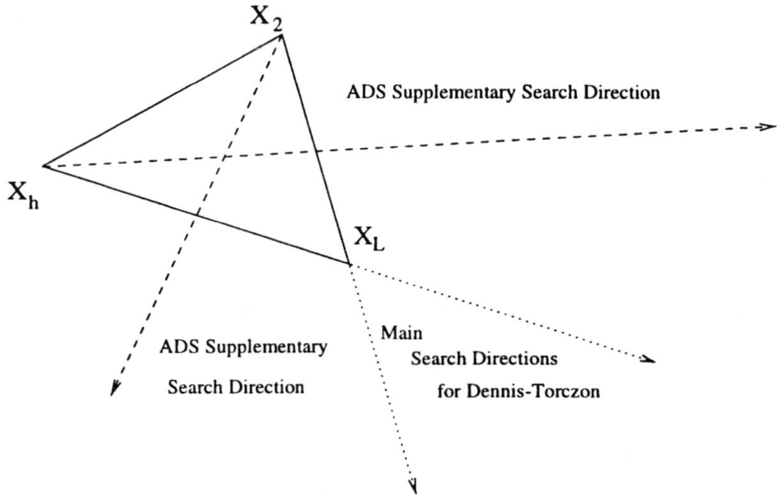


*Figure 14.8.* Example of ADS supplementary search directions for the parallel algorithm.

Replace $X_h$ by $X_r$ ;
**end if**
**else**
**call** the-Adaptive-Search;
**end if**
**End While.**

The **Adaptive Search** will determine, according to the progress of the algorithm, whether or not a contraction should be made. It may be useful to explore, in order of their importance, some supplementary search directions. If a better point $X_{AS}$ is found, $X_h$ is replaced by $X_{AS}$ and the main procedure continues in the While loop. Otherwise, a contraction or, in the worst case, a reduction of the polytope is made.

**Remarks.**
The AS fails only when it cannot find a better point in any of the supplementary search directions (their number may be fixed by the user). In this case a contraction is justified.
The user has to fix a certain number of supplementary searches to be performed over the maximum authorized. Then the algorithm will explore them in their order of importance ($[X_{h-1}, X_0]$ first, then $[X_{h-2}, X_0]$, ...).

**ADS parallel**

Step 1: Perform the Dennis-Torczon Reflection step.
Step 2: If the Reflection is a success perform next step otherwise go to Step 4.
Step 3: Perform the Dennis-Torczon Expansion step. If the expansion is a success update the polytope with the expanded vertices. Update otherwise the polytope with the reflected vertices. Go to step 6.
Step 4: Call the Adaptive Search. If the AS is a success update the polytope and go to step 6. Otherwise continue with next step.
Step 5: Perform the Dennis-Torczon Contraction step. Update the polytope and continue with next step.
Step 6: Test convergence. If it has not converged return to step 1.

## 3. Preliminary Numerical Experiments

We have implemented our own versions of [1], [2] and ADS in Fortran 77, and tested some constrained problems by using the following reformulation:

The problem to

$$\text{Minimize } f \text{ over } C$$

is replaced by the problem to

$$\text{Minimize } \phi_C,$$

where $\phi_C(X) = f(X)$ if $X \in C$ and $\phi_C(X) = +\infty$ if $X \notin C$.

This was enough for our test-functions since the feasible region has a nonempty interior.

**Choice of initial polytope**: Starting from a user-specified initial point $X$ we automatically construct $N$ additional points in the following manner:

$X_i = X + e_i$ where $e_i$ is the unit vector of $\mathbb{R}^n$ $e_i = (0, 0, ..., 1, 0, ..., 0)$.

This gives us the initial polytope $\mathbf{S} = \{X_1, X_2, ..., X_{N+1}\}$ from which we can start the moving polytope algorithms.

Constructing the initial polytope in this way ensures that it will be non-degenerate (i.e. the $N+1$ vertices span $\mathbb{R}^n$). This condition is required by all three algorithms.

**Choice of the geometrical coefficients:**

We obtained the best results with the following values:

- for the reflection coefficient: $\alpha = 1$

- for the expansion coefficient : $\beta = \frac{1}{2}$

- for the contraction coefficient : $\gamma = 2$.

**Stopping Conditions**: The Moving Polytope algorithms avoid the use of gradients. Theoretical convergence is rather difficult to establish for these kinds of problems. Dennis and Torczon give a convergence theorem for their algorithm but only for continuously differentiable problems. We know that for continuously differentiable problems, gradients methods are most suitable (both in terms of function evaluations and guaranteed convergence). In practice, implementations of Moving Polytope algorithms test numerical convergence by using some of the following tests:

- **Stopping test 1**:

$$\sum_{i=1,N+1} \frac{(f(X_i) - f_m)^2}{N+1} \leq \epsilon \text{ where } f_m = \sum_{i=1,N+1} \frac{f(X_i)}{N+1}$$

- **Stopping test 2**:

$$|f(X_i) - f(X_L)| \leq \epsilon_1 \text{ for each i and}$$
$$\frac{1}{N+1} \sum_{i=1,N+1} ||X_i^{k+1} - X_i^k||^2 \leq \epsilon_2$$

where $\epsilon_1$ and $\epsilon_2$ are user-specified tolerances.

- **Stopping test 3**:

$$\text{Max}\{ \frac{1}{\Delta} ||X_i^{k+1} - X_L^k|| \text{ , for , } i = 1, N+1\} \leq \epsilon,$$

where $\Delta = Max\{1, ||X_L^k||\}$.

In our case we have used the first test with $\epsilon = 10^{-3}$ most of the time, since the test-functions are not noisy.

**Test- Functions**

(i) **F01**: Rosembrock Function
$F01(X_1, X_2) = 100(X_2 - X_1^2) + (1 - X_1)^2$.
The minimum (local) is at $X_{local} = (1,1)$ for $F01 = 0$
Starting point $(X_1, X_2) = (-1.2, 1)$.

(ii) **F02**:
$F02(X_1, X_2, X_3) = (1 - X_1)^2 + (2 - X_2)^2 + (3 - X_3)^2$.
The minimum (local) is at $X_{local} = (1,2,3)$ for $F01 = 0$
Starting point $(X_1, X_2, X_3) = (0,0,0)$.

(iii) **F03**: An unconstrained (global) optimization
$F03(X) = X^4 - 14X^3 + 60X^2 - 70X$.
This is a global optimization problem having
a local minimum at $X_{local} = 5.9572$ for $F03 = 11.9576$
and the global minimum is attained at $X = 0.7808$ for $F02 = -24.3696$
Starting point $X = 0$.

(iv) **F04**: A constrained (global) optimization
$F04(X_1, X_2) = \frac{1}{3}(X_1^3 + X_2^3) - X_1^2 - X_2^2 + 8/3$
subject to: $X_1 \geq 0$ $X_2 \geq 0$ and$(r \leq 0.5$ or $r \geq 1)$
where $r = (X_1^2 + X_2^2)^{\frac{1}{2}}$ is the radius from the origin.
This is a global optimization problem having
a local minimum at $X_{local} = (\sqrt{\frac{1}{8}}, \sqrt{\frac{1}{8}})$ for $F04 = 2.4313$
and the global minimum is attained at $X = (2,2)$ for $F04 = 0$
Starting point $(X_1, X_2) = (0,0)$.

(v) **F05**: An unconstrained (global) optimization
$F05(X_1, X_2) = (-13 + X_1 - 2X_2 + 5X_2^2 - X_2^3)^2 + (-29 + X_1 - 14X_2 + X_2^2 + X_2^3)^2$.
This is a global optimization problem having
a local minimum at $X_{local} = (11.4128, -0.89681)$ for $F05 = 48.98425$
and the global minimum is attained at $X = (5,4)$ for $F05 = 0$
Starting point $(X_1, X_2) = (0,0)$.

| Test-Function | Nelder-Mead | Dennis-Torczon | ADS parallel |
|---|---|---|---|
| F01 | 200 | 600 | 240 |
| F02 | 106 | 200 | 160 |
| F03 | 88 | 88 | 88 |

Table 14.1 : Number of Function Evaluations for Local Optimization.

| Test-Function | Nelder-Mead | ADS sequential |
|---|---|---|
| F03 | 10% | 10% |
| F04 | 15% | 30% |
| F05 | 10% | 35% |

Table 14.2 : Local Minimum Improvement.

The aim of this work was just to show that our approach will not increase the number of function evaluations in a prohibitive way while at the same time it may give a better minimum than the Dennis-Torczon and Nelder-Mead algorithms. As we can see in these two tables, the ADS procedure is rather useful for improving both situations: better percentage of convergence to local minimum and some possible escapes from local minimum to a better one.

In the first, the results are the average of 10 runs for each algorithm. We start 5 times from the worst initial point given in this section, with $\epsilon = 10^{-3}, ..., 10^{-7}$. The other runs are the 5 best over some "easy" points with $\epsilon = 10^{-5}$.

In the second table, we have considered that the algorithm succeeded only in the sense that, with $\epsilon = 10^{-3}$ , we find (or not) the global minimum from 10 starting points close to the attraction region of the local minimum. One could argue that the success rate is small, but recall that the method is originally designed for local minimization.

In the two tables one can see that the results for $F03$ are the same for all the algorithms. This is because the basis for geometrical operations are the same, since for this test-function a polytope is of the form $\{X_L, X_h\}$

and therefore $X_0 = X_L$.

## 4. Conclusions

We have presented some results for three algorithms based on the Moving Polytope Method. They all start from an initial polytope with vertices spanning $\mathbb{R}^n$ and perform some geometrical operations on the vertices in order to find the region of attraction of a minimum.

The proposed algorithm is the Adaptive Direct Search (ADS) algorithm. It combines the original sequential Nelder-Meald algorithm and the Dennis-Torczon parallel algorithm, adding a procedure which provides not only the basic search directions of these two algorithms, but also the possibility to perform additional searches based on the progress of the approach. These supplementary searches are justified by the fact that, in many situations, contracting a polytope can be nonprofitable. In some cases, the supplementary searches perform also an escape from the regions of attraction of local minima, improving thus the final solution.

## Acknowledgments

I gratefully acknowledge J. Dennis and V. Torczon for an earlier draft of their work, and J.-B. Hiriart-Urruty who asked me to work on the subject. I would also like to thank two colleagues, A. McCoy and K. Bouyoucef, for their helpful comments during the preparation of this paper.

# References

[1] J. E. Dennis and V. Torczon, Direct Search Methods on Parallel Computers, *SIAM Journal on Optimization* 1, 1991 pp 448-474.

[2] J. A. Nelder and R. Mead , A Simplex method for function minimization, *Comput. Journal* 7, 1965, pp 308-313.

[3] W. Spendley, G. R. Himsworth, Sequential applications of simplex designs in *optimization and evolutionary operation*, Technometrics, 4, pp 441-461, 1962.

[4] B. Hamma, La méthode du Polytope Mouvant: application à l'Optimisation Globale, *Séminaire d'Analyse Numérique* vol. 1990-91, Univ. Toulouse, 1991.

# Chapter 15

# RELATIONAL COMPLEMENTARITY PROBLEM

G. Isac[1], M. M. Kostreva[2], & M. Polyashuk[3]

[1] *Department of Mathematics and Computer Science*
*Royal Military College of Canada*
*P. O. Box 17000 STN Forces*
*Kingston, Ontario*
*Canada K7K 7B4*

[2] *Department of Mathematical Sciences*
*Clemson University*
*Clemson, South Carolina 29 634-1907*
*U.S.A.*

[3] *Department of Mathematics*
*Q. Northeastern Illinois University*
*R. Chicago, Il 60 625*
*U.S.A.*

**Abstract** The concept of binary relation is applied to generalize the complementarity problem. Dual relations are introduced as an analogue to dual cones in the generalized complementarity problem. A concept of linearization is introduced by considering the minimal linear relation stronger than the given relation. Geometric characteristics are studied as are the interconnections between a binary relation, its linearization and its dual. Existence of solutions for the new type of complementarity problem is investigated. In particular we consider the complementarity problem associated to a relation defined by the union of a family of pointed closed convex cones.



*A. Migdalas et al. (eds.), From Local to Global Optimization*, 327–339.

**Keywords: Complementarity problem, binary relation, cones, linearization.**

## 1. Introduction

In recent years the Complementarity Problem [9] and several generalizations [8], [11], [12], [21] have been studied and have developed into an important class of applied mathematical problems. There exists a body of theoretical results, algorithms for computing solutions and many applications for the complementarity problem. On the other hand, recent studies of binary relations in multidimensional spaces have produced interesting theoretical and practical results [3]. The concept of a binary relation provides a generalization of Pareto optimization as well as the value function (utility function) approach. The object of this paper is to show connections between these theories which been recently discovered, and applied to produce new results for complementarity problems.

A related problem, the generalized order complementarity problem [2], [11], [12] has been studied. It was known for some time that under certain conditions on the functions involved, there exists a solution to the linear generalized order complementarity problem. Recently, more extensive results have been obtained. Results for the infinite dimensional version of the generalized order complementarity problem are presented in [12]. Isac and Kostreva proposed a variant of the generalized order complementarity problem in which the mappings are point to set mappings [11]. Through all the progress and generalization of concepts, certain features of the notion of ordering have mot yet been explored. It would seem to be of importance to the complementarity problem, and other related problems as well, to study some abstract notions of ordering. In particular, we will study the complementarity problem with respect to the order of the union of a family of pointed closed convex cones. Let us now introduce the mathematical setting.

## 2. Formulation

To begin the development, we introduce some definitions.

**Definition 1** *A binary relation $\mathcal{R}$ is defined by $\mathcal{R} \subseteq \mathbf{R}^n \times \mathbf{R}^n$ . We say that $x\mathcal{R}y$ if and only if $(x,y) \in \mathcal{R}$. A binary relation $\mathcal{R}$ is reflexive if and only if $x\mathcal{R}x$ for all $x \in \mathbf{R}^n$.*

**Definition 2** *An upper section of a binary relation $\mathcal{R}$ at point $z$ is defined as*

$$\mathcal{R}_x = \{y \in \mathbf{R}^n : x\,\mathcal{R}\,y\}.$$

Any binary relation $\mathcal{R}$ defines a point to set mapping from $\mathbf{R}^n$ into $\mathbf{R}^n$ by $x \xrightarrow{\Psi} \mathcal{R}_x$. Remark that $Gr(\Psi) = \{(x,y) : y \in \mathcal{R}_x\}$ is exactly $\mathcal{R}$. Conversely, if $T : \mathbf{R}^n \to 2^{\mathbf{R}^n}$ is a point to set mapping then its graph $Gr(T) = \{(x,y) : y \in T(x)\}$ defines a binary relation in $\mathbf{R}^n \times \mathbf{R}^n$.
If $\mathcal{R}$ is a convex set we say that the binary relation is *convex*. Convex binary relations were considered in [1].
The upper section will be interpreted as the set of points which are not worse than $x$ in the case of a reflexive binary relation. Denote by $<,>$ the Euclidean inner product.

**Definition 3** *Given a binary relation $\mathcal{R}$ its dual $\mathcal{R}^*$ is defined by*

$$x\mathcal{R}^*y \Longleftrightarrow \langle y, (z-x)\rangle \geq 0 \; for\ all\ x \in \mathcal{R}_x.$$

Now we define the *Relational Complementarity Problem*:

Let a binary relation $\mathcal{R}$, be given. Let $f : \mathbf{R}^n \to \mathbf{R}^n$ be given mapping. Suppose $a \in \mathbf{R}^n$ is given. *The relational complementarity problem* is to find $x \in \mathcal{R}_a$ such that $y = f(x) \in (\mathcal{R}^*)_a$ and $\langle y, (x-a)\rangle = 0$.

**Remark** If $K \subset \mathbf{R}^n$ is a pointed convex cone and $\mathcal{R}$ is $x\mathcal{R}y \Longleftrightarrow y - x \in \mathbf{K}$, then for $a = 0$ the relational complementarity problem is the classical nonlinear complementarity problem.

**Definition 4** *Given $\Phi : \mathbf{R}^n \to \mathbf{R}^n$, a relation $\mathcal{R} \subseteq \mathbf{R}^n \times \mathbf{R}^n$ is said to be $\Phi$ - invariant if and only if $x\mathcal{R}y \Longleftrightarrow \Phi(x)\mathcal{R}\Phi(y)$, for all $x, y \in \mathbf{R}^n$.*

**Definition 5** *A relation $\mathcal{R} \subseteq \mathbf{R}^n \times \mathbf{R}^n$ is a linear relation if it is $\Phi$ -invariant for any $\Phi \in \{\Phi : \mathbf{R}^n \to \mathbf{R}^n : \Phi(x) = \alpha x + \beta, \alpha \geq 0, \beta \in \mathbf{R}^n\}$.*

**Definition 6** *For a set $S \subseteq \mathbf{R}^n$, define $\mathcal{H}_x(S) = \{\lambda(y-x) : y \in S, \lambda \geq 0\}$.*

**Definition 7** *For relation $\mathcal{R} \subseteq \mathbf{R}^n \times \mathbf{R}^n$ define its linearization $\mathcal{L}(\mathcal{R})$ by*

$$[\mathcal{L}(\mathcal{R})]_x = \mathcal{H}_x(\mathcal{R}_x).$$

**Theorem 1** For any binary relation $\mathcal{R} \subseteq \mathbf{R}^n \times \mathbf{R}^n$, and for any $x \in \mathbf{R}^n$ such that $\mathcal{R}_x$ is non-empty we have that the upper section $(\mathcal{R}^*)_x$ of the dual relation $\mathcal{R}^*$, is a convex cone with its vertex at the origin.

**Proof** Let $x \in \mathbf{R}^n$ be an arbitrary element such that $\mathcal{R}_x = \{y \in \mathbf{R}^n : x\mathcal{R}y\}$. is non-empty. By definition we have

$(\mathcal{R}^*)_x = \{y \in \mathbf{R}^n : \langle y, (z-x)\rangle \geq 0 \; for \; all \; z \in \mathcal{R}_x\}$. It is easy to show that the set $(\mathcal{R}^*)_x$ is a convex cone with its vertex at the origin, i.e., for every $y, y_1, y_2 \in (\mathcal{R}^*)_x$ and every $\lambda \geq 0$ we have $\lambda y \in (\mathcal{R}^*)_x$ and $y_1 + y_2 \in (\mathcal{R}^*)_x$. ■

**Corollary 1** *For any reflexive binary relation $\mathcal{R} \subseteq \mathbf{R}^n \times \mathbf{R}^n$ we have that for any $x \in \mathbf{R}^n$, $(\mathcal{R}^*)_x$ is a convex cone with its vertex at the origin.*

Given a binary relation $\mathcal{R} \subseteq \mathbf{R}^n \times \mathbf{R}^n$, we denote by

$$Pr^1(\mathcal{R}) = \{x \in \mathbf{R}^n : there\ exists\ y \in \mathbf{R}^n\ with\ (x,y) \in \mathcal{R}\}$$

**Corollary 2** *For any binary relation $\phi \neq \mathcal{R} \subseteq \mathbf{R}^n \times \mathbf{R}^n$, if we denote by $A = Pr^1(\mathcal{R})$ and $B = \bigcup_{x \in A} (\mathcal{R}^*)_x$, we have that $\mathcal{R}^* \subseteq A \times B$, where $B$ is a union of convex cones, each with its vertex at the origin.*

**Proof** This follows from the fact that $\mathcal{R}^*$ is the graph of the point to set mapping $\Psi^*$ defined by $x \xrightarrow{\Psi^*} (\mathcal{R}^*)_x$. ■

**Theorem 2** *For any binary relation $\mathcal{R} \subseteq \mathbf{R}^n \times \mathbf{R}^n$, its dual $\mathcal{R}^*$ coincides with the dual of its linearization $\mathcal{L}(\mathcal{R})$, i.e., $\mathcal{R}^* = [\mathcal{L}(\mathcal{R})]^*$.*

**Proof** The statement of the theorem is equivalent to the following: for any $x \in \mathbf{R}^n$, $(\mathcal{R}^*)_x = ([\mathcal{L}(\mathcal{R})]^*)_x$. Let $z \in ([\mathcal{L}(\mathcal{R})]^*)_x$. By *Definition 3*, $\langle z, (w-z)\rangle \geq 0$, for all $w \in [\mathcal{L}(\mathcal{R})]_x = \{\lambda(y-x) : y \in \mathcal{R}_x, \lambda \geq 0\}$. Since $\mathcal{R}_x \subseteq [\mathcal{L}(\mathcal{R})]_x$, for all $w \in \mathbf{R}^n$, then $\langle z, (w-x)\rangle \geq 0$, for all $w \in \mathcal{R}_x$ and therefore $x \in (\mathcal{R}^*)_x$. Let $z \in (\mathcal{R}^*)_x$, then $\langle z, (y-x)\rangle \geq 0$ for all $y \in \mathcal{R}_x$. This implies that for all $\lambda \geq 0$, $\langle \lambda z, (y-x)\rangle \geq 0$, for all $y \in \mathcal{R}_x$. By *definitions 6 and 7*, the proof follows. ■

# 3. Relational complementarity problem. Existence of solutions

We do not know of any existence results for the general relational complementarity problem. It is hard to solve this general problem. So, we will consider an interesting special case. This is the case of a binary relation defined by a *conical set, not necessarily convex.*
If $D \subset \mathbf{R}^n$ is non-empty subset, we denote by $\mathbf{K}(D)$ the *smallest closed convex cone which contains the set* $D$. Let $D \subset \mathbf{R}^n$ be non-empty subset. Then $D$ is a *pointed conical set* if the following properties are satisfied:

(1) *for all* $x \in D$ *and* $\lambda \in \mathbf{R}_+$, $\lambda x \in D$,

(2) $\mathbf{K}(D) \bigcap [-\mathbf{K}(D)] = \{0\}$.

We say that a non-empty set $B$ is a *base of a pointed conical set* $D$, if there exists a linear functional $\Psi$ such that $\Psi(x) > 0$ for all $x \in D \setminus \{0\}$ and $B = \Psi^{-1}(1) \bigcap D$. The following particular cases of $D$ are interesting for practical problems. The set $D$ is supposed to be closed and pointed.

I. $D = \bigcup_{i \in I} \mathbf{K}_i$, where for every $i \in I$, $\mathbf{K}_i$ is a pointed convex cone.

II. $D = \bigcup_{i \in I} \mathbf{K}_i$, where for every $i \in I$, $\mathbf{K}_i$ is a polyhedral cone not necessarily convex.

III. Recall that $B$ is *star-shaped with respect to a convex set* $A \subset B$, i.e., $x \in B$ if and only if $\lambda x + (1 - \lambda)y \in B$ for all $\lambda \in [0, 1]$ and all $y \in A$. An interesting case occurs when $D$ is a pointed conical set with a base $B$ which is *star-shaped* with respect to a convex subset.

IV. Let $\epsilon > 0$ be a real number, eventually very small. We say that a non-empty subset $B$ is $\epsilon$ - *convex* if and only if, whenever $[x, y[ \subset conv(B) \setminus B$ we have that $||y - x|| < \epsilon$. We recall that $[x, y[= \{\lambda x + (1 - \lambda)y : \lambda \in ]0, 1]\}$. In this last example suppose that $D$ is a pointed conical set with an $\epsilon$-*convex base* $B$.

The situations presented in examples (I) - (IV) can arise when $D$ is close to a pointed convex $\mathbf{K} \subset \mathbf{R}^n$ and *represents a tolerance* for certain types of (numerical) errors in the vector $x$ relative to the classical complementarity problem $CP(f, \mathbf{K})$, which requires $x \in \mathbf{K}$. Also, all the above

mentioned examples may develop in sensitivity studies related to complementarity problems.
If $D$ is a pointed conical set, the *dual* of $D$, is by definition,

$$D^* = \{y \in \mathbf{R}^n : \langle y, x\rangle \geq 0\, for\ all\ x \in D\}.$$

Let $D$ be a non-empty pointed conical set. Consider the binary relation

$$x\mathcal{R}(D)y \Longleftrightarrow y - x \in D.$$

Given $f : \mathbf{R}^n \to \mathbf{R}^n$, a function and taking $a = 0$, the *Relational Complementarity Problem is*:

$$RCP(f, \mathcal{R}(D)) : \begin{cases} find\ x_* \in \mathcal{R}(D)_0\ such\ that \\ y_* = f(x_*) \in \left(\mathcal{R}(D)^0\right)_0\ and\ \langle y_*, x_*\rangle = 0 \end{cases}$$

We observe that the problem $RCP(f, \mathcal{R}(D))$ is equivalent to the following problem:

$$RCP(f, D) : \begin{cases} find\ x_* \in D\ such\ that \\ y_* = f(x_*) \in D^*\ and\ y_*, x_* = 0 \end{cases}$$

**Theorem 3** *The problem $RCP(f, D)$ is equivalent to the following variational inequality*:

$$VI(f, D) : \begin{cases} find\ x_* \in D\ such\ that \\ \langle f(x_*), y - x_*\rangle \geq 0\ for\ all\ y \in D. \end{cases}$$

**Proof** Suppose that $x_*$ solves the problem $VI(f, D)$. Since $y = 0 \in D$ it follows that $\langle f(x_*), -x_*\rangle \geq 0$, or

$$\langle f(x_*, x_*)\rangle \leq 0 \tag{1}$$

Since $D$ is a cone we have that $2x_* \in D$ and because $x_*$ is a solution of the problem $VI(f, D)$ we have

$$\langle f(x_*), x_*\rangle = \langle f(x_*), 2x_* - x_*\rangle \geq 0 \tag{2}$$

From (1) and (2) we deduce that $\langle f(x_*), x_*\rangle = 0$. Let $y \in D$. We have

$$\langle f(x_*), y\rangle = \langle f(x_*), y - x_* + x_*\rangle = \langle f(x_*), y - x_*\rangle + \langle f(x_*), x_*\rangle \geq 0.$$

Hence, $f(x_*) \in D^*$ and we have that $x_*$ solves the problem $RCP(f, D)$. Conversely, suppose that $x_*$ solves the problem $RCP(f, D)$. We have, $x_* \in D$, $f(x_*) \in D^*$ and $\langle f(x_*), x_*\rangle = 0$. Thus, for every $y \in D$ we have

$$\langle f(x_*), y\rangle \geq 0\ for\ all\ y \in D \tag{3}$$

and

$$\langle f(x_*) - x_*\rangle = 0 \tag{4}$$

From (3) and (4) we obtain $\langle f(x_*), y - x_* \rangle \geq 0$ for all $y \in D$ which means that $x_*$ solves the problem $VI(f, D)$. ■

By *Theorem 3* we have that the problem $RCP(f, D)$ can be solved applying the theory of variational inequalities with respect to a nonconvex sets. Regrettably, the theory of variational inequalities with respect to nonconvex sets is not well developed. Perhaps, some results can be obtained using the mathematical tools developed in [5], [6], [16] and [17]. It seems that the methods developed in [1] and [18] are not applicable since the set $D$ is star-shaped with respect to the origin only and not star-shaped with respect to a ball.
The possibility of applying some special fixed point theorems [20] can also be considered. We will use, in this paper another approach to study the solvability of $RCP(f, D)$.
Let $\mathbf{K} \subset \mathbf{R}^n$ be a closed pointed convex cone and $f : \mathbf{R}^n \to \mathbf{R}^n$ a mapping. Denote by $\mathbf{K}^*$ the dual cone of $\mathbf{K}$. The following notion is fundamental for our main result. Denote by $NCP(f, \mathbf{K})$ the classical nonlinear complementarity problem associated to $f$ and $\mathbf{K}$.

**Definition 7 [4], [10]** *The family of elements $\{x^r\}_{r>0} \subset \mathbf{K}$ is an exceptional family of elements for $f$, with respect to $\mathbf{K}$, if and only if, for every real number $r > 0$ there exists a real number $\mu_r > 0$, such that the vector $u_r = f(x^r) + \mu_r x^r$ satisfies the following conditions:*

(i) $u_r \in \mathbf{K}^*$,

(ii) $\langle u_r, x^r \rangle = 0$,

(iii) $||x^r|| \to +\infty$ as $r \to +\infty$.

**Theorem 4** *For any continuous function $f : \mathbf{R}^n \to \mathbf{R}^n$, there exists, either a solution for the problem $NCP(f, \mathbf{K})$, or an exceptional family of elements for $f$ with respect to $\mathbf{K}$.*

This result is due to G. Isac, V. Bulavski and V. Kalashnikov [10], [4].

Let $D \subset \mathbf{R}^n$ be a closed pointed conical set. Evidently, $D$ is supposed to be non-empty. If $D$ is not convex, we say that $D_r = \mathbf{K}(D) \setminus D$ is the *residual set of $D$ with respect to* $\mathbf{K}(D)$. Clearly $0 \notin D_r$ and $D_r$ is empty if $D$ is a closed pointed convex cone.

The next theorem is the main result of this paper.

**Theorem 5** *Let $D \subset \mathbf{R}^n$ be a closed pointed conical set (not necessarily convex) and $f : \mathbf{R}^n \to \mathbf{R}^n$ be a continuous function. If there exists $\rho > 0$ such that the following assumptions are satisfied:*

(1) *for every $x \in \mathbf{K}(D)$ with $||x|| = \rho$, there exists $y \in \mathbf{K}(D)$ such that $||y|| < \rho$ and $\langle f(x), x - y \rangle \geq 0$,*

(2) *for every $x \in D_r(\leq \rho) = \{x \in D_r : ||x|| \leq \rho\}$ there exists $y \in \mathbf{K}(D)$ with $||y|| < ||x||$ such that $\langle f(x), x - y \rangle \geq 0$,*

*then the problem $RCP(f, D)$ has a solution $x^*$. Moreover, $||x^*|| \leq \rho$.*

**Proof** Consider an arbitrary $\epsilon > 0$ and define $f_\epsilon(x) = f(x) + \epsilon x$ for any $x \in \mathbf{R}^n$ . The function $f_\epsilon$ satisfies the following properties:

(i) *$f_\epsilon$ is a continuous function,*

(ii) *For every $x \in \mathbf{K}(D)$ with $||x|| = \rho$, there exists $y \in \mathbf{K}(D)$ such that $||y|| < \rho$ and $\langle f_\epsilon(x), x - y \rangle > 0$,*

(iii) *For every $x \in D_r(\leq \rho)$, there exists $y \in \mathbf{K}(D)$ with $||y|| < ||x||$ such that $\langle f_\epsilon(x), x - y \rangle > 0$.*

Indeed, let $x \in \mathbf{K}(D)$ with $||x|| = \rho$ (resp. $x \in D_r(\leq \rho)$) then by (1) (resp. (2)) there exists $y \in \mathbf{K}(D)$ such that $||y|| < \rho$ (resp. $||y|| < ||x||$) and $\langle f(x), x - y \rangle \geq 0$.For such $x$ and $y$ we have

$$\langle f_\epsilon(x), x - y \rangle = \langle f(x) + \epsilon x, x - y \rangle =$$

$$\langle f(x), x - y \rangle + \epsilon \left[ ||x||^2 - \langle x, y \rangle \right] \geq \epsilon \left[ ||x|| \right] - ||x|| ||y|| > 0$$

Therefore, (ii) and (iii) are satisfied. For all $x \in \mathbf{K}$ with $||x|| > \rho$ denote by $T_\rho(x)$ the radial projection onto $S_\rho^+ = \{x \in \mathbf{K}(D) : ||x|| = \rho\}$, i.e., $T_\rho(x) = \frac{\rho x}{||x||}$. Consider the function

$$g_\epsilon(x) = \begin{cases} f_\epsilon(x), \; if \; ||x|| < \rho \\ f_\epsilon(T_\rho(x)) + ||x - T_\rho(x)|| x, \; if \; ||x|| > \rho \end{cases}$$

For any $x \in \mathbf{K}(D)$ with $||x|| > \rho$ there exists $\lambda_x > 0$ such that $x = \lambda_x T_\rho(x)$. Since $f_\epsilon$ satisfies (ii), we have that, for $T_\rho(x)$ there exists $u_p^x \in \mathbf{K}(D)$ with $||u_\rho^x|| < \rho$ such that

$$\langle T_\rho(x) - u_\rho^x, f_\epsilon(T_\rho(x)) \rangle > 0.$$

Using this fact, observe that $\langle x - \lambda_x u_\rho^x, g_\epsilon(x)\rangle > 0$. If for a given $x$ with $||x|| > \rho$ we let $y = \lambda_x u_\rho^x$, we obtain that the function $g_\epsilon$ satisfies the following property:

(iv) *for every* $x \in \mathbf{K}(D)$ *with* $||x|| > \rho$, *there exists* $y \in \mathbf{K}(D)$ *with* $||y|| < ||x||$ *such that* $\langle x - y, g_\epsilon(x)\rangle > 0$.

For any $\epsilon > 0$, the function $g_\epsilon$ is without exceptional families of elements with respect to $\mathbf{K}(D)$. Indeed, suppose that $g_\epsilon$ has an exceptional family of elements $\{x^r\}_{r>0} \subset \mathbf{K}(D)$.
We have

$$u_r = g_\epsilon(x^r) + \mu_r x^r \in (\mathbf{K}(D))^*, \ for\ all\ r > 0,$$

$$\langle x^r, u^r\rangle = 0\ for\ all\ r > 0\ and$$

$$||x^r|| \to +\infty\ as\ r \to +\infty.$$

Take $r > 0$ such that $||x^r|| > \rho$. Since $g_\epsilon$ satisfies (iv), there exists $y_r \in \mathbf{K}(D)$ such that
$||y_r|| < ||x^r||$ and $\langle x^r - y_r, g_\epsilon(x^r)\rangle > 0$.
We have

$$\begin{aligned} 0 < \langle x^r - y_r, g_\epsilon(x^r)\rangle &= \langle x^r - y_r, u_r - \mu_r x^r\rangle \\ &= \langle x^r - y_r, u_r\rangle - \mu_r ||x^r||^2 + \mu_r\langle y_r, x^r\rangle \\ &\le -\mu_r ||x^r|| \, [||x^r|| - ||y_r||] < 0 \end{aligned}$$

which is impossible.
Applying *Theorem 4* we obtain that for any $\epsilon > 0$, the classical problem $NCP(g_\epsilon, \mathbf{K}(D))$ has a solution $x_\epsilon^*$. Since $g_\epsilon$ satisfies (iv) it is impossible to have $||x_\epsilon^*|| > \rho$. Hence, $||x_\epsilon^*|| \le \rho$, which implies that $g_\epsilon(x_\epsilon^*) = f_\epsilon(x_\epsilon^*)$.
Therefore, we obtained the following result:
for any $\epsilon > 0$, the problem $NCP(g_\epsilon, \mathbf{K}(D))$ has a solution $x_\epsilon^*$ such that $||x_\epsilon^*|| \le \rho$.
Now, considering the fact that $f_\epsilon$ satisfies (iii) we have that $x_\epsilon^* \in \{x \in D : ||x|| \le \rho\} = D_\rho$.
Considering the sequence $\epsilon_n = \frac{1}{n}, n = 1, 2, \ldots$; we obtain a sequence of solutions $\left\{x_{\frac{1}{n}}^*\right\}_{n=1}^{\infty}$ such that $x_{\frac{1}{n}}^* \in D_\rho$, and for every $n \in N$, $x_{\frac{1}{n}}^*$ is a solution of the problem $NCP(f_\epsilon, \mathbf{K}(D))$. Since $D_\rho$ is compact the sequence $\left\{x_{\frac{1}{n}}^*\right\}_{n=1}^{\infty}$ has a convergent subsequence $\left\{x^*_{\frac{1}{n_k}}\right\}_{k=1}^{\infty}$. We have that $x^* = \lim\limits_{k\to\infty} x^*_{\frac{1}{n_k}}$ is an element of $D$. Considering the definition of $f_\epsilon$ and the fact $(\mathbf{K}(D))^* \subset D^*$ we obtain that $x^*$ is a solution of the problem $RCP(f, D)$ and $||x^*|| \le \rho$. The proof is complete. ■

**Conclusions** This paper introduces a new generalization of the complementarity problem which replaces the ordering of a convex cone by that of a general binary relation. By doing this, several significant insights have been found:

1. applications to approximation which occur quite naturally in the use of generalized complementarity problems may be handled by means of the new relational complementarity problem.
2. the existence of solutions for relational complementarity problems is non-trivial to prove, due to the complexity and generality of the sets involved.
3. an interesting special case may be solved (and existence proved) by appeal to the techniques relating to the absence of an exceptional family of elements, under weak additional assumptions such as continuity and a form of coercivity.

The findings we present here and the methods used should be further developed, and combined with other new techniques to make further progress on this interesting problem. In particular, it would be of interest to combine the duality theory with the present relational complementarity problem.

# References


[1] J. M. Borwein: Automatic continuity and openness of convex relations. *Proc. Amer. Math. Soc , 99, Nr 1 (1987), 49-55.*

[2] J. M. Borwein and A. H. Dempster: The linear order complementarity problem. *Math. Operations Research, Vol. 14 (1989), 534-558.*

[3] B. Borzenko: Approximational approach to multicriteria problems, in: *Methodology and Software for Interactive Decision Support, (A. Lewandowki and I. Stanchev, Eds.), Lecture Notes in Economics and Mathematical Systems, Nr. 337, Springer-Verlag, 1989.*

[4] V. Bulavski, G. Isac and V. Kalashnikov: Application of topological degree to complementarity problem, In: *Multilevel Optimization: Algorithms and Application, (A. Migdalaset al (Eds.) Kluwer Academic Publishers, (1998), 333-358.*

[5] A. Carla: *On a separation approach to variational inequalities, in: Variational Inequalities and Network Equilibrium Problems, (F. Giannessi and A. Maugeri, (Eds.)), Plenum Press, New York and London (1995), 1-7.*

[6] F. Giannessi: Separation of sets and gap functions for quasi-variational inequalities, in: *Variational Inequalities and Network Equilibrium Problems, (F. Giannessi and A Maugeri, (Eds.)), Plenum Press, New York and London (1995), 101-121.*

[7] D. Goeleven: Noncoercive hemivariational inequality approach to constrained problems for star-shaped admissible sets, *J. Global Opt., 9 (1996), 121-140.*

[8] G. Habetler and A. Price: Existence theory for generalized nonlinear complementarity problems. *J. Opt. Th. Appl. 7, (1971), 223-239.*

[9] G. Isac: Complementarity Problems, Lecture Notes in Mathematics 1528, Springer-Verlag, New York (1992).

[10] G. Isac, V. Bulavski and V. Kalashnikov: Exceptional families, topological degree and complementarity problems, *J. Global Opt., 10 (1997), 207-225.*

[11] G. Isac and M. M. Kostreva: Kneser's theorem and the multivalued generalized order complementarity problem, Appl. Math. Letters 4 Nr. 6, (1991), 81-85.

[12] G. Isac and M. M. Kostreva: The generalized order complementarity problem, *J. Opt. Theory Appl. 71 Nr. 3 (1991), 517-534.*

[13] G. Isac, M. M. Kostreva and M. Wiecek: Multiple objective approximation of feasible but unsolvable linear complementarity problems, *J. Opt. Theory Appl. 86 Nr. 2 (1995), 389-405.*

[14] J. Jahn: Mathematical Vector Optimization in Partially Ordered Linear Spaces, Verlag Peter Lang, Frankfurt am Main, Germany (1986).

[15] M. M. Kostreva and M. M. Wiecek: Linear complementarity problems and multiple objective programming, Math. Programming 60 Nr. 3 (1993), 349-359.

[16] C. Marco and M Giandomenico: On the duality theory for finite dimensional variational inequalities, in: *Variational Inequalities and Network Equilibrium Problems, (F. Giannessi and A. Maugeri (Eds.)), Plenum Press, New York and London (1995), 21-31.*

[17] U. Mosco: Dual variational inequalities, *J. Math. Anal. Appl. 40 (1972), 202-206.*

[18] Z. Naniewicz: Hemivariational inequality approach to constrained problem for star-shaped admissible sets, *J. Opt. Theory Appl. 83 Nr. 1 (1994), 97-112.*

[19] P. D. Panagiotopoulos: Hemivariational Inequalities. Applications in Mechanics and Engineering, *Springer-Verlag, Berlin, Heidelberg (1993).*

[20] J. P. Penot: Fixed point theorems without convexity, *Analyse nonconvexe [1977-Pau], Bull. Soc. Math. France, Memoire 60 (1979), 129-152.*

[21] R. Saigal: Extension of the generalized complementarity problem, *Math. Operations Research, 1 (1976), 260-266.*

[22] Y. Sawaragi, H. Nakayama and T. Tanino: Theory of Multiobjective Optimization, *Academic Press, Orlando (1985).*

[23] Z. You and S. Zhu: A constructive Approach for proving the existence of solutions of variational inequalities with set-valued maps, *Kexue Tongbao, 31 (1986), 1153-1156.*